This book belongs to:

A ROOM WITHOUT BOOKS IS
LIKE A BODY WITHOUT A SOUL.

—Cicero

BALUS

The Aeroplane in Papua New Guinea

Volume I: The Early Years

G-ACGG

BALUS

The Aeroplane in Papua New Guinea

Volume 1: The Early Years

James Sinclair

Designed & published by
Robert Brown & Associates (Aust) Pty. Ltd
P.O. Box 29,
Bathurst, N.S.W. 2795 Australia

Distributed in PNG by
Gordon & Gotch (PNG) Pty. Ltd
P.O. Box 107, Boroko, PNG

Design by Libby Grundy
Cover illustration by Tony Ward
Printed in Singapore by Toppan Printing Company

National Library of Australia Cataloguing-in-Publication data

Sinclair, James, 1928- .
Balus, the aeroplane in Papua New Guinea. Volume 1, The early years.

Bibliography.
Includes index.
ISBN 0 949267 50 3.

1. Aeronautics - Papua New Guinea - History. I. Title.

387.7'0995'3

Contents

This book is dedicated to

Wing-Commander R.H.M. Gibbes, DSO, DFC and Bar

Foreword

In 1978, I published a history of civil aviation in Papua New Guinea from its beginnings in 1922 to the outbreak of the Pacific war. This book was entitled *Wings of Gold*.

The present history tells the story of post-Pacific war civil aviation in Papua New Guinea to 1982, the year in which the Goroka-based airline, Talair, celebrated its 30th anniversary.

Talair is the oldest purely commercial air operation in PNG, born in the early post-war years of the old colonial era, and continuing into the present. For most of its existence, Talair has been under the direction of R.D. Buchanan, and it is through his generosity that I have been able to undertake the very considerable research that was necessary before this book could be written. Without the support of Dennis Buchanan, I could not have extended my account, originally intended to be a history of Talair only, to cover the entire aviation scene. I have had a completely free hand, and have not been subjected to any direction or control.

The subject I have attempted is vast, and consequently the book that has resulted is too large to be published in a single volume. It will, therefore, appear in two volumes, of which this is the first. Once again, I have to acknowledge the assistance of Dennis Buchanan, who has subsidised publication, and enabled the book to be issued at a reasonable cost.

The main title of this work is *Balus: The Aeroplane in Papua New Guinea*. Volume One is titled, *The Early Years,* and covers the period 1944 to 1960. The operations of all companies are examined. This period saw the birth of Territory Airlines, the company that was in later years renamed, Talair.

Volume Two, to be published in 1987, will be titled *The Rise of Talair,* and covers the period from 1961 to 1982. This volume will be considerably larger than Volume One. Once again, all operations, not just those of Talair, are covered.

I have made every effort to ensure the accuracy of my account, but with so complex a subject, it would indeed be surprising if mistakes have not been made. I ask the reader's indulgence if I have erred.

A great many people and organisations have helped me, with information, photographs, documents, publications, and personal memoirs. I acknowledge these elsewhere. I can only say that the book could not have been written without the help so generously given by the aviation community.

I commenced researching this history in June, 1982.

James Sinclair

Acknowledgements

The extent of my indebtedness to Dennis Buchanan has already been made apparent.

The Department of Aviation and the Civil Aviation Agency, through Jim Schofield, has given every assistance, and supported my request for special access to records of the old Department of Civil Aviation held by Australian Archives. The officers of the archives were most helpful.

Qantas, Ansett and Talair provided much appreciated free air travel during the research phase of this work. Qantas also allowed me full access to their photographic files.

Aviation historian, Geoff Goodall, kindly made available his unique collection of aviation files, aircraft and company histories. Bob Fulton, then of Talair, turned over to me the material he collected on Territory Airlines/Talair history, and gave much practical assistance besides.

John Behr provided me with a copy of his unpublished manuscript on Qantas in New Guinea. Max Meyers allowed me to use the material collected by Vic Ambrose on the operations of Missionary Aviation Fellowship in Papua New Guinea and West Irian. The Aviation Department of Summer Institute of Linguistics has been particularly helpful.

Bruce Hoy, Curator of the Air Museum, Port Moresby, saved me an enormous amount of tedious work, and time, by making available his extremely comprehensive collection of photostats on aviation matters from the various PNG newspapers, and from that mine of information on PNG matters, *Pacific Islands Monthly*. Gus Smales, then editor of PIM, allowed me the use of photographs from the magazine's files. Wayne Grant, managing director of the *Post-Courier,* allowed me to photocopy relevant pages of back issues — and provided the photo-copier.

Al Bovelt, aviation historian, continued the support he gave me whilst I was writing *Wings of Gold*. Ray Thurecht loaned me the material collected by the late Vince Sanders on South Pacific Aero Clubs. Tim Bowden, of the ABC, sent me copies of tapes he made whilst preparing his radio documentary series, *Taim Bilong Masta*.

If this work is judged by the aviation community to be of value, then it will be largely because of the co-operation I have received from individual pilots, engineers and aviation figures, who answered my constant letters and telephone calls, submitted to tape-recorded interviews, provided photograpahs and in some cases travelled considerable distances to my home, thus saving me valuable time.

I am particularly grateful to the late Bill Forgan-Smith and the late Tom Drury, and to Tom Deegan, Sam Jamieson, Len McNeil, Phil Oakley, John Simler, Dick Burt, Dennis Douglas, Hugh Birch, Bill Sherwood, Tony Chadim, Bobby Gibbes, Fred Fox, Laurie Crowley, Ron Firns, Cliff Jackson, Ray Harris, Vic Smith, Dave Robertson, Bill Johns, Dick Glassey, Paul Raasch, Geoff Masters, Sten-Erik Molker, Graham Syphers, Bryan McCook, Richard Leahy, Bryan Grey and Ralph Conley.

The names of others who helped me greatly appear below, in alphabetical order:

Juin Arthur, Jan Balenzuela, Ron Ballard, Jim Baptista, Kel Barclay, Peter Barter, Willis Baughman, John Bayles, Ivan Bennett, Phil Bennett, Ross Biddulph, Ted Booker, J.N.B. Boyd, Bruce Brearley, Tom Briggs, Bill Burns, John Cale, Arthur Callard, Brother Larry Camilleri, Dave Campbell, Ross Christensen, Vic Cox, John Cruikshank, Doug Cummings, Ken Davenport, Dick Davis, William Decker, Roger Dodson, John Downie, Ian Downs, Bill Dowsett, Pat Dunch, Eddie Van Ede, Doug Elphinstone, Jim Entz, Gerry Fallscheer, Bruce Fawcett, Bob Fawkes, Ray Feltman, Wren Fennamore, A.S. Fisher, Gerd Franck, John Freeman, Frank Furniss, Fraser Gibson, Jim Glindemann, Frank Goossens, Joan Gray, Robin Gray, Wes Guy,

Graham Hawthorne, Bertie Heath, Tom Howes, Lloyd Hurrell, Ken James, Norm Janke, Bruce Jephcott, Doug Jones, Jeff Joynton-Smith, Tony Karas, John Kessey, Frank King, Alf Lane, Tom Lee, Dave Lourey, John Mabry, Dave Maclure, Peter Manser, Dick Mant, Jasper Maskeleyne, Jock McKee, York Mendoza, Stewart Middlemiss, Phil Miller, Brian Mitchell, Ken Montagu, Harold Morton, John Naylor, Bill Neale, Frank Newell, Adrian Nisbet, Paun Nonggor'r, Bill Nye, Napoleon Onsem, Dennis O'Malley, Max Paynter, Noel Peacock, Tony Peeters, Jim Perry, Brian Peters, Fred Phillips, Lionel Purkiss, Ian Ralfe, John Regan, Bert Ritchie, Ron Roach, Ron Roberts, Chris Robinson, Harold Rudd, John Seaton, M. Seymour, Mal Shannon, Elwin Smith, Mal Smith, Jim Spark, John Squires, Jack Stammer, Phil Stedman, Bill Taylor, Lionel Thrift, Bill Ticehurst, Ross Treadgold, John Tracey, Bill Vink, Brian Walker, Bert Wickham, Noel Wicks, John Wielens, Ken Wiggers, Roger Wilson, George Woodward, Ron Yates and John Zehnder.

Author's Note

An essential part of aviation history is obviously air accidents, and in this work I have tried to record the major crashes that have occurred in PNG since the war. No serious account of PNG aviation can possibly avoid discussion of air crashes, but it needs to be emphasised that air travel in PNG, as in every other country in the world, is vastly safer than road travel.

On any basis of comparison, one has a a much greater chance of being killed or injured in a motor vehicle than in an aeroplane. We have become so accustomed over the years to the road carnage in accounts in newspapers and on television that we have become blasé, our feelings calloused by the never-ending toll of horror. Multiple roads deaths in single accidents are so commonplace that they rarely rate more than a mention in the news, and are forgotten next day by the general public. But the slightest accident involving an aeroplane is NEWS!

I ask the reader to bear the above in mind when reading this book. The use of aeroplanes in PNG is so intensive that, statistically, accidents are bound to happen. On the basis of numbers of accidents for kilometres travelled, however, the PNG record is good.

QANTAS

1

The Beginning

The Pacific War

In January, 1942, Japanese Forces invaded Rabaul and in the space of a few weeks occupied those portions of the Territory of New Guinea known as the Bismarck Archipelago, the Admiralty Islands, the Solomon Islands, and the coastal and subcoastal portions of the island of New Guinea described as the Sepik, Madang and Morobe administrative districts.

'By August, 1942, they had invaded the coastal portion of the Territory of Papua as far south as the Buna-Gona-Sananada-Kokoda areas, crossed the Owen Stanley Range and penetrated to within thirty miles [forty-eight kilometres] of Port Moresby, the capital of that Territory.

'Port Moresby became a major base from which the military operations of the Allied Forces were directed against the Japanese. Hundreds of thousands of soldiers, sailors and airmen halted here on their way to the forward areas to the north. Places now famous such as the Kokoda Trail, Buna, Gona, Sanananda, Kokoda, Milne Bay, Wau, Mubo, Salamaua, Lae and others became bloody battlefields. The Territory of New Guinea became one of the major battlegrounds of the war in the Pacific; and the war of attrition which developed with increasing tempo and intensity was catastrophic in its devastation of commercial and industrial establishments, and its effects on the native people inhabiting the areas where fighting occurred.

'The devastatation caused by modern weapons is so well known as scarcely to merit emphasis other than perhaps to say that in those areas where land fighting occurred and in those attacked from the air while occupied by the enemy no pre-war installation or establishment remained when the 'cease fire' was sounded. To the devastation caused by active combat has to be added the damage caused by the construction of military establishments, camps, aerodromes etc, ancillary to active combat. Lastly, there has to be mentioned in particular the damage done to the property of the indigenous population — their garden lands, villages, homes and canoes; and to the suffering which modern war inevitably brought to a simple people who hitherto knew nothing more lethal than a spear, arrow, shotgun and rifle. Thousands lost their all in the ebb and flow of advancing armies, and in consequence were forced to subsist on what could be found in the jungle...

'It is not known how many natives lost their lives through causes directly attributable to the war, but when the disruption of their daily existence is remembered it will be surprising if the number is not high.

Owing to military exigencies, it became necessary early in 1942 to place under the military commander, control of the Territory of Papua and that portion of the Territory of New Guinea not then occupied by the enemy. So, on 12 February of that year, the Administrators of Papua and New Guinea were relieved from exercising the duties of their offices, and the Judges and civil services of the Territories were suspended by National Security Regulations until such time as the suspension should be removed by proclamation.

'This resulted in a cessation of commerce and industry. The resources and establishments of the two administrations passed to the Army; all fit males of military age were drafted into the Forces; while males who were unfit or above military age were repatriated to Australia.

'A military unit called the Australian New Guinea Administrative Unit was then formed, and henceforth administered the Territories according to the Australian ideal as far as was commensurate with military needs...'

From *Territory of Papua Report for the Period 30 October, 1945, to 30 June, 1946.*

When the Pacific war engulfed Papua New Guinea, the towns of Lae and

1. Lockheed Lodestar VH-CAA, operated by Qantas for U.S. Army, Dobodura, December, 1942. Hudson Fysh (hat) and Sergeant Bird. *Photo: Qantas.*

Salamaua, the principal centres of the aviation industry, were totally destroyed. Most of the aeroplanes of the Morobe Goldfields fleet were wiped out. The aviation industry, one of the most efficient in the world, collapsed overnight.

When the Japanese struck, internal services were being provided in PNG by the industry leader, Guinea Airways Limited; by that company's principal competitor, Mandated Airlines Limited, and by Stephens Aviation Limited and the small operators like Kevin and Keith Parer and Stan Johnson. Many earlier organisations and individuals had fallen by the wayside, or had been absorbed by other companies. The most significant of these operators had been Holdens Air Transport and Pacific Aerial Transport. The chief aircraft types employed by the companies at the time of the Japanese invasion were Junkers G31 and W34; Ford Tri-Motor; Stinson Reliant; Boeing 40-H-4; and a number of de Havilland types: DH50A, DH60 Moth of several different models, DH66 Hercules, DH83 Fox Moth, and — most important of all — the DH84 Dragon.

An aerial service between Australia and PNG was begun by W.R. Carpenter Airlines on 30 May, 1938, under a contract with the Australian Government, expiring in May, 1943. The terms of the contract provided for the payment by the government to the company of a subsidy of one shilling per mile for one return flight between Sydney and Rabaul per week, an amount of approximately £9,654 per annum. DH86B 'Express' four-engined aircraft were initially employed on the route.

The achievements of the PNG aviation industry in heavy air-lifting during the early 1930s had amazed the fledgling aviation world. The development of the Morobe Goldfield and the exploitation using giant dredges of the rich gravels of the Bulolo valley were entirely due to aviation, and in particular to just four aeroplanes, the big three-motor Junkers G31 freighters operated by Guinea Airways. But perhaps the finest achievement of the pilots, engineers, aeroplanes and administrators of that vanished era — the 1930s seem so far away, yet fifty years is but a blink of time — was the evacuation of the white women and children of PNG before the Japanese menace and the later evacuation of most of the civilian population at the last possible hour, a task carried out with the magnificent aid of Qantas Empire Airways, Australian National Airways, and the Royal Australian Air Force.*

Of the four Junkers G31 freighters, three were destroyed by Japanese aerial attack on 21 January, 1942. The fourth was impressed into the RAAF, as were most of the handful of surviving civil machines. Many of the pilots, engineers and administrators joined the armed forces; some were killed in action.

The W.R. Carpenter Airlines service between Sydney and Rabaul was suspended when the Japanese attacked. Prior to this, the company's DH86 aircraft had been taken over by the RAAF, and Carpenters had maintained the service using two second-hand, but modern and efficient, Lockheed 14H Super Electras. After the suspension, the two Lockheeds were employed under RAAF direction in the evacuation of the civilian population. One was lost. The surviving Carpenters Lockheed, VH-ADT *Coronia* was eventually purchased by Qantas.

Qantas' involvement with PNG aviation began in September, 1939, with the outbreak of the Second World War. The company had been operating within Australia since November, 1922, and from December, 1934, as an overseas operator in partnership with the great British concern, Imperial Airways, on the Australia-England route, employing DH86 aircraft and then the splendid Short C-Class Empire flying boats. After war was declared, five of the Qantas Empire boats came under RAAF control. Qantas' flying staff were already members of the RAAF Reserve — as were the staffs of the other civilian airlines. In September, an RAAF Flying Boat Squadron, No 11, was formed and hastily equipped with two of the Qantas Empire boats, and two Seagull Mark V amphibians. The squadron was under the command of Flight-Lieutenant Jim Alexander. Fifteen Qantas captains, first officers, radio officers, engineers and ground staff joined the squadron, which also contained sixteen permanent Air Force members. The squadron left for Port Moresby on 25 September, 1939, their mission 'to institute a comprehensive coast watching and communication organisation; to investigate the establishment of advanced bases, and to extend reconnaisance and intelligence'. This squadron, so curiously equipped with the only machines then available, its personnel from such differing walks of life, quickly became known as 'Alexander's Ragtime Band'.

During the remainder of the war years, Qantas aircraft, and those of ANA and

*For further details of the war period, see the author's *Wings of Gold*. Part of the ANA war story is told in *Wings of Tomorrow* Clive Turnbull, F.H. Johnson Co, Sydney, 1945. *Front-Line Airline* E. Bennett- Bremner, Angus & Robertson, Sydney, 1944, tells of Qantas' role in the war. The official history of the RAAF contains a mass of material on the work of civil operators in PNG during the war. See also *Catalina Chronicle*, David Vincent, Catalina National Committee, Adelaide, 1981, and *Qantas Empire Airways (Western Operations Division) Indian Ocean Service 1943-1946*, Barry Pattinson and Geoff Goodall, Aviation Historical Society of Australia, 1979.

Guinea Airways, were employed on charter flights and special operations to and within PNG on behalf of the military authorities and both the Australian and United States governments. The most cursory account of these operations would require another book. Particularly notable was the evacuation of seventy-seven men and one woman from Mount Hagen to Horn Island in two ex-Qantas DH86 aircraft in May, 1942. This operation, hazardous in the extreme, was commanded by Qantas captain Orm Denny, and flown by Qantas crews. Two well-known Papua New Guinea identities, pilot Tommy O'Dea and Catholic missionary Father John Glover, were with the rescue party.

The most intensive use of impressed and chartered civil aircraft was made in November-December, 1942, during the bloody Buna-Sananda-Gona campaign. This was the day of the 'Bully Beef Bombers', the unarmed Douglas and Lockheed transports which supplied the forward troops by airdrop under conditions of great peril, and later evacuated the wounded to Townsville.

The Lockheed Lodestar transports used in the Buna and other campaigns were originally Dutch-owned, flown out to Australia with other surviving aircraft from the Netherlands East Indies just before the Dutch surrendered to the Japanese. In time of war, modern transport aircraft were almost literally worth their weight in gold. The Lodestars were therefore purchased by the United States Government, and late in 1942 they were allocated to Australian companies by the Allied Directorate of Air Transport (ADAT). Qantas got three, ANA five, and Guinea Airways two (plus two Lockheed 14's).

The agreement between the directorate and the three airline companies called for the Lockheeds to be operated on a charter basis for ADAT. Four eventually crashed, and the remainder were over a period of time returned to the United States Army Air Force. Qantas used Lodestars to inaugurate a chartered supply service from Brisbane to Townsville and on to Port Moresby. One, VH-CAB, crashed near Ward's Strip on 26 November, 1943, killing the pilot, Captain W.G. Campbell, and his three crew and fourteen passengers. The Lockheeds were replaced by DC3 and C47 transports, operated under the same contract conditions.

To assist civil aviation companies, the Department of Air began to release volunteer pilots and navigators to the Reserve as the tempo of war in PNG began to ebb. Ninety pilots and eight navigators eventually joined the companies, and some of these flew in PNG after the war.

During all this intense activity, the single surviving Carpenter Airlines Lockheed 14, VH-ADT *Caronia* was employed with company pilots and groundstaff under charter to the American authorities, to the Australian Army, to Ansett Airways on charter work for the American Army, and to ANA. This followed a brief and unsuccessful attempt to resume the contract airmail service to PNG suspended when the Japanese invaded.

The outbreak of war had almost entirely destroyed the Carpenter aviation operation (the New Guinea internal company, Mandated Airlines, was Carpenter-owned). The Lockheed 14 was all that was left, and Carpenters were anxious to resume their Sydney-Papua New Guinea service as soon as they could. Carpenters took the view that the service had been suspended, not terminated, by the Japanese invasion, and protested vigorously when their Airline Licence No. 1 was cancelled by DCA.

The company did not give up. They were determined to revive the service; the airline licence may have been cancelled, but they held a legal contract with the Australian Government which did not expire until May, 1943. There was no provision in the contract for variations in the event of war.

In January, 1943, the company made a formal approach to the Director-General of Civil Aviation, A.B. Corbett, for approval to resume the PNG service, between Sydney and Port Moresby (Rabaul being in the hands of the Japanese). The original contract still had four months to run. Carpenters proposed a twice-weekly service, employing their single Lockheed 14. Corbett referred the matter to Australian Army authorities; they supported the application, on the grounds that additional services to Port Moresby would be of considerable value to them.

Corbett recommended to the Minister for Civil Aviation, A.S. Drakeford, that Carpenter's application be granted. The company was by now something of a nuisance to DCA, with a valid contract, only one aeroplane, no route facilities and a single base, at Sydney. In time of war, no serviceable aircraft could be allowed

2. Captain Orm Denny, November, 1942.
Photo: Qantas

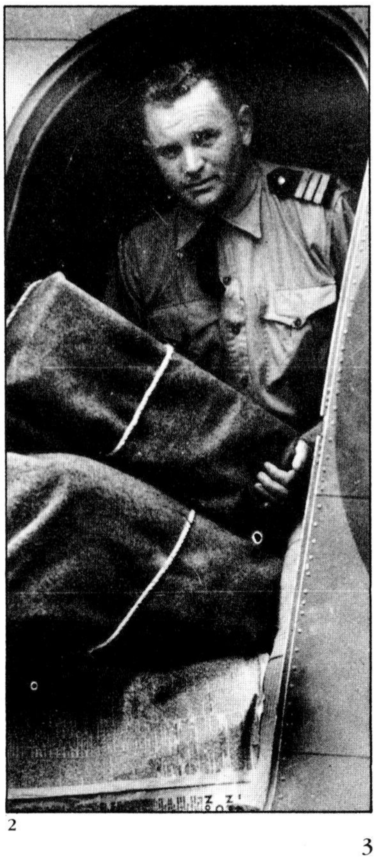

2

to remain idle. The government could certainly have taken over the Lockheed and allocated it to one of the operating companies and this course was considered. But Carpenters had undoubtedly suffered heavily, and the Director-General was loath to penalise them further. A resumption of the PNG service would be a nice way out of the problem. Ample military loading was available, and the aircraft was suitable for the proposed service. Corbett recommended that W.R. Carpenter & Co. be allowed to resume the contract service once weekly at the original subsidy rate of one shilling per mile, with the right to run an additional weekly trip without subsidy. Mails would have to be carried on all flights in accordance with the contract, but if two weekly services were run, the company would not be required to carry mails exceeding a total of about 18 tonnes per quarter.

Minister Drakeford gave his consent. Carpenter's airline licence was renewed, for a period of twelve months, from 9 February, 1943, the planned date of resumption of the service. On 10 February, the Director-General issued his Circular No.48, distributed widely to civil and military authorities, including those of the United States. This gave full details of the service. Under the National Security Regulations, authority had been delegated to the Director- General to grant priority reservations on aircraft within Australia, and this authority was retained for loadings from Sydney to Port Moresby. On the return flight to Sydney, loadings would be controlled by the Army Movements Control Officer at Port Moresby.

There was an immediate reaction to the circular. The RAAF was particularly concerned. Air Vice Marshall G. Jones, chief of the air staff, wrote to Minister Drakeford. He pointed out that all air transport then operating between Australia and New Guinea was under the control of the Director of Air Transport, 5th US Air Force, from his Brisbane headquarters. 'It would be unsound to pay subsidy for and to operate a regular civil air service parallel with but independent of the present air transport control which is responsible for meeting the requirements of all Allied Forces in New Guinea,' Jones wrote. Aircraft then flying to and from New Guinea were sufficient in number to meet present requirements. Jones also raised the question of Commonwealth liability, should the Lockheed be destroyed by enemy action.

A.B. Corbett made an immediate reply, expressing his surprise that an additional aircraft on the Sydney-Moresby route should be considered unnecessary. At the very least, over 18 tonnes of mail per quarter would be carried, free, under the contract with Carpenters, easing the load on military air transport. Moreover, Army authorities had assured him that they were unable to get all the air transport they needed to and from Port Moresby. He continued: 'Carpenter's aircraft leaves Cairns and Moresby strictly under control of the RAAF officers. In that respect, it operates under the same arrangements as Qantas Flying Boats or ANA and Qantas aircraft operating for USA'.

This RAAF opposition killed the Carpenter service. The inaugural flight to Port Moresby was made as scheduled, on 9 February, 1943. Success depended on RAAF cooperation, particularly the provision of petrol and facilities at Ward's Strip, at company expense. On one occasion, a flight was delayed at Moresby because of the refusal of the Air Force to supply the necessary fuel. Carpenters sought the assistance of Colonel Elsmore, Director of Air Transport, Allied Air Forces, and then despatched a telegram to the Director-General of Civil Aviation, claiming that Elsmore had authorised the continuation of the service. Upon the receipt of this assurance, Corbett allowed the company to keep flying.

The service, though, was doomed. Not only the RAAF but the American Air Force authorities were, rather naturally, not keen to have a single civil aircraft operating into a war zone. The inevitable occurred. On 1 March, 1943, DCA advised Carpenters by telegram: 'Early discontinuance your service to Moresby expected.' The last flight was made on 15 March with the return of the Lockheed to Sydney.

For most of the remainder of 1943, the Lockheed was employed, under direction, by Ansett, and then by ANA. In 1944 the aircraft was leased by Qantas.

One must admire the tenacity of W.R. Carpenter & Co. The management of the company refused to accept the apparently inevitable. On 15 September, 1944, Carpenters again applied for permission to resume the New Guinea service, to Finschhafen via Moresby, noting that conditions in the Territory had 'considerably changed' and early resumption of civilian and business activities 'can shortly be foreseen'.

Indeed, PNG was now a backwater in the Pacific war. Considerable Japanese forces were still bottled up in a number of localities and Rabaul was still a fortress in the hands of the enemy — but a beseiged fortress, pinched off and largely neutralised. The tide of war had rolled on, towards the home islands of Japan. Allied forces in New Guinea — mainly Australian by now — were engaged in dangerous, bloody and probably unnecessary mopping-up operations, but there was no longer any possibility that the Japanese could regain the initiative. The end was certain: it was simply a matter of time.

The Department of Civil Aviation had a delicate role to play as the war drew to its close. During the war years, the national emergency demanded the suspension of many civil liberties taken for granted in times of peace. Where necessary, military requirements took priority in most fields, including civil aviation. But the Department was keen to restore normal services as rapidly as the military situation permitted. On 7 July, 1944, the Director-General — who was now D. McVey — advised the Department of the Army that his Minister, Drakeford, wanted a civil airline service to be established immediately between Brisbane, Port Moresby and Finschhafen, to be operated by Qantas. The Commander-in-Chief of the Army, General Blamey, was 'strongly opposed' to the idea, and the matter was reluctantly dropped.

When, two months later, W.R. Carpenter & Co made their application, the Director-General supported the company. 'This Department's view is that it is in Australia's peace-time interest that services be resumed as soon as military considerations permit'. But PNG was still under military control, and the question was once again referred to the Department of the Army.

This was the final attempt by W.R. Carpenter & Co to resume their contract service to PNG. Negotiations had been under way for some time with Qantas, for the acquisition of the air transport interests of Carpenters, including their interest in the Sydney-Rabaul service, aircraft, plant, spares and buildings. The takeover was supported by DCA.

The deal was finalised on 16 October, 1944. For the sum of £44,000 — actually, 40,000 Qantas shares — Qantas took over the Carpenter aviation interests, including the faithful old Lockheed 14 with six spare engines, but not including Mandated Airlines which was a separate company, though Carpenter- owned. In a letter to Director-General McVey, Hudson Fysh, the Qantas managing director, formally advised the completion of the transfer, and applied for permission to operate the PNG service on the Carpenter licence. He concluded: 'It is understood that permission of the Military Authorities to operate a civilian service into PNG is still withheld, but it would be of of invaluable assistance to us in planning if you were able to advise permission for us to operate on Carpenter's licence when the Military Authorities allow implementation of our plans.'

The Department of Civil Aviation had previously indicated to W.R. Carpenter & Co that their airline licence would be reissued once the authorities agreed to the re-establishment of a civil service to PNG and McVey now advised Fysh that the undertaking could be regarded as applicable to Qantas, in view of the takeover.

The way was thus clear for the resumption of the PNG service from the standpoint of the civil authorities. General Blamey had other ideas. His decision was that an 'operational necessity still exists for the prohibition of civil air lines into New Guinea'. Qantas sought political support, and in the meantime submitted details to DCA of a proposed schedule of three flights per week between Sydney and Finschhafen — then one of the largest military bases in PNG — with calls at Brisbane, Townsville and Port Moresby, and Cairns and Salamaua as possible optional stops. Extension of the route to Rabaul via Madang would be made as circumstances permitted. It was proposed to use either Douglas DC3 or Lockheed Lodestar aircraft. The possibility of a flying-boat service between Sydney, Townsville, Port Moresby and perhaps Finschhafen was also mentioned.

In anticipation of a strong demand by civil operators for suitable transport aircraft when peacetime services were resumed, the Commonwealth Government had acquired twelve C47 aircraft — the military transport version of the DC3 — under Lend-Lease arrangements from the Americans in 1944. The aircraft were in good condition, but were taken over on an as-is basis. Six of the C47's were allocated to Australian National Airways, and three each to Qantas and Guinea Airways (who were still operating within Australia). Designed for the carriage of large and heavy items of cargo, the conversion of the C47's to DC3 standard involved major structural changes to the fuselage to remove the large military cargo doors, the reworking of the floor supporting structure, and the installation of toilet and buffet compartments and front and rear baggage compartments. The existing military radio equipment was unsuitable for airline use, and new Bendix equipment was obtained by DCA. The C47 interiors were not soundproofed, and interior trim, ventilation and cabin seating had to be provided. The work of converting the aircraft under DCA supervision was done by the companies themselves — by Qantas at Archerfield, Guinea Airways at Parafield, and ANA at Essendon and Mascot.

The first C47 to be converted, by ANA, was completed in September, 1944, and was registered as VH-AEP. The first Qantas conversion, VH-AEU, was completed in February, and was initially operated by Guinea Airways.

While the work of conversion was proceeding, negotiations with the military authorities continued. Qantas was 'terribly anxious' to get the service started, wrote acting general manager H.H. Harman on 13 January, 1945. On 10 February, Director-General McVey advised Qantas that the C47's would be available for use 'in some four or five weeks'.

On 19 February, the Secretary of the Department of the Army advised McVey that General Blamey had given 'further consideration to the military questions involved in the light of the changed operational situation in New Guinea'. He now considered that there was no military objection to the establishment of a civil air service, but as far as Lae only, not to Finschhafen.

One further hurdle had to be overcome. In individual cases of the return to PNG of civilians where no objection was raised by the Army, the decision lay with the Minister for External Territories. The entire question of the return of civilians was under consideration by Cabinet. Already planters, businessmen, missionaries and individuals compulsorily evacuated from the Territories when the Pacific war began, were starting to clamour for permission to return, and as 1945 progressed the uproar increased. The decision was made to allow Qantas to commence the service, but under stringent conditions, outlined by J.R. Halligan, Secretary of the Department of External Territories in a letter to the Director-General of Civil Aviation, dated 27 March: 'With regard to the carriage of passengers by this service to the Territory, it is advised that civilians may not enter the Territories of Papua or New Guinea unless approval for their entry has been given by the General Officer Commanding, First Australian Army, and they are in possession of a permit issued by this Department. The Company, therefore, should not accept as a passenger any person who is not in possession of a permit to enter Papua or New Guinea.'

All bookings to PNG were under the priority control of the District Superintendent, Department of Civil Aviation, Mascot. Civilian passengers had to have a valid permit to enter, and an income tax clearance certificate. Passengers returning to Australia were subject to customs formalities, and to quarantine requirements.

The Department of Civil Aviation's attitude was that Qantas had taken over the Carpenter airline licence to operate the new service, but not the contract. A new contract, with no provision for the payment of a subsidy, would have to be negotiated. The Sydney-Papua New Guinea service was envisaged by DCA as the first sector of a contemplated service to the Philippines and Hong Kong.

There were some additional problems to be ironed out, particularly the question of Qantas' right to carry passengers within Australia on the Sydney- Port Moresby section — for example, between Sydney and Brisbane. Australian National Airways, the internal operator, protested vigorously, holding that it would be unfair competition, a position which DCA was inclined to support. Captain Ivan Holyman, of ANA, was determined to protect his airline's interests — a stand instantly appreciated by all in the aviation industry, in that day and this.

In March, 1945, the Qantas operations manager, Captain Lester Brain, and the traffic manager for Queensland, Harold Hindwood, briefly visited PNG to make preliminary arrangements for the commencement of the new service. A converted C47, VH-AFA, was expected to be available for a test flight on 19 March, and another a fortnight later. The service would terminate at Lae, where Harold Hindwood would be in charge of the extremely modest Qantas headquarters.

The Qantas service from Sydney to Lae began on 2 April, 1945, with VH-AFA, on a once-weekly basis. The round trip occupied three days. The route followed was from Sydney to Brisbane, Rockhampton, and Townsville, where an overnight stop was made. On the second day, the aircraft flew to Lae via Cairns and Port Moresby. The return to Sydney was made in one long haul, calling at all stopping places. The service was not subsidised, but payment was made for the carriage of mails at the rate of .055 pence per pound-mile. The adult return fare, Sydney-Sydney, was £66/12/-.

The service, which was immediately successful, was increased when the second C47 became available to two flights per week from 27 April, and to three from 16 July. It became known as the 'Bird of Paradise' service, and it was one of the longest scheduled DC3 routes anywhere in the world. During the first twelve months, the converted C47's flew more than 650,000 miles [1,046,110 kilometres], carried 7,000 passengers, forty-three tons [forty-four tonnes] of mail and about the same quantity of freight.

One of the original four Qantas captains on the Bird of Paradise service was D.M. Elphinstone, who had flown in New Guinea pre-war. Early in 1939, young Doug Elphinstone, with a new commercial pilot's licence, had been trying to scratch a living giving joyflights at Mascot. One day he met the well-known Guinea Airways pilot, Ken Garden — later killed in the crash of a Liberator bomber. Garden was about to leave Guinea Airways, and he suggested to Elphinstone that he go and see the company manager, Eric Chater, who was in Sydney at the time, about a job in New Guinea. In 1983, Elphinstone remembered those days, vividly:

'It sounded like a hell of a good idea, so I went and saw Eric Chater, and the next thing I knew I was up to Port Moresby and across in a Ford Tri-Motor to Lae with Tommy O'Dea. I was then taken for a cruise to do a few trips around the place, and then sent up to Wau to fly the top 'dromes, in Fox Moths. Most of the time I spent on the top 'dromes — Slate Creek, Watut, Sonia, Roamer and the rest — and I finished up on Fords. I flew the Stinson Reliant a bit, too.

'One thing sticks in my memory as a lucky escape. I was flying a DH60 Moth, and went in to see Charlie Duchatel, who was mining in the Watut. His place was about a mile from the strip, so I stopped the engine and walked up to see him. The way to stop a Gipsy Moth was to cut the switches, then open the throttle. You had to be careful to shut the throttle before re-starting. That's what I did — swung the prop, and started with the throttle wide open. As soon as the engine fired I realised what I'd done. I raced for the cockpit as she started to roll, but only got as far as the wingtip. I grabbed that wingtip hard, and around and around she went, weather-cocking around. I had petrol for three hours, and I sure couldn't run around in circles holding that damned wingtip for three hours!. The only thing to do was to straighten her up, and run like hell for the cockpit. That's what I did. I made it, just managed to grab the side of the cockpit and pull off the throttle...'

In 1941, Doug Elphinstone left Guinea Airways and tried to join the Royal Australian Air Force. He was rejected on medical grounds, and went to Canada to join Ferry Command. Later, he joined Qantas, flying the chartered Empire flying boats on the New Guinea run. He was a logical choice for the new Bird of Paradise service.

'There were four of us captains, four first officers, four radio operators, four pursers. This was when we were flying two services per week. Arthur Jacobsen was one of the other captains...can't remember the others now. One captain would fly the service up to Lae and down to Townsville, and stop off there until the following service when he would fly the thing back to Sydney and north again, to relieve the second captain. We were on the go all the time...'

Elphinstone left Qantas in September, 1945, but he was not yet done with Papua New Guinea.

Those who flew the grand old DC3's during the early days of the Bird of Paradise service will never forget the experience. The DC3, king of propeller-driven airliners and one of the few truly immortal aeroplanes in aviation history, was slow, quite noisy, and rather cramped. Only limited supplies of food and drink could be carried, and the passenger baggage allowance was but 16 kilograms. After the long hours in the air, ears pounded by the roar of the Pratt and Whitney engines, one would emerge on the ground stiff and travel- weary, but never in any doubt that one had been flying in a real aeroplane.

The civil air link between Australia and PNG was thus restored, before the Second World War came to its end.

3

4

5

6

7

3. A Qantas Empire Airways flying-boat being serviced, Rose Bay, 1939. *Photo: Qantas*
4. Qantas DH86 being serviced at Archerfield, 1936. This aircraft was used by Qantas in the Mt Hagen evacuations of May, 1942. *Photo: Qantas*
5. USAAF cargo plane dropping supplies to the 37th Army Division on the Bougainville airstrip. *Photo:* Pacific Islands Monthly
6. Lockheed 10A Electra, the first American-built aircraft owned by Qantas. This aircraft assisted in the New Guinea evacuation at Dobodura in December, 1942. *Photo: Qantas.*
7. Qantas flying-boat taking off from Groote Eylandt, 1939. *Photo: Qantas*

The Provisional Administration

In July, 1945, the *Papua-New Guinea Provisional Administration Act 1945* was passed by the Commonwealth of Australia. It was proclaimed to commence on 30 October. On that date, civil administration was restored to Papua, and to that portion of New Guinea lying south of the Markham River. The transfer from military to civil control was progressively extended to the rest of the mainland, and to Manus, New Britain, New Ireland and Bougainville, by 24 June, 1946.

The Act provided for a single administration for the two former Territories of Papua and New Guinea, and for the appointment of an Administrator. Of necessity, Port Moresby would be the seat of government; it was the only important centre in PNG to survive the war.

Australian policy in regard to PNG had taken an abrupt change. The events of the Pacific war had had a profound effect on Australians. There were few families who did not have a son, a brother, an uncle, a father, or a friend serving in the armed forces in the New Guinea theatre during the war. Servicemen knew of the debt they owed to the native people of PNG; to the carriers who maintained supply lines to forward areas, to the bearers who brought out the sick and wounded — the 'Fuzzy Wuzzy Angels' in the sentimental phrase of the time. A Labor government had been in power in Australia for most of the war years, and as the tempo of the conflict in PNG slackened and the end was in sight, a broad policy for future administration was evolved.

The principal emphasis of the new policy was outlined in a speech by the Minister for External Territories, E.J. Ward, in July, 1945: 'This government is not satisfied that sufficient interest had been taken in the Territories prior to the Japanese invasion or that adequate funds had been provided for their development, and the advancement of the native inhabitants...the government regards it as its bounden duty to further to the utmost the advancement of the natives and considers that can be achieved only by providing facilities for better health, better education, and for a greater participation by the natives in the wealth of the country and eventually its government.'

The man chosen as Administrator, responsible for the implementation of Labor's plans, was Colonel J.K. Murray, who, as a member of the Australian Army's Directorate of Research and Civil Affairs headed by the enigmatic A.A. Conlon, had helped to formulate the new policies. Murray arrived in Port Moresby on 23 October, 1945. Former Professor of Agriculture at the University of Queensland, Murray had limited experience in administration, and he faced a heart-breaking task. In August, a party of officials had briefly visited Port Moresby to make preliminary arrangements for the restoration of civil government on 30 October, and to determine in principle the method of transfer from military to civil control of the supplies, equipment and services that would be required. But no firm plans for the administration of PNG had been formed.

Much of PNG was a shambles. The country was jammed with military installations, services and equipment, little of it of use to a civil government. PNG had in effect been a vast military encampment during the war. Everything had been subordinated to the military need; everything had been geared to the exigencies of campaign demands. Roads, bridges, wharves, airstrips, services, all had been developed to suit military requirements, usually built to last for a limited period.

There were other difficulties facing Colonel Murray and his Provisional Administration. Most of the essential records of the previous administrations had been destroyed during the war. Many experienced pre-war officers had been killed, or failed to return to PNG. The economy of the country was shattered, although wherever possible the military Production Control Board had kept rubber and copra plantations going, for war purposes. The goldmining industry of the Morobe Goldfield, so vital to the economy of the Mandated Territory during the 1930s had been crippled. Every major settlement, apart from Port Moresby, was in ruins. In the words of the first post-war Territory of Papua Report: 'Where to start was indeed a problem. The Army had everything, the civil authority nothing. It was decided therefore to deal with essentials according to priority...arrangements were made for the Australian New Guinea Administrative Unit [ANGAU] personnel to remain at their posts for a reasonable time or until replacement by civil officials...'

ANGAU establishments and installations at outstations were taken over by the Provisional Administration as staffing permitted. Hospitals were taken over, and district vessels. The ANGAU Marine Section continued to operate coastal shipping, and the ANGAU Signals Section remained responsible for telecommunications. Army, Navy and Air Force postal services were continued, and power stations, water and refrigeration plants remained under military control, until the arrival of civil technical staff. Military buildings were taken over as they were vacated; most had been built for wartime needs using flimsy materials, designed for a brief life of two or three years. By arrangement, the Army supplied the Provisional Administration with stocks of food, motor transport (generally in poor condition), medical requirements and general supplies. The nucleus of a Commonwealth War Disposals Commission was established to facilitate the transfer of items required by the Provisional Administration.

To establish some sort of order from chaos was primarily the responsibility of the Administrator, Colonel Murray. He very rapidly concluded that the rehabilitation of such a war-shattered country was going to depend largely on reliable air transport services.

This need was well appreciated by the Department of External Territories and by the Department of Civil Aviation. In August, 1945, DCA was asked by External Territories to establish services 'as early as possible' between Port Moresby and Milne Bay, Dobodura and Wau, in preparation for the early resumption of civil administration.

Very little information was available to DCA on facilities and conditions in PNG, still at that date entirely under military control. There was no DCA officer in the Territory. Reliable information would be required before any attempt to establish civil services could be made. Val Augenson, Chief Inspector of Aerodromes, was therefore despatched to PNG, to investigate and make recommendations.

Augenson made a thorough inspection, visiting as many districts as weather and the military situation permitted. He found that both the RAAF and the United States Army Air Force were transporting supplies required by ANGAU for the administration of the parts of PNG under their control. This Air Force support was on a considerable scale, and would be difficult to replace when it was, inevitably, withdrawn. The RAAF was sending in three C47's per fortnight to Goroka (spelled 'Garoka' then, and for some time thereafter); occasional C47's to Wau and Kokoda, and a weekly C47 to Aiyura, which airdropped supplies. RAAF and USAAF C47's serviced Mount Hagen each week, and the Americans also supplied Arona.

Augenson found that ANGAU was maintaining a number of small outstations and aerodromes, but because of the lack of suitable light military aircraft, these centres were increasingly difficult to supply. The larger aerodromes were in varying stages of condition, depending upon the extent of their military use. Many new military aerodromes had been constructed. Those in use were maintained by the RAAF, sometimes with Army help.

Augenson concluded that a number of aerodromes were suitable for use by civil aircraft, and in his report he considered and described these. This was the first up-to-date information gained by DCA.

The situation at Port Moresby was satisfactory. There were three parallel strips at Jacksons; No.3 was the best, 2400 metres long by 46 metres wide, constructed of gravel overlaid by pierced steel matting. RAAF radio and meteorological services, field strip lighting and a control tower were available. All buildings suitable for civil air transport use were occupied by the RAAF; Qantas was still trying to acquire land upon which to erect a terminal building. An alternative runway was available at Higgins Field, Cape York Peninsula should Port Moresby be closed because of bad weather.

Ward's Strip base was occupied by a RAAF aircraft repair squadron. In the inner harbour a flying boat base was also maintained, equipped with a good concrete slipway, with tractors, gear and tender and air-sea rescue vessels. The slipway was capable of taking aircraft up to the size of the ponderous Short Sunderland.

There was a fine airstrip at Finschhafen, 1828 metres long with 152-metre overruns at each end. It was constructed of coral covered by steel matting. It possessed field runway lights, and was controlled and maintained by the USAAF, which was expected to withdraw in the near future. The RAAF would then assume control.

Facilities at Finschhafen were among the best then existing in PNG. There was a control tower, elaborately equipped; an operations building; a transport, signals and meteorological building 45.7 metres long and another big general operations building. There were nine huge Butler hangars, each 97.5 metres by 40.2 metres, and ample fuel storage. It was expected that the Americans would leave all of their magnificent equipment at Finschhafen, apart from the Butler hangars. The RAAF was trying to acquire four of those.

Lae strip was 1500 metres long and 39.6 metres wide, gravelled, and suitable for all-weather use. Electric field lights were installed, and the RAAF maintained their usual meteorological, radio and control tower facilities. The old Guinea Airways hangar had somehow escaped destruction during the war, and on its roof was erected a Pundit beacon, pyramid-shaped and outlined in neon strips, operated by a coded timer, which continually flashed the recognition letters, 'LY'.

Dobodura, near the huge Oro Bay base complex, had been developed into a major Air Force base, with lavish facilities. It was 27.4 kilometres from Oro Bay, and already the twenty-five bridges and culverts on the connecting road were deteriorating. No.4 Strip, 3050 metres long and 183 metres wide, was capable of taking the largest existing civil transport types. It would be a useful alternative when bad weather closed Lae.

Of the military airstrips at Milne Bay, Gurney was the best. Sealed with bitumen, Gurney was 1800 metres by 36.6 metres with all usual facilities. There was a flying boat base at Gabuna Bay, with slipway, jetty, and crash- launches. A Pundit beacon was installed on the jetty, flashing the letters, 'MB'.

The Japanese had held on at Wewak until May, 1945, and at Rabaul until August. Both towns suffered severe damage, and Augenson was not able to visit them. Airstrips had existed at both places before the war, and the Japanese had constructed others.

Madang strip was built of coral gravel, 1300 metres by 61 metres again with RAAF control facilities. Madang was still being supplied by regular services by Sunderland and Martin Mariner flying boats, and a jetty, mooring buoys and a rescue launch were available. A Pundit beacon flashed the letters, 'MG'. From the point of view of weather, Augenson rated Madang as one of the best coastal air bases in PNG.

At Nadzab, sitting all on its own in the broad grassed vastness of the Markham valley, was a complex of airstrips; a huge collection of discarded military aircraft was eventually concentrated there. No.1 Strip, 2200 metres by 30.5 metres sealed with bitumen and with 304.8 metres of steel matting at each end, was the best for civilian use, but there were no facilities or buildings of any kind at Nadzab, and the connecting road with Lae, about forty kilometres away, was rapidly washing out. Nadzab was a problem, then, as far as its civilian potential was concerned, and was still a problem when PNG became an independent State.

Augenson did not visit the famous strip at Wau, one of the first to be built in the Mandated Territory, and notorious for its grade of one in twelve. The strip was 1021 metres long and 121.9 wide, all-weather, and had of course been used by heavy aircraft since the beginning of the 1930s and by military aeroplanes during the conflict. The Provisional Administration wanted a priority air service to Wau, but the official DCA view was that Wau strip was not really suitable for modern aircraft; it was suggested that the strip was potentially dangerous, and the Department preferred Bulolo. Bulolo, however, had not been maintained since 1942, when the town was burned and the dredges and machine shops partly dismantled to deny their use to the enemy. The great mining company, Bulolo Gold Dredging Limited, was already in the process of returning to Bulolo, and would without doubt quickly repair the strip. The Administration was anxious to see the gold mining industry restored. During the latter stages of the war, a road had been constructed by Army engineers between Labu crossing, near Lae, Bulolo and Wau. It was to prove of critical importance to the reconstruction of the industry. At the time of Augenson's visit, it was closed beyond Mumeng through landslides, a problem that was to continue for many years.

Before the war, Salamaua strip had been extensively used to service Wau and Bulolo, mainly by light aircraft. Salamaua had been the administrative headquarters of the huge Morobe District. The strip had been heavily bombed, and would have to be reconstructed before it could be used by aircraft. There was also a flying boat alighting area on the west side of the Salamaua Isthmus. But Salamaua had had its day. The future belonged to Lae, soon to become the principal centre of aviation in PNG.

There were, of course, a great many other strips of varying quality in PNG. In the Highlands, there were strips at Mount Hagen — 1097 metres — Goroka (or Bena, as it was also known) — a main runway 1828 metres long and another, 670 metres — Arona — 838 metres — Aiyura, near Kainantu, which the Air Force considered too dangerous to use — and Ogelbeng. Small strips existed at many other places. Of the light aircraft strips in the Waria valley, only Garaina was still operational. The Kokoda strip was used regularly by C47 transports, lifting out rubber for military needs. There was a good strip at Saidor, but it was remote from any likely civilian settlement.

Other serviceable strips were located at Maprik, Hayfield, Daru, Kikori, Awar, Annenberg, Josephstall, Dumpu, Gusap, Kaiapit, Tsili Tsili, Abau, Yule Island, Goaribari Island, Kerema, Tadji, Kiriwina, Goodenough Island, Momote, Torokina, Jacquinot Bay, Emirau, Hood Point and other places.

It was an imposing line-up, but many of the strips were in isolated locations, of indifferent quality, and with no facilities. There was enough to make a start.

Whilst at Port Moresby, Augenson saw the harrassed Colonel Murray, who again called for urgent action to establish civil air services. 'He considered that the first requirement was an air service between Moresby-Samarai, and Moresby- Daru which could make calls at intermediate places. He therefore favoured the use of seaplanes, flying boats or amphibious aircraft. Later such service he thought should be continued right round the coast to Wewak. As regards the inland districts, the first requirements would be flying supplies in by C47 aircraft to Goroka and Hagen, and supply dropping on the strips at Aiyura, Kainantu, and Kerowagi from Lae, and flying supplies into Kokoda and Dobodura from Port Moresby'.

In his report to DCA, Augenson made a number of remarkably pertinent comments and shrewd suggestions. He correctly foretold that Port Moresby and Lae would become the chief business and administrative centres for the south and north coasts of mainland New Guinea. Port Moresby should become the base for marine aircraft services for all of PNG and Lae for land aircraft services. Because of its manifest advantages, Lae — at the gateway to the Highlands and the Morobe Goldfield, with good facilities including land for a township on the terrace overlooking the pre-war settlement and for an 'extensive aerodrome' at Malahang — should become the headquarters of a new DCA region for PNG. He recommended the early appointment of a district superintendent to represent DCA in PNG at Lae, and the appointment of inspectors of airports, radio and aircraft. These men should be attached to existing RAAF operational bases, so that they could operate facilities without interruption after the Air Force withdrew.

Augenson favoured the use of multi-engined aircraft in PNG. Because of the need for both passenger and freight services, he recommended that the RAAF C47 aircraft then being employed for the transfer of ANGAU stores be either transferred to DCA, or sold to civil operators, to ensure the continuity of services. These C47's should not be modified, except perhaps for the provision of moveable passenger seats. The existing big cargo doors would be invaluable for loading bulky freight. He foresaw the continuing need for airdropping supplies to remote locations during the early stages of rehabilitation, and suggested that properly designed wooden ejection chutes be installed.

Undoubtedly, many aerodromes would have to be improved to make them safe for civil operations, and Augenson recommended that a Civil Airfield Construction Unit be formed, with specialised air-borne equipment, including light bulldozers and graders.

To meet existing requirements and prepare the way for the resumption of civil services, Augenson made a number of suggestions. The RAAF should make a Catalina available to a civil operator based at Port Moresby for the services required by Colonel Murray. A C47 should be leased to a civil operator to be based at Lae for Highlands and other flights. A DH84 Dragon or Beechcraft 18 should be made available through Disposals, ex-RAAF, under similar conditions and based at Lae to service Wau and Bulolo and smaller centres.

A great deal of confusion in these early post-war years would have been avoided had Augenson's recommendations been adopted across the board. They were not.

Following Augenson's return from PNG, the Minister for External Territories, E.J. Ward, issued a press statement, dated 26 October, 1945. This formally announced the restoration of civil administration on 30 October, and stated that it was not yet possible to authorise the return of civilians generally to PNG because of the difficulties associated with the supply of stores, accommodation and transport.

'Air transport was vital to the life of PNG,' said Ward. He continued: 'Pending certain decisions in regard to the future operation of air services within the Territory, it has been decided that airline licences will be issued to private operators under conditions determined by Civil Aviation authorities, and applications for such services should be addressed to the Director- General...'

The Director-General was already holding a number of informal applications from individuals and organisations anxious to commence civil air transport services in PNG. No decisions had been made. There were a number of difficulties. Existing DCA policy was not to issue licences for new services save under exceptional circumstances. The whole question of the development of a post-war civil aviation policy was still being fought out by the politicians. The *Australian National Airlines Act* had been passed on 2 August, 1945. This led eventually to the formation of Trans-Australia Airlines, and the adoption of Australia's Two Airline Policy. At issue was the burning question: Who should control domestic air transport between the States, the government, or private enterprise? Three major companies — ANA, Guinea Airways and MacRobertson-Miller Airlines — combined to challenge the Act in the High Court, and the ensuing legal fight went as far as the Privy Council.

All this was in the future when the relatively minor matter of the resumption of internal services in PNG was being considered, but it meant that civil aviation policy was in a state of flux, making it difficult to come to decisions.

The question of the performance of ex-military aircraft likely to be employed by civilian operators — in the absence of anything better — was also causing the Department concern. Little attention had been paid to aircraft performance in PNG operations pre-war. Services then had been operated with the very minimum of DCA control, using whatever types of aircraft that were available. This situation could not be allowed to continue. The ultimate object would be to allow the use of multi-engined aircraft only for passenger services, within PNG as well as Australia. Some latitude would undoubtedly have to be allowed, however, during the early post-war period.

DCA engineers began a close investigation of the single-engined performance of ex-military machines. Multi-engined aircraft could only be considered safer than single-engined if they had a reasonable performance with one engine shut down.

The first new approach, to the Department of External Territories, for approval to commence internal operations in PNG was by the well-known Goldfields identity, C.A.M. ('Boss') Adelskold, on 21 September, 1945. Adelskold advised that arrangements were 'well under way' for the formation of a small company of ex-New Guinea residents for the operation of services from the coast to the interior, using DH84 aircraft. He asked whether DCA would place restrictions on the use of the DH84, and what controls the Department would exercise over operations. Adelskold was referred to DCA, which was not then in a position to make a reply.

On 2 October, M.A. (Joe) Taylor wrote to the Minister for External Territories, outlining his plans to form a 'New Guinea Co-Operative Air Transport' company, with a nominal capital of £10,000 and a working capital of £6000. It was intended to purchase from Disposals a DH83 Fox Moth and a DH84 Dragon. Joe Taylor estimated that each aeroplane would be operated for 1,000 hours per year. Pilots would be paid £125 per month, and engineers half that amount.

The DH84 was the 'only machine available that will fulfill New Guinea requirements,' Joe Taylor submitted. One of the pioneers of aviation in Papua New Guinea, he continued: 'I have been directly associated with the aircraft industry for the past twenty-five years as a licenced engineer, 1927-1928, and 1930 to 1941, I was employed by airlines in New Guinea — the immense power and control exerted by the Companies led me to the conclusion that something of the nature of my proposal is essential to the well-being of PNG residents.'

Joe Taylor's proposals, too, were referred to DCA.

Also in October, W.R. Carpenter & Co wrote from Sydney, on behalf of Mandated Airlines, applying for 'such licences as may be necessary to enable us to operate our airline services between Moresby, Wau, Lae etc and other places in Papua, as soon as same is approved by Military or Civil Authorities'. MAL intended to use DH84 Dragons on these services.

In October, too, Guinea Airways applied through its manager, G.H. Archibald, to DCA for approval to resume its pre-war operations in PNG. 'We made application to the Department of External Territories some months ago,' Archibald wrote. Guinea Airways proposed to employ DC3 aircraft, out of its old base at Lae. As in the case of the other applicants, DCA could do nothing but advise that most of the Territory of New Guinea was still under military control. Performance requirements of aircraft were still being examined. No decision had yet been reached in regard to the DH84.

Following E.J. Ward's announcement of 26 October, a detailed circular was prepared and sent to all persons who had expressed an interest in operating PNG services. Applicants were asked to provide information on routes to be flown; aircraft to be employed with full details of their equipment, including radio; personnel, their numbers, licences and qualifications; facilities to be provided, and spares carried. Operations manuals showing full maintenance proposals were to be submitted. Prior to commencing operations, successful applicants would have to make proving flights, to the satisfaction of the Department. Attached to the circular was a list of aerodromes considered to be suitable for civilian operations, based on the information provided by Augenson, supplemented by submissions from DCA civil engineer, J.A. McCormack.

The Minister for Air, Arthur Drakeford, made a statement explaining the reason for the calling of applications: 'The departure from the policy of not issuing, at present, airline licences for new air services is due to the urgent necessity of providing air transport in areas where civil administration is to be re-established...in order to allow prospective applicants the opportunity of investigating conditions and of ascertaining the requirements of DCA, examination of applications would not be commenced before 30 November, 1945.'

In November, a letter was sent to all applicants. Services were needed to replace those currently being provided by the RAAF, and for the routes required by the Administration. Also required, for general public use, was a weekly return trip linking Daru, Kikori, Kerema, Kairuku, Port Moresby, Kapakapa, Abau, Samarai, Misima, Milne Bay, Baniara, Tufi, Dobodura and Salamaua, with trips as required to the D'Entrecasteaux Islands, the Trobriands and Woodlark. Many of these locations could only be serviced by flying boat in 1945.

Time went by. No decisions were made on the applications received, despite Drakeford's statement. The Australian National Airlines Commission was established by the Act. The attention of the commission, when set up, would be first directed to the vexed question of interstate services. When that aspect was finalised, the matter of services operating to and within Australian Territories — which included PNG — would be considered. This was going to take time.

Pressures mounted. The Minister for External Territories wanted immediate action. The Administrator, Colonel Murray, in a letter to the Minister in November, 1945, emphasised the 'very great importance' of an early establishment of internal air services — of which Ward was well aware. Murray's own movements within PNG were being seriously hampered. 'I am afraid that there is not much possibility of getting assistance of any notable character from the RAAF,' he complained. Murray suggested the formation of a government-owned Territory Air Service. 'At present I cannot get to Daru and back by boat in less than seven to eleven days, depending largely on the weather ... similar conditions exist with regard to Samarai, Misima and the various island groups ... reversion to the old state of affairs of leisurely movement by boat will cripple the expeditious handling of government business...'

In March, 1946, J.R. Halligan, Secretary of External Territories, again wrote to the Director-General of Civil Aviation: 'Air services provided by the RAAF are diminishing rapidly...the position is becoming critical...it is essential civil aviation be restored at the earliest possible moment...most urgent...'

W.R. Carpenter & Co again took up the cudgels on behalf of Mandated Airlines. In February, 1946, the company applied for permits to enter PNG for five men, to 'operate an air service'. They were: A.T. Collins, pilot and engineer-in- charge; J.R. Rose, pilot; E. Serafini, engineer; and two apprentice engineers, R.B. Fuller and B.A. Brearley. The acting Director-General, the veteran Captain E.C. Johnston, replied, saying that the date for the examination of applications for airline licences had not yet been fixed. This could not be done until 'the performance requirements of aircraft are finally determined'. Later that month, MAL sent a telegram to DCA: 'Please grant necessary permission to operate Dragons in New Guinea'. Again the Department replied: 'No decision yet made on limitations of DH84...'

Further pressure developed in February, from Guinea Airways. This company now had a close association with ANA, and was operating successfully within South Australia, its home State. But Guinea Airways had been the giant of pre-war aviation in PNG, and quite naturally wanted to recapture this position. If any one organisation could be credited with the creation of the PNG aviation industry, that organisation was surely Guinea Airways. It was not to be — Guinea Airways was eventually refused permission to resume its PNG operations.

However, in February, the company, sick of waiting for a decision, despatched Captain H.T. Hammond to PNG, to make preparations for a resumption of operations. 'Bunny' Hammond was absolutely the right man for the job. Before the war, he had been in turn chief pilot for Holdens Air Transport Service, and Guinea Airways. The company planned to establish Hammond at Lae, to operate airline services using one of the three converted C47 transports that had been allocated in 1944. A Fox Moth would be used to service the smaller strips.

Hammond inspected the Wau 'drome that was viewed with disfavour by DCA. He wrote to Colonel Murray, saying that the strip was in 'better condition than it has ever been' and quite suitable for DC3 operations. His personal landings at Wau had totalled 'some 2,000'. Guinea Airways was 'most anxious to assist in the rehabilitation of New Guinea', and Hammond asked the Administrator to 'expedite this permission'.

In March, 1946, Guinea Airways advised DCA that Hammond had returned to PNG after reporting his findings. It was intended to send him a Fox Moth. Once again, the company asked for permission to resume operations with the C47.

Meanwhile, the DCA investigation into the single-engined performance of aircraft likely to be available to civilian operators had been completed. DCA engineers found that the performance of some of these aircraft was so poor that to reduce the hazards of operating them, their use would be limited by classifying airline services into three categories, based primarily on density of traffic: Major, Intermediate and Minor. Aircraft considered satisfactory for major services were Douglas DC4, DC3 and C47, Lockheed 18 Lodestar, Beechcraft Dl8S and DH104 Dove. For intermediate services, the Lockheed 10A and 10B, Avro Anson and Douglas DC2 could be employed, under certain limitations. Minor services could be flown by all types, other than single-engined, eligible at maximum permissable weight consistent with Certificate of Airworthiness and the requirements of the airline licence. Aircraft types apart from those mentioned, could only be employed if specific approval was given by DCA. The ban on single-engine machines appeared to rule out the Fox Moth, and also the Dragon, which had a quite unsatisfactory performance on one engine. Indeed, DCA at first were inclined to limit the use of the DH84 to the carriage of freight only. However, it was clear that a special dispensation would have to be made if civil services were to commence, and permission was finally given to employ both the Fox Moth and the Dragon.

In June, 1946, DCA despatched a circular to all airline licence applicants, advising that the way was now clear for applications to be considered. Licences would be issued over specific routes, not for general operations within PNG.

Mandated Airlines, however, was already flying in PNG. The post-war civil aviation story had begun.

8

9

10

8. Madang from the air. The Coastwatchers' Light can be seen right centre foreground. *Photo: Qantas*
9. DC3 — VH-AEY - was loaned to Qantas by the Commonwealth Government and used in PNG in 1946. *Photo: Qantas*
10. Rabaul township as seen from the Volcanological Observatory. The crater in the centre background of the photo erupted in 1937. *Photo: Qantas*

Mandated Airlines

Late in 1945, W.R. Carpenter & Co. were offered the chance of purchasing a batch of twenty-four DH84 Dragons from the Commonwealth Disposals Commission. They were in storage at the RAAF base at Tocumwal, in varying stages of airworthiness, but going cheap. Carpenters eventually put half of them into service with MAL in PNG. The first two were handed over on 31 January, 1946.

This coup enabled MAL to steal a march on all its competitors. The takeover of the Dragons was handled by MAL's manager, Arthur T. Collins, at the W.R. Carpenter premises in Camperdown, Sydney.

Arthur Collins was a veteran of New Guinea aviation. He was employed by Guinea Airways at Lae in the early 1930s as a licenced engineer, and in 1933 purchased a Simmonds Spartan biplane in which to gain his private pilot's licence. His tutor was another well-known engineer-pilot, W.C. Gatenby. Later, Collins joined MAL as an engineer, and after he had obtained his commercial licence, as a pilot. He played a distinguished part in evacuation of civilians after the Japanese invasion. Collins became foreman and manager of the W.R. Carpenter Airlines workshop at Camperdown, and was responsible for the maintenance of the two Lockheed 14 Super Electras. As previously recounted, one of these aircraft crashed, killing both pilots — Joss Crisp and Clive Bernard — and the other was eventually transferred to Qantas.

Soon after he took over the Camperdown position, Collins accepted as an indentured apprentice in aviation and general engineering a lad called Brearley. Bruce Brearley, born in October, 1925, had been fascinated by aircraft from his childhood days. As soon as he was able to toddle, Brearley would sneak out from the family home at Brighton-Le-Sands, cross a sewage farm and hang on the fence that surrounded Mascot aerodrome, watching in awe the comings and goings of the beautiful aeroplanes of that day. Repeated hidings from concerned parents had no effect — he kept on visiting Mascot. Brearley wanted to spend his life around aeroplanes, and in 1942 he eagerly seized the chance offered by Arthur Collins to become an aircraft engineer. He attended technical college at nights, and during the day worked on the maintenance of the Lockheed 14, *Caronia*. When Qantas acquired *Caronia* in the takeover of the Carpenter aviation interests, Bruce Brearley went with the deal.

Arthur Collins remained with Carpenters. Brearley found the experience of working for a large organisation like Qantas not to his taste, and he soon returned to Carpenters.

With no aircraft available, Brearley spent his time working on the repair and maintenance of the crushing plants used by the company in the production of coconut oil, a major activity of W.R. Carpenter & Co. He continued his aviation engineering studies at technical college, but could see his ambition to work around aeroplanes slipping away. Then came a stroke of good fortune, which Brearley recalled in 1984, shortly before his retirement from Ansett Airways, after a lifetime in aviation: 'About the end of 1945, Arthur Collins came and said to me, "You started on aircraft, and you've only done general engineering since we sold to Qantas. How would you like to come to New Guinea with me, and work on Dragons?" I said, 'Wonderful! When do we start!'.

'Carpenters had bought twelve I think it was, may have been thirteen, Dragons from Air Force Disposals. They were going to get MAL started again, in Lae. The aeroplanes were all down at Tocumwal, all in Air Force colours and fitted out for radio and navigational training. Carpenters paid 150 quid for the first one I remember, and fifty each for the rest!

'Arthur, another pilot called Johnny Rose and I went down to Tocumwal in

11. The wreck of the *Tenyo Maru* was a famous landmark near Lae airstrip until it slid off the reef and sank in the late 1970s. *Photo: Mal Shannon*

11

February, and flew back the first two, to de Havilland's at Mascot, where the big Qantas complex is now. We pulled the engines out and they were fully overhauled. The aeroplanes were completely stripped down, everything new put in, re-bagged. Then they were given civil registrations, VH-AOP and VH-AOQ. All this, and getting ready to go up to New Guinea, took weeks.

'It was late March before we were ready to start. We flew up the coast side by side, seven of us in the two Dragons full of spares, our gear and other stuff. Arthur flew one kite, and Johnny Rose the other. There was another apprentice, Ron Fuller; Ernie Serafini, engineer, and Bill Marshall and Wally Coote, who were going to do the paperwork and look after traffic. They didn't stay with us long, though.

'We had to stop to refuell all the way up the coast. We only took about thirty gallons a side in the Dragons, enough for about four hours, at most. So we stopped all the way up — Brisbane, Rocky, Townsville, Iron Range, Cooktown, Horn Island. We didn't stop at Daru. Flew in to Lae and landed, one after another.

'Lae was a real mess. The strip was full of Air Force, a big control tower, just a box stuck right on top of long wooden poles, and all round the balcony a great big sign, IS YOUR IFF ON? We didn't know what this meant, and of course we had no radio in the Dragon. Found out later that it meant IDENTIFICATION FRIEND OR FOE. Lucky they didn't shoot us down.

'We landed, and a great mass of troops of all shapes and sizes came screaming out in boots and shorts. They'd never seen anything like the Dragons. One bloke raced up to our plane and poked it in the side. The plywood fuselage wobbled on the wheels, and he said, "Christ, it's wood!"

'The big old hangar from pre-war was still standing with the Pundit beacon on top, and the *Tenyo Maru* was sticking up out of the water at the end of the strip. We were the first to start up at Lae. Didn't move into the hangar at first, we were camped on the other side of the strip, where the Air Force stores section was. Then we had to move out to the CTC — the Civilian Transit Camp, at the old AWAS Barracks that Ma Stewart made into the Hotel Cecil. What a place. There were only about fourteen civilians in Lae then. BGD was just getting cracking, and Bertie Heath was their beach man at Lae. A few other old timers were back...

'So we settled down, and started flying...'

12. Doug Elphinstone. *Photo: Doug Elphinstone*

12

The first large group of pre-war residents to return to PNG arrived on the M.V. *Ormiston* in March, 1946. Others quickly followed and as the civilian population increased there was plenty of work for the two MAL Dragons. Two more, VH-AOS and VH-AOP, joined the little fleet in May. Two DH82 Tiger Moths, VH-ASW and VH-AQS, were acquired in April. They were dirt cheap. A good Tiger could be purchased from Disposals for as little as sixty pounds, then.

Arthur Collins and John Rose flew long and hard with the first two Dragons, to Salamaua, Wau, Bulolo, Finschhafen, Kainantu, Aiyura, Goroka, Chimbu, Mount Hagen, Madang and Wewak. The first MAL schedule of freight and passenger rates was issued on 24 April. These applied only when an aeroplane happened to be going to the desired destination, or when enough cargo was offered to make up a full load. If not, an aeroplane had to be chartered. Charter rates were listed as £12/10/- per hour for Dragons, with a capacity of 544.32 kilograms and £5 for a Tiger Moth. The capacity of a Tiger was given, optimistically, as 226.8 kilograms.

Another pilot, R. Devenish-Meares, and an engineer, C.A. Haydon, were granted permits to enter PNG in May, to join MAL. The Secretary for External Affairs, J.R. Halligan, advised DCA of the issue of the permits, but asked a pertinent question: 'Has MAL been licenced to operate?'

MAL did not in fact have an airline licence. None had yet been issued for PNG. However, on 2 May, W.R. Carpenter & Co again asked for the issue of a licence, based on the employment of DH84's. 'We are at present operating at Lae with a view to establishing Mandated Airlines Ltd as we were pre-war...'

Not only was MAL operating a charter, passenger and freight service for the general public — small as it then was — but also a regular bi-monthly, subsidised service for the Provisional Administration, quite without the knowledge or concurrence of DCA.

MAL made no effort to conceal any aspect of their operations from DCA. In July, W.R. Carpenter & Co submitted a full schedule of the services it proposed to operate to support the May application for an airline licence. Among these, DCA officers were intrigued to notice, were details of 'Service One, Lae/Madang/Hansa Bay/Angoram/Wewak. At present service is being operated on £50 per month subsidy for bi-monthly service. First Class mail matter carried free, Second Class mail matter ordinary freight rate, and priority for Government passengers and freight'.

DCA representation in PNG was very slim in mid-1946. There were but five officers in the Territory, three at Port Moresby and two at Lae. DCA was in the process of assuming control of the aerodromes at Port Moresby and Lae from the RAAF at this time, and had established a regional control flying centre at Port Moresby, jointly operated with the RAAF. The DCA men in PNG were far too busy to worry about what MAL were doing. No one had reported the MAL subsidised service to headquarters in Melbourne.

The Director-General sought the advice of the Department of External Territories, who had issued the entry permits that had allowed Arthur Collins and his team to enter PNG and begin the service. In due course, J.R. Halligan received an explanation from Colonel Murray, which he sent on to the Director- General: 'When Civil Administration was restored...the shipping and air services that were operated for Services needs gradually disappeared without anything to replace them. The work of the Government in rehabilitating the Territory had to proceed...in this transitional period between the change from Military Occupation communications were bad and something had to be done to meet a pressing need.

'Mandated Airlines Ltd introduced some Dragon aircraft to the Territory and with these machines commenced aerial transportation on a charter basis. To meet the needs of the Administration the Company was approached during May to make two trips per month from Lae to Wewak via Madang each way...'

The arrangement with MAL called for the carriage of Administration passengers and freight on a priority basis at ten per cent less than scheduled rates. Twin-engined aircraft had to be employed, and the agreement was subject to cancellation by either party on one month's notice. The service began late in May, and flights were thereafter made on the second and fourth Fridays of each month. The Administrator pleaded: 'To date it has worked satisfactorily. There is need of this service and for others as well, and it is desired that nothing be done to discontinue it unless it is replaced by something better.'

Colonel Murray went on to note that a seaplane service linking Port Moresby with stations west to Daru, and east to Samarai, the islands and the north coast to Tufi was still required urgently. A service linking Rabaul with faraway Bougainville was required.

The Administrator's plea could not be ignored. The MAL subsidised service was allowed to continue. The fact was that RAAF services had almost disappeared, and MAL was still the only civilian operator. However, fresh applications for airline licences had been received from Qantas and Guinea Airways in response to the DCA circular of June, 1946. Qantas had also been approached by the great Australasian Petroleum Company, with a firm proposition that Qantas should operate charter services for them in Papua, using a Walrus amphibian. APC had recently been granted permission from the Commonwealth Government to resume their pre-war activities. Qantas was willing to operate the required services, if DCA would allow them to.

The Guinea Airways airline licence application was curious in that it was submitted on that company's behalf by Australian National Airways. Should Guinea Airways be refused a licence, then ANA would submit its own application. Captain Bunny Hammond was still at Lae, watching Guinea Airways' interests — without the Fox Moth the company had intended to base there — but it was beginning to look as though Guinea Airways would not be resuming its operations in PNG. Qantas Empire Airways Ltd, a private company, was about to become an Australian Government-owned airline with the purchase of the old fifty per cent Imperial Airways' interest in Qantas from the British Overseas Airways Corporation (BOAC), announced in December, 1946, and the subsequent purchase of local private share holding. In 1947, Qantas became the designated operator of Australia's overseas air services. It would be Qantas that would inherit the Guinea Airways mantle in PNG.

In October, 1946, DCA set up an inter-departmental committee to study the development of new domestic air services, and to make recommendations. In the meantime, official policy was to 'not encourage' the establishment of new services, which once again stalled the issue of PNG licences. MAL continued to operate.

In support of its application for a licence, the company had submitted a proposed Flight Manual. In this was outlined the MAL policy: 'The prime factor in the success or otherwise of and the ultimate aim of all air transport services is SAFETY. Other considerations are subordinate...It is the policy of the Company that the relative importance in flying operations is:-

1. Safety
2. Passenger comfort
3. Maintenance of Schedule
4. Economy

and no attempt is to be made to achieve one object to the detriment of any other...

'All flying operations are under the direct control and authority of the Flight Superintendent...'

The manual laid down rules governing the conduct of pilots, for breach of which the penalty would be 'summary dismissal'. These included carelessness or inattention to duty, flying while under the influence of liquor or 'habit forming drugs', and allowing the carriage of passengers in excess of the number authorised in the aircraft's Certificate of Airworthiness (C of A).

MAL's rules to ensure the competency of its pilots drew upon the experience gained in its pre-war operations in PNG, and substantially similar rules were adopted by other early post-war operators. Before any pilot could be employed on a regular service, he had to have a complete knowledge of the route. In addition, he was required to traverse the route at least four times and land at all intermediate and emergency strips, under the guidance of an experienced pilot. This 'route familiarisation' or 'checking-out' procedure was to be followed by all other air transport companies in PNG, although there were minor differences in details.

The Manual continued: 'All pilots shall do everything possible to increase their technical knowledge...the general standard of flying is to be kept at the highest pitch...all pilots are to undergo a practical test under the supervision of the Flight Superintendent during each three-month period...any pilot failing to prove his ability during a periodical flight test will forthwith be suspended from flying on regular services...'

13. Dick Burt, 1961. *Photo: Sam Jamieson*

13

One clause severely warned pilots against the temptation to perform stunts in the air. 'No stunt flying of any nature whatsoever is to be performed...climbing turns and steep climbs are classified as stunt flying...'

This was a timely warning. The great majority of the early post-war pilots flying in PNG were young men, filled with the spirit and optimism of youth. They were, moreover, RAAF-trained, many in high-performance, single-seat fighters. The job of piloting a service aircraft was relatively straightforward. Commercial flying was more complex, requiring a modified technique, a knowledge of flight engineering and flight control, and of air navigation legislation. Commercial pilots needed to know the practical operation of aircraft radio communication and navigation equipment — even though such aids were at first non-existent in PNG. To remedy these deficiencies in the skills of many service pilots aspiring to careers in civil aviation, DCA had in 1945 begun a series of intensive training courses that were attended, eventually, by hundreds of pilots, some of whom flew in PNG.

In August, 1946, a man who was the archetype of the dashing ex-RAAF pilot joined MAL. Richard Cecil Turner Burt was, however, older than most — but as bold as any. Born in Hobart in 1909, Dick Burt gained his private pilot's licence before joining the Air Force on the outbreak of war. All Dick Burt wanted to do was fly combat missions, in any type of aircraft the Air Force wanted to give him, fighters or bombers, singles or twins. It was all the same to Burt. However, in those critical days at the beginning of the war the RAAF badly needed to train pilots, as many as possible. Burt was already a pilot, and he was considered too old for combat flying. To his chagrin, he was posted to Narrandera in New South Wales, as an instructor. He was still disgusted when he spoke of those days, in 1983: 'I put in a year down there. I hated every minute it. Didn't like being an elementary flying instructor, not cut out for it. Didn't have the patience, for a start. Then I went to Evans Head, and I put in another year there, commanding the Towing Flight. Then down to Sale, and from there to No. 5 Aircraft Depot, outside Wagga. I was a test pilot there, testing aircraft after they had been repaired or re-engined. Flew Lockheed Hudsons, Beauforts, Beaufighters, Kittyhawks, Wirraways, Spitfires. I piled up quite a bit of time on the Lockheeds. Then I went to Bankstown, to the Test and Ferry Flight...'

It was at this time that the decision was made to release a number of volunteer pilots for airline service. Qantas wanted men experienced on Lockheeds to fly the machines leased from the Americans. Dick Burt was offered a job with Qantas. Convinced by now that he was not going to be allowed to fly in combat — he never stopped making application for transfers — he accepted. He joined Qantas in December, 1943, and commenced to fly Lodestars on the PNG courier run.

'I was an almost instant captain. I had a lot of time up on Hudsons, and the Lodestar was just a longer version, but better, more stable. Had tail brakes instead of differential brakes. The Hudson used to sort out the men from the boys, with that very short fuselage. Had to be handled properly. They could bite you very quickly if anything went wrong, if you didn't know what you were about. The Lodestars were actually owned by the US Government. Qantas supplied the captain and did the maintenance. The second pilot and radio operator were Air Force sergeant-pilots and sergeant-radio operators. After a while we got two brand new C47's. Lovely aeroplanes...'

Dick Burt was a man of independent mind, who did not take kindly to the rules and regulations that hedge and sometimes make miserable the lives of mankind in this modern world. One of the regulations governing the use by Qantas of the Lodestar and C47 aircraft during the war was that all passenger seats were allocated by the American authorities. In June, 1945, Burt gave an unauthorised lift to a RAAF squadron-leader, from Darwin to Sydney, to oblige a friend. Unluckily for Dick Burt, an American courier was travelling on the aircraft, and he reported this breach of the rules. Burt was suspended by Qantas, and he thereupon resigned. He joined the famous Dutch airline, Koninklijke Nederlandsche Indische Luchtraart Maatachappij, for obvious reasons always referred to as KNILM, the company that had originally owned the Lockheed Lodestars that had been flown out of the Netherlands East Indies when the Japanese invaded.

Dick Burt flew Douglas C47 transports for KNILM until July, 1946. Later that month, he visited the headquarters of W.R. Carpenter & Co in O'Connell Street, Sydney. Arthur Collins was there, having travelled to Australia to fly back another newly-overhauled Dragon, VH-ARI. Says Burt: 'I asked about a job, and Arthur Collins said, "Yes, flying Dragons. Can you fly a Dragon?" I said I could. I flew them at 5 Aircraft Depot at Wagga. They had them there as navigation kites. When I had nothing to do, I flew them. So I got the job. Arthur wanted me to fly back another Dragon that was ready to go, VH-ARJ. But I had just bought a Tiger Moth from Disposals, for sixty-five quid. It was at Western Junction, just outside Launceston. I said I would have to fly it back, first. DCA weren't very happy about me flying a Tiger across Bass Strait, they said I could go any time or any how after they heard I was an ex-Qantas captain.

'So I flew up to Lae in ARJ with a load of passengers, a week after Arthur Collins had gone back in ARI...'

No pilot enjoys being involved in an aircraft accident. Quite apart from all other considerations, wounded pride and a dented ego result. Dick Burt had his full share of both qualities, in common with many men who become professional pilots. Pride in one's ability, and in one's profession, are hallmarks of the successful aviator. It was galling, then, to Burt when he had the first major aircraft accident to occur in PNG since the war, in the just-delivered Dragon, VH-ARI. A great deal was to flow from this crash.

Scant weeks after his arrival in PNG, Burt was sent to Wewak to make a series of charter flights to Maprik and Angoram. His Dragon, ARI, was in splendid condition, having operated for only seventy-eight hours since its flight from Sydney to Lae. The aircraft had in fact only flown a total of 780 hours since manufacture. Port and starboard engines had done 785 hours (engines were commonly tested on test-beds after being removed from the aircraft, which accounts for the difference in hours). Prior to departure from Lae on 30 August, 1946, a Certificate of Safety had been issued covering VH-ARI, signed by engineers Arthur Collins (airframe) and Cliff Haydon (engines). This Certificate was issued at 7.30 a.m. It expired twenty-four hours later.

Refuelling arrangements at Wewak — and everywhere else in PNG apart from Lae and Port Moresby — were primitive in the extreme. Vacuum Oil Company was the supplier of aircraft fuel, and this was delivered to aerodromes in 200-litre drums. A hand pump was used to fill the aircraft's tanks. Prior to refuelling, the pilot was supposed to pump out about 9 litres of fuel from the drum onto the ground to see if water — greatest enemy of the internal combustion engine — was present. The only two Vacuum water-trap funnels owned by MAL were in constant use, at Lae and Moresby. Repeated requests by Arthur Collins to Vacuum to improve outstation fuelling facilities had not produced results. Pilots operating away from base would, if prudent, drain the aircraft's tank sumps or filters to make quite sure that the fuel was uncontaminated before taking off.

Discussing this accident years later, Dick Burt said: 'I drained both tanks for water prior to take-off, but apparently there was still water in one. I had five passengers on board. I took off without trouble, but the instant we were off the starboard engine began misfiring, and it lost power fast. Because of the country, I commenced a left turn, into the live engine, to try to reach the beach, but I couldn't make it. I put her down on a bit of a clearing in a sacsac swamp. The starboard wing hit a coconut palm, and we slewed right round. I banged my head somewhere, and got a cut scalp. The passengers were all right, though. She burst into flames — was a write-off.'

There was of course no accident investigation branch of DCA in PNG at that time, and the Department sent A.H. Affleck, Senior Flying Inspector, and A.R. Murray, Aircraft Surveyor, to the scene from Melbourne.

Arthur Affleck was a man of vast experience of the aviation industry. He gained his wings as a Civil Aviation Cadet, trained by the RAAF, in 1924. Only three cadets had been chosen from hundreds of applicants. Another of the three was Orm Denny. In 1925, Affleck became a commercial pilot, with the long- vanished Australian Aerial Services Ltd, flying DH4 and DH9C biplanes between Melbourne and Hay. He joined Qantas in 1927, and was the original pilot of the famous Flying Doctor Service. He flew in New Guinea during 1933 as a pilot for Ray Parer's Pacific Aerial Transport company. In 1936, Affleck joined what was then the Civil Aviation Branch of the Defence Department.

'Rube' Murray was to become a legend in post-war civil aviation circles in PNG. He, too, had spent years in the industry, and was an expert on crash investigations.

These two seasoned professionals found that the crash of ARI was the direct result of a failure of the starboard engine, which in turn was 'probably' caused by the presence of water in the fuel.

Affleck then proceeded to make a number of comments about PNG air operations in his accident report, which was closely studied by the Director-General and his senior staff. Affleck noted that VH-ARI was not covered by a valid Certificate of Safety at the time of the crash, since it had expired at 7.30 a.m., before Burt took off. MAL was operating a scheduled service from Lae to Wewak and return without an airline licence, thereby contravening Regulation 79 of the Air Navigation Regulations. This service, however, was sponsored and subsidised by the Provisional Administration. MAL was also operating on a charter basis to other parts of PNG in contravention of the Regulations.

14. Arthur Affleck. *Photo: Arthur Affleck*

14

Dick Burt came under fire in the Report. Affleck, a mild-mannered man, was plainly nettled by the rather nonchalant attitude of Burt, who seldom bothered to mask his low opinion of officialdom and all its works. Burt 'exhibited amazing indifference towards any reasonable safety precautions,' Affleck wrote.

Despite his findings, Arthur Affleck recommended that no legal action be taken against MAL. 'If DCA is unable to take immediate steps to control and licence PNG aerodromes,' he wrote, 'some interim authority must be given to companies such as this to permit them to carry on charter operations without contravening Regulations. None of the aerodromes in New Guinea are licenced, nor do they come within the control of the Department. This also applies to Lae and Moresby, which are used by the regular Qantas service, although still RAAF 'dromes...The only DCA staff in New Guinea is that of D.A.W. at Lae and Port Moresby...we have no real knowledge of local flying activities in regard to pilot technique, aircraft maintenance...suitability of routes flow, condition and suitability of aerodromes, weather conditions, loadings etc...

'DH84 are virtually single-engine aircraft. Performance on one engine is very bad. They carry no radio — those available are too heavy. The only radio navigation facilities available are at Moresby and Lae...meteorological facilities are very thin...

'The terrain throughout New Guinea is such that air transport offers the only feasible means of supplying settlers in centres away from the coast...it is essential for the good of the country that every assistance should be given to reputable air transport undertakings to enable them to establish as complete a network of air communications as is possible throughout the Territory. To enable this to be done the first essential is to...build up a framework of reasonable safety requirements which will permit the operation of air services, whether charter or regular.

'The history of aviation in New Guinea provides adequate proof that those responsible for the actual operation of the services have been fully aware of the limitations and restrictions which must be applied to operations, but it is contended that these are and must be of a different nature to those in operation within Australia. The nature of the terrain, the type of weather conditions, the lack of navigational facilities and the nature of the aerodromes which it is possible to prepare, together with their location, make for a type of flying which is unknown in Australia but which has been perfected in New Guinea over a period of twenty years. If, now, the same restrictions are applied to New Guinea as are in operation within Australia, the majority of services throughout the Territory must cease, and the settlers in inland districts will be forced to leave...

'I believe a survey should be made of the Territory to decide what air services are required, and then decide the organisation and equipment necessary to meet the needs of those services...'

Affleck recommended that the attention of operators, pilots and engineers should be drawn to the necessity for a strict adherence to proper refuelling procedures, particularly when aircraft were being flown away from main bases. Essential aerodromes should be licenced without delay. Conditions for the issue of airline licences should be reviewed, and made less stringent than those in force within Australia. All air services throughout PNG, whether on a regular or charter basis, should be licenced.

Out of Dick Burt's crash, therefore, had come far-sighted recommendations that were to have an immense influence on the development of civil aviation in PNG. For Arthur Affleck's recommendations were accepted by C.S. Wiggins, in charge of air navigation safety standards within DCA. In a submission to the new Director-General, Air Marshall Richard (later Sir Richard) Williams — ex-Chief of Air Staff, RAAF, and often regarded as the father of the Air Force — Wiggins agreed with Affleck's many comments. DCA knew virtually nothing of what was happening in civil aviation in PNG, particularly in regard to the safety and maintenance aspects, air routes, and the condition and suitability of aerodromes. Meteorological facilities were known to be poor, and radio communications were non-existent except in Port Moresby and Lae. Since the aircraft being flown by MAL — still the only operator — carried no radio, and no suitable lightweight equipment was in sight, advice of hazardous weather conditions could not be passed to a pilot once he had taken off.

Wiggins continued: 'It is all the more necessary that a proper point-to-point communication service be provided in order that complete information as to whether a flight may be safely undertaken is available prior to the commencement of a flight...

'I believe that a survey should be made of the Territory to decide what air services are required and then decide the organisation and equipment necessary to meet the needs of those services.'

In mid-November, 1946, the Director-General signified his agreement that a survey would have to be made. In a minute, he wrote: 'The District Superintendent will be down in two or three weeks, and perhaps we can discuss details then...'

15. MAL DH84 Dragon taking off from Kokoda, November, 1946. *Photo: Sam Jamieson*
16. MAL DH84 Dragon, Kokoda, November, 1946. *Photo: Sam Jamieson*

15

16

The Department Of Civil Aviation

It will be recalled that in his report following his visit to PNG in 1945, DCA officer, Val Augenson, recommended the early appointment of a district superintendent to represent the Department and control the development of post- war civil aviation in PNG. The appointment was made in October, 1946. On 16th of that month, the new district superintendent, J.S. Arthur, arrived in Port Moresby. He was accompanied by George Woodward, senior clerk.

Although Lae was the logical place for the headquarters of DCA in PNG from the point of view of civil aviation, it had been decided in Melbourne that Arthur would have to be based at Port Moresby, since this was the headquarters of the Provisional Administration, the place where the Administrator and the heads of all government departments were located. Because of the sorry state of communications within PNG at that time, the advantages of Port Moresby were felt to outweigh those of Lae, for the district superintendent would need to be in constant touch with the men who were governing the Territory.

John Arthur was a remarkable man. He was to guide the development of civil aviation in PNG for the next ten years, and during the process would make some enemies, but many friends. He was a little above average height, of medium build, balding, rather brusque of manner, and direct of speech. He wore his authority — which was considerable — easily, but was ever prepared to jump on those who strayed too far outside the rules, for the limits he set were generally very reasonable. He was an administrator of rare ability, not a pilot when he came to PNG — although he soon obtained his private licence — and thus regarded with some reservations by those who held that only a man with a full grounding in civil aviation could hope to properly understand the problems of the industry.

Had John Arthur insisted on the letter of the law during that decade, the difficulties of the air transport operators, already considerable, would have been immeasurably increased. He was hard headed, but he was understanding. There are those who still today curse the memory of John Arthur — he died in April, 1979 — but there are many others who consider he has never been given his full due for what he did for civil aviation in PNG from October, 1946, to February, 1956.

John Arthur had little to build on when he came to PNG. Prior to his appointment, PNG came under the nominal control of the Regional Supervisor of Airways Operations at Cairns, Len Coleman. Coleman was also responsible for North Queensland. DCA was drawn into PNG after the Qantas Bird of Paradise service began in April, 1945. The RAAF was then operating the airports at Port Moresby and Lae, and providing necessary radio and meteorological services. But the RAAF were in the process of withdrawing. DCA therefore arranged to take over control of the Aeradio stations at Port Moresby and Lae from the RAAF, in April, 1946.

Flight Control Officer Charles A. Rowles was sent to Port Moresby with an operator, R. Bonney, to work with the Air Force operators and then take over the station. Conditions at Port Moresby were grim, as Rowles outlined in a report to Coleman, dated 17 April: 'Rain started on 11 April, and continued to early morning on the 13th. During this period, twenty-five inches [635 millimetres] of rain fell at Jacksons...on the morning of the 12th, Mr Bonney and myself had to wade through water up to five feet deep [about 1.5 metres] to get to the tower...Water was driving right through the radio room...Aeradio functioned on and off until noon when the last control line broke down...

'Needless to say, we had a torrid time of it. Our quarters were in a shocking state. We had been living on rice for two days...the fact that I had no transport made it impossible for me to inspect the Strip properly...with the state of communications here, repeated failures of equipment etc, it must be apparent that flying in this area is a very hazardous undertaking...'

There was a critical shortage of accommodation and transport at Port Moresby. DCA had been late on the scene; the Provisional Administration had taken over the best of what was available. More seriously, much of the RAAF radio equipment and power units that were to be taken over were in poor condition, or arranged in an unsuitable layout, and there was an urgent need for accommodation for radio, electrical and mechanical technicians to carry out the required work. Similar problems existed at Lae, where C.A. Mabbott was stationed.

On 30 May, 1946, the Director-General wrote to the Secretary for External Territories, warning that unless 'immediate action' was taken to overcome DCA's accommodation and transport problems at Port Moresby and Lae, it would be necessary to suspend the Qantas Bird of Paradise service. The extension of the service to Rabaul, which was being planned, would be impossible, too, unless adequate facilities were made available there. He continued: 'The Qantas air service is dependent, for its safe operation, on the provision of certain meteorological, communication and navigation facilities. While the RAAF was properly manned it was possible to rely upon them for the provision of these services. However, as the RAAF commenced to reduce its activities in New Guinea...it became necessary for the Department of Civil Aviation to place its own staff in New Guinea...due to the real need to keep the Qantas service in operation it was not possible to plan and prepare this move into New Guinea in a manner that would ensure that the installations and conditions would be satisfactory...'

DCA's problems were compounded when the RAAF advised that it would cease to provide radio communications for meteorological services to and within PNG from 30 June, 1946.

It was obvious that individual officers at Port Moresby and Lae were in no position to negotiate with the Provisional Administration for improved facilities. A senior DCA officer was required on the spot, pending the arrival of a district superintendent. T.A. Housely, Supervising Airways Engineer, visited PNG in July and August to look into the situation. J.G. Walker had by this time taken over from Rowles as OIC DCA Port Moresby.

Housely wrote in his report following this visit: 'Whilst the various air strips are being maintained by the RAAF, the most pressing requirement is the establishment, as soon as possible, of reliable airways communication, navigation and control facilities. It will be necessary to have an Airways Engineer in New Guinea continuously for some months at least...we can ill afford an Engineer for such a purpose, but I am convinced of the necessity for such action, and the most suitable man for the position is Mr F.S. Maynard...'

Frank Maynard had been a radio engineer with Amalgamated Wireless (Australasia) Ltd in the Mandated Territory in the 1930s and fought there during the war. He was currently Airways Engineer at Cairns, and had already visited PNG with Len Coleman. Frank Maynard was sent to PNG as senior representative of DCA, and things began — but very slowly — to improve. After a time, Maynard went to Rabaul to prepare for the extension of the Bird of Paradise service to that centre.

One of the first of the DCA radio technicians to be sent to Port Moresby was Geoff Masters. He had served in PNG during the war, as a motion picture projectionist in the Army Cinema Corps, and after taking his discharge completed a radio technicians course at the Melbourne Technical College. At the age of twenty-one, he joined DCA as a radio technician, and in 1946 was sent to Port Moresby. Masters remembers: 'I was met at Jacksons by one John Walker, OIC of the DCA station. He took me to the RAAF office — he was sharing the RAAF setup. There I met senior technician Aub King, who had arrived a couple of weeks before. Our first job was to reconstruct the J6924 AWA transmitter at the bottom of Three Mile hill, opposite the entry to Murray Barracks which were our quarters. The transmitter had been under water in the big April floods. From the start, I could see that attempting to rebuild it was a dead loss...different technicians worked on the damn thing for years...

'After working with Aub on that thing for a month or so, I got a telegram from Frank Maynard at Rabaul, telling me to go to Lae. DCA was taking over there, from the RAAF. Soon after I arrived there, old George McGee turned up. It was very hard at first, nobody wanted to give up anything. Air Force people were reluctant to leave, although they knew they had to, but were hanging on to the death knock...

'Bernie O'Connor arrived, and the three of us started to rearrange the station. Stuff was all over the place, Air Force had receivers here, transmitters there, and we started to get everything together at what was then the Operations Base at Lae. One long building, with a big sign, BASE OPERATIONS US ARMY AIR FORCE. An old tower stood alongside it — we rebuilt this later on...'

After John Arthur arrived in PNG, Frank Maynard retired from DCA and settled on a plantation, outside Rabaul. The RAAF was almost out of PNG flying operations by this time. One C47 was still being operated between Townsville and Port Moresby, Lae, Finschhafen, Rabaul and return, and another flew Lae- Goroka-Mount Hagen and return, once weekly. The RAAF was preparing to withdraw and it was suggested that they be taken over by Qantas, as an extension of the Australia-PNG service. The Bird of Paradise service was extended to Rabaul, on 15 January, 1947. Finschhafen was included as an optional stopping place, from 29 January.

Arthur quickly began to build up a DCA organisation in PNG, although at this time the Territory was still a part of the North East District, which also included North Queensland. Engineer E.H. Booker arrived in January, 1947, and during that month John Arthur was able to advise that he was now in a position to accommodate 'any number' of staff at Port Moresby, Lae and Finschhafen. Construction of accommodation for married and single staff was under way at Rabaul.

Problems multiplied for John Arthur from the day he first set foot in Port Moresby. As well as his difficulties with staff and equipment, he was faced with the fact that MAL was still operating the subsidised service for the Provisional Administration without a licence. The service was 'illegal', Arthur wrote, since DCA was the only authority that could enter into such contracts. Far more serious was the presence of a number of Avro Ansons in PNG, carrying passengers bound for sales at Lae and Rabaul being held by the Disposals Commission.

It is now history that war disposals fortunes were made by a number of speculators in PNG in the 1945-47 period, and there is no doubt that the way in which many valuable assets were sold for a song was scandalous. Vast quantities of magnificent equipment that would have been invaluable to the civil administration were sold off for ridiculous amounts, or destroyed by United States Army units, because no arrangements for the transfer to civil authority had been made. Officers of the Disposal Commission in PNG often had little appreciation of the needs of the Provisional Administration. They wanted to get the job over, and get out. Many allegations of corruption and dishonesty floated around PNG.

One Australian purchased 900 vehicles from the Army at Lae for thirty-eight shillings each — at a time when new vehicles were literally impossible to buy. One well-known Goldfields identity bought the rights to the military telephone cables in the Markham valley for £350, and sold the cables as scrap for £20,000. He purchased fifty trucks for £25. In the words of the Australian author and observer, Osmar White: 'Until the early 1950's, the only really profitable post-war business conducted in New Guinea was buying and selling junk'.

Buyers from all over Australia were attracted like flies to honey to the long series of Disposals sales in different parts of PNG. Many came by air, some travelling with Qantas but many chartering aircraft, which were invariably Avro Ansons, also purchased cheaply from Disposals.

Between 18 and 22 November, 1946, five flights were made by chartered and privately owned Ansons from Australia to PNG. None of these aircraft was fitted with radio, none carried emergency rations or ditching equipment. Only one of the Anson pilots had engineering qualifications. No engineers were available in PNG to issue these aircraft with the mandatory Certificates of Safety. In most cases, the Ansons landed in Port Moresby from Higgins Field before advice of their coming was received. No route details were sent, and no indication of the amount of fuel carried was given. In the event of a ditching during the flight from Cape York Peninsula to PNG, it would be impossible to take any affective search and rescue action. Reporting these flights, Charles Rowles, Flight Control Officer, wrote: 'I will hesitate to clear aircraft for the Coral Sea crossing unless some steps are taken to provide a measure of safety, in the event of a forced landing, for the passengers, who pay large sums for their passages. I would be pleased to receive advice as to Departmental Policy with regard to these flights...'

Almost inevitably, one of the Ansons, VH-AVZ, got into trouble. On 22 November, pilot V.J. Madsen took off from Higgins Field on the final leg to Port Moresby. He had three passengers on board. Twenty minutes out, he noticed that

17. Mr and Mrs John Arthur, 1953. *Photo: Juin Arthur*

17

the brake pressure gauge was reading NIL. The Anson had no air compressor, so the passengers took it in turns to build up pressure by using the foot pump, but, as soon as pumping stopped, the pressure dropped rapidly. On landing, the port brake failed, the Anson swung to starboard and left the steel matting and plunged into the deep drainage ditch. The main plane was fractured on the port side, an aileron and flap were damaged, and the undercarriage broken. Pilot and passengers escaped uninjured.

It was only after this incident that DCA Port Moresby was advised that VH-AVZ had experienced brake trouble at Townsville, before making the flight to PNG.

John Arthur had only been on the job for a matter of weeks when this incident occurred. He took immediate action, advising DCA headquarters, Melbourne, by telegram: 'Will assemble all Ansons at Moresby for return to mainland. Have arranged for Catalina escort Moresby to Higgins Field. Aircraft will fly formation to Higgins.'

Henceforth, approval was given to fly Ansons to Disposals sales only if two- way radio, rubber boats, emergency rations, first aid equipment, Mae Wests, and Verey pistols and cartridges were on board.

In December, 1946, John Arthur himself brought up the question of the need for a thorough survey of civil aviation requirements in PNG, following an interview with the Administrator, Colonel Murray. Planning for the survey recommended by Affleck and Wiggins was under way at Melbourne, but no action had yet followed. Communication between headquarters and the new district supervisor at Port Moresby was still far from satisfactory, however, as the following letter from John Arthur showed: 'It is advised that QEA Ltd are operating a DH84, VH-AMN, from Lae to airfields up the Markham Valley to Wau, and to Madang and Wewak. This aircraft arrived on or about 18/11/46. No notification was received from QEA of their intention to operate this aircraft on a charter service from Lae... 'Qantas agents here say the Company will be stationing a C47 at Lae. No official notification has been received...'

Arthur asked for 'urgent advice' as to whether Qantas had signified their intentions to DCA headquarters, and if so, for full details. He was clearly affronted that such a development could occur without his knowledge. Qantas had in fact been granted permission to operate in PNG on a charter basis only, but nobody had thought to notify the new district supervisor at Port Moresby.

18. Jackson's tower and met/aeradio, c1947. *Photo: Ted Booker*
19. George Woodward in Jackson's tower, c1947. *Photo: Ted Booker*

18

19

2
The Entry Of Qantas
Qantas

Enough has been said in this account to indicate the depth of Colonel Murray's desire to have adequate air transport services established in PNG. The coming of MAL had eased the situation, but there was a limit to what could be achieved with a handful of Dragons. From the beginning, Murray wanted Qantas, with its far greater resources, in PNG.

Early in 1946, the managing director of Qantas, Hudson Fysh, visited Port Moresby at the invitation of the Administrator, together with his general manager, H.H. Harman, and assistant operations manager, Captain Orm Denny. This was the first step towards Qantas' entry into PNG. Following the discussions, both Colonel Murray and Hudson Fysh approached DCA for approval to proceed, but as previously explained the civil aviation policy was still in a state of confusion, and no decision could be made.

Bulolo Gold Dredging Ltd was beginning its move back to Bulolo, and although sensible to the potential advantages offered by the existing military road from the Labu crossing near Lae, foresaw the need for heavy air transport. The BGD manager, J.D. Simpson, approached Qantas, and the company applied for permission to operate the new and advanced Bristol Freighter on the Lae-Bulolo run, and for general air transport within PNG. Once again, nothing was done. Aircraft performance standards were still being determined. Finally, approval to operate the DH84 on charter operations in PNG was given.

In September, 1946, Qantas circulated a notice among its pilots, calling for applicants for a job in PNG, flying the Administrator around in the course of his official duties. The aircraft to be employed would be a DH84 Dragon, VH- AMN. The pilot selected was L.J. McNeil.

Len McNeil was born in March, 1922, and grew up on the family farm in the Tweed River district of northern New South Wales. Like so many of the early postwar pilots, he was trained in the RAAF, on Tiger Moths. He completed his flying training at Point Cook in January, 1943. The group commander who signed McNeil's log book was Flight Lieutenant W.G. Campbell, who as noted in Chapter 1, The Pacific War, was the pilot of Lockheed Lodestar VH-CAB which crashed near Ward's Strip in November of that year, killing all on board.

Len McNeil was one of the pilots who volunteered to leave the RAAF and join civil transport operations as the tempo of war eased. The RAAF had more pilots than it required for the conduct of the operations then being undertaken, and McNeil could see that his chances of combat flying were, at best, limited. Civilian pilots were making 'Bully Beef Bomber' flights in PNG, and when Qantas called for pilots, McNeil applied, hoping to be employed on this exciting work. He joined Qantas in July, 1943.

McNeil flew DH86 and Lodestar aircraft for some time, and did make two flights to Port Moresby, as second pilot to Orm Denny. In October, he was endorsed on a Fox Moth, VH-UZD, an aircraft which he was to later — to his embarrassment — 'prang' at Tapini — and became a pilot with the Royal Flying Doctor Service for Qantas, based at Charleville. In October, 1944, he was endorsed on the DH84 and in January, 1946, on the DC3.

Len McNeil was still with the Flying Doctor service when the call for a Dragon pilot for PNG was made. With his extensive outback experience on Fox Moths and Dragons, McNeil was an obvious choice for the PNG position.

The DH84 that Qantas acquired for the PNG job, VH-AMN, had originally been owned by the noted aviator, K.A. Virtue. It was a deluxe passenger-type aircraft, beautifully upholstered and trimmed, with carpets and curtains and a tiny washbasin, a fit carriage for an Administrator. McNeil left Sydney for PNG on 20 November, 1946, and arrived in Port Moresby three days later. He was met by

Captain R.O. Mant, who was a flight captain on the Bird of Paradise service.

Dick Mant was a true veteran of New Guinea flying. He was the original pilot for the W.R. Carpenter company that eventually became Mandated Airlines. From 1933, Mant flew Fox Moths and Dragons for Carpenters in PNG, and when the company won the airmail service contract in 1938 he was appointed Flight Superintendent. Joe Taylor was then the MAL chief engineer. Dick Mant was the captain when the DH86B Express airliner, *Carmania* inaugurated the Australia-PNG service in May, 1938. There was no more experienced PNG pilot than Dick Mant. He joined Qantas during the war.

Because the Qantas base was at Lae, where Harold Hindwood was manager, it had been decided that McNeil would operate from that centre, flying across to Port Moresby as required by Colonel Murray. McNeil was new to PNG, and Mant flew with him in VH-AMN, to show him the route.

The DH84 was a great aeroplane in its day, but that day was in the past. It first flew in November, 1932. Designed by the de Havilland team led by A.E. Hagg to a joint military and civil specification, the Dragon was a biplane, of simple wood and fabric construction, with fixed undercarriage and folding wings. The engines were Gipsy Majors of 130 horsepower each, driving two-bladed fixed-pitch wooden propellers. These had to be swung by hand — there were no self-starters. Normal seating was for six passengers, and the single pilot sat in a narrow cabin in the nose. A total of eighty-seven Dragons were built in Australia by de Havilland's at Bankstown for service with the RAAF during the war; VH-AMN was one of these aeroplanes. The Dragon had a normal cruise speed of 175.4 kph and a fuel consumption of only 59.1 litres per hour, yet it was capable of lifting a disposable load almost equal to its own weight. It was an instant success, and Dragons served with airlines in Britain, Australia, New Zealand, Canada, East Africa, Egypt, and of course PNG.

When peace came, the Dragon and the Avro Anson were the only readily-available aeroplanes suitable for commercial purposes, and they both played a vital part in early post-war operations in PNG and Australia. But this was only because of the lack of a suitable modern, low-cost alternative. On one engine, all that the Dragon could really do was maintain a slow, powered glide. It could not maintain height, much less climb, on one engine, and it was thus a dangerous aeroplane in inexperienced hands, particularly in the high mountains of PNG. Yet most of the pilots who flew it remember the Dragon with great affection. It was a survivor from an earlier era of aviation.

VH-AMN was a heavy Dragon, with its special equipment and deluxe fittings, plus a full load of spares and personal effects, and Len McNeil and Dick Mant had a hard time of it on the flight to Lae. McNeil remembers it well: 'AMN had folding wings, which did effect its performance, because these things ride along on a cushion of air, and if you break the cushion the air streams through. I found that trying to get through the Kokoda Gap with Dick was very hard. We had to circle to gain height to get through...Dick said, "Christ, we'll have to do something about this. It's got to perform, you have to get 10,000 feet [3048 metres]."

'Later, this became a criterion when we started to get more Dragons. Qantas bought them from the Air Force, from different operators around Australia, anywhere they could get one. They'd do them up, change the dihedral and what have you, but the criterion was, fully loaded down at Archerfield it had to make 10,000 feet in one hour. Otherwise, it was no good to us in New Guinea, because when you left Lae with a full load it would take about an hour to get to the Highlands. If you had 10,000 feet then, at least you could find a gap, dive through it with the speed you had, get under the cloud, and then into the valleys. If you had an engine loss it was a downhill power-assisted glide. The maintenance, though, was good. Qantas maintenance has always been good...'

They finally succeeded in gaining sufficient height to clear the gap, and reached Lae without incident. The entire Qantas staff was there to greet them — Harold Hindwood, traffic officer Ken Wightman, and engineers Keith Vickerman and the veteran Joe Taylor, who was then with Qantas.

Qantas facilities at Lae were virtually non-existent. There was a small office, but most of the aircraft maintenance work was carried out in the open air. But there was plenty of work available, more than MAL could comfortably handle, and McNeil was soon flying very long hours, after Dick Mant had showed him around the Highlands aerodromes. To increase the payload of the Dragon, all of the fancy fittings were stripped out, reducing the aeroplane to bare bones. As it turned out, Colonel Murray seldom flew in AMN.

The legal flying limit for pilots then was 100 hours in thirty days, with a maximum of 1,000 hours in any one year. Few pilots in PNG bothered too much about limits in those free-and-easy days, when DCA controls were so lax. Len McNeil's philosophy was simple: He wanted all the business he could get for Qantas: 'I flew thirty days a month, if I could get the loadings. Pointless sitting on the ground, there was nothing much to do in Lae then! I note from my logbook that 123 hours a month was not uncommon. I was given a tremendous amount of latitude by Qantas. Looking back, I wonder that they didn't exercise more control. But all Qantas pilots who went to New Guinea were volunteers, and Qantas trusted us.

'One of the greatest problems we had was in getting sufficient spares. We had some hellish battles with a certain chief engineer at headquarters. I don't think he realised the amount of flying those little aeroplanes and their Gipsy Majors did. I remember one thing in particular. The exhaust pipe was held on with brass nuts, and we used to go through them. They only cost about 2/6 and they were made of brass so they couldn't be overtightened on steel fittings — the thread would strip first. He couldn't understand why I was forever writing down for these brass nuts.

'The Gipsy Majors were worked very hard — were very lifesaving machines. Never lost a Gipsy in New Guinea...'

There was, naturally, keen competition between MAL and the fledgling Qantas operation. Len McNeil was the first to acknowledge the skill of the MAL pilot, John Rose. Rose, too, was RAAF-trained.

'He was my greatest opposition up there, I can tell you. Trying to get business away from MAL was tough if Johnny Rose was handling it, because he never turned back. I remember once I was a passenger with John in an MAL Dragon, AYB I think it was. I had the day off, and he was doing a trip to Kokoda and Popondetta. I'd never been there, and when he offered me a seat I jumped at it. He had another passenger on board, an engineer called Fred Starr.

'Well, we took off and headed out across the Huon Gulf in the general direction of Salamaua. Then, bang, the starboard engine went. I wasn't too worried, we were only a few minutes out of Lae, and John was a good pilot. We could get back in a powered glide on one engine. I said, "This'll test your skill and aplomb, old son..."

'Then, bang, the port engine cut. Must have been water in the fuel. This was another kettle of fish! I said to John, "How are we going?" He said, "Not a hope in Hades..."

'We hit the water with a tremendous splash, right alongside the *Tenyo Maru*. My seat belt snapped and I was hurled forward. My head went right through the top of the cabin, right between two metal spars. The hole was big enough to let me climb out as the Dragon settled in the water. I helped Fred Starr out, and looked around for Johnny. There he was, still sitting in the cockpit with his head just above water, fishing around for his camera. We persuaded him to join us on the upper wing...'

At this dramatic juncture, a local fisherman chanced along in his canoe, and rescued the stranded aviators.

McNeil remembers another incident involving John Rose: 'Johnny was up in the Highlands one day, after he had joined Qantas. He reckoned that he was forced down by bad weather, did about four feet [over one metre] off the end of the lower wing of a DH84. Came back to Lae with this yarn...I was smiling while he was putting in his report...I said, "You lying bastard, no way in the world would you hit a tree in bad weather...where were you shooting-up?" Finally he confessed, it was up at Goroka, he was shooting the strip up and his wingtip hit a forty-four gallon [200-litre] drum. I said to him, "I knew it, you're too good to hit a bloody tree..." '

Len McNeil had a shrewd head for business, and during his frequent flights to the Highlands and the outstations, he noticed the eager desire of the native people for the sort of goods that were sold in trade stores before the war — mirrors, knives, axes, tinned meat, coloured calico, rice, sugar, tobacco. Such goods were now hard to get in the bush, and many of the people had money to spend; war damage payments from a grateful Australian Government. Both Arthur Collins and John Rose were keen entrepreneurs. Collins, in partnership with the MAL agent at Wewak, Jerry Stewart, was rumoured to have paid just £1,000 for the salvage rights to the huge quantities of war materials in the Aitape district. John Rose owned trade stores. McNeil recalls: 'Johnny Rose put me onto this. "Get yourself a trader's licence," he said, "you are legal then, you can take stuff along on your trips and flog it to the people in the hills. There's money in it!" So I got a licence. Used to throw a suitcase or two into the Dragon on top of the cargo and sell it...later on, Orm Denny came to me and said, "Len! You're making more money out of this than Qantas is. What's it to be, are you going to be a pilot or a trader?" So I turned in my trader's licence...'

Henceforth, he was 'Trader' McNeil to the aviation men at Lae.

It was probably inevitable that accidents would happen in those days, when the aeroplanes in use were so underpowered, the pilots generally so inexperienced in PNG operations, and DCA control almost non-existent. In January, 1947, came the first post-war fatality, the first of the great many that were to occur over the next forty years.

D.G. Tapsall was an ex-RAAF pilot, who after the war joined the staff of the Commonwealth Bank at Port Moresby. Like so many young men who gained their wings during the war, the itch to fly was in Doug Tapsall's blood, and when Arthur Collins offered him a job flying Dragons, he eagerly accepted. He joined MAL in December, 1946.

On 29 January, 1947, Doug Tapsall took off from Lae in Dragon VH-AOS with a load of freight for Kerowagi. He never made it. He crashed in bad weather into Mount Kerigomna, and was killed instantly. AOS, like all the Dragons, carried no radio. No one will ever know what happened during Tapsall's last flight, but pilot error was presumed to be the cause — a convenient phrase often used to explain a fatal accident when no other explanation fits. Certainly, Tapsall had very little experience in PNG flying — where second chances are seldom allowed.

Tapsall was killed on impact, but he would have had little chance of survival had he received injuries and lived. For pilots of light aircraft took off then and were on their own, sometimes away from base for days. With no radio, a pilot could not advise base if he was delayed somewhere due to weather, or with mechanical trouble. Nor could he transmit a Mayday signal. Nobody worried unduly if an aeroplane was a day late. No search-and-rescue organisation had yet been formed by DCA in PNG. The only service provided was by the RAAF. It was late on 30 January before it was concluded that Doug Tapsall was down somewhere.

An air search began. The handful of MAL pilots, and Len McNeil joined in, and the RAAF search-and-rescue Catalina from Port Moresby. The wreckage was eventually spotted on the harsh slopes of Kerigomna, but it was not until 4 February that the ground parties brought out Tapsall's body. Tapsall was little known to the pilots. He had not been with MAL for long, but they all threw in to procure and erect a cross over his grave. Says Len McNeil: 'You learned the hard way up there. The thing was to keep air all round you, don't let your wings hit the sides of those valleys, keep a bit under you, and make sure you had enough room to do a steep turn, probably with the nose down, because you had no horizon in those valleys...'

20. Fred Barlogie. *Photo: Bill Burns*

20

Arthur Collins still flew Dragons. He dressed for comfort in the air, as the Goldfields pilots used to do in the old days. A white shirt, white shorts and sandals. Three weeks after Doug Tapsall crashed, Arthur Collins came in to Lae in VH-APL, and during his landing run, the starboard undercarriage leg collapsed. The lower wing had failed. The starboard propeller was splintered into shreads, but Collins escaped injury.

Rube Murray was now based in PNG, as Senior Aircraft Surveyor. He investigated this accident, and found that the cause of the failure of the lower wing was metal fatigue of the round-section bracing strut between the upper and lower wings.

VH-APL was one of the eighty-seven Dragons manufactured by de Havilland Aircraft at Bankstown during the war. The original design specified a strut of streamlined section, but wartime supply difficulties forced de Havilland to use a strut of round section, with a wooden streamlined fairing attached. The round-section strut was considered to be perfectly safe, but the company admitted that the original streamlined strut was 'more satisfactory'. The company was now supplying the original streamlined strut against all spares orders.

Rube Murray recommended that all DH84 aircraft in PNG be immediately grounded. This was not done — it would have crippled air operations — but the struts of all DH84's were closely inspected to ensure that the aircraft were safe to operate. As a result of these inspections, DCA concluded that the general condition of the MAL Dragons was not satisfactory, and the company was reprimanded.

Then came another crash, this time with the loss of four lives. The aircraft was an Avro Anson, VH-AKH. On 1 March the Anson left Lae on a flight to Bulolo. In command was R.G. Peters. On board were passengers P. Wheatherstone, aircraft engineer, F.J. Purvis, and B. Turner. When the Anson failed to reach its destination, a search was mounted. The wreckage was spotted in the high, cold country of the Upper Wampit valley, but it took several days for a ground party to reach the site.

The Anson was being operated by a newly-established company, Guinea Air Traders.

21

21. Qantas traffic building at Lae, July, 1946. *Photo: Qantas*

22. Captain Len McNeil pictured in later years at Cairns, stands by an old Dakota in which he flew during his time with Qantas in New Guinea. *Photo: Len McNeil*

22

23

24

23. Qantas Tiger Moth. *Photo: Qantas*
24. Qantas Dragon. *Photo: Roger Wilson*
25. Dragon ditched at Lae. *Photo: Jim Glindemann*
26. Qantas Dragon. *Photo: Roger Wilson*

25

26

Guinea Air Traders

During 1946, a thriving business in the sale of surplus Avro Ansons for civil transport purposes developed within Australia. Ansons, and Dragons, were both cheap, and readily available. Both were obsolete types in 1946, but both were widely employed for a number of years. These were the aircraft that re- established internal operations in PNG, and in many parts of Australia, after the war.

The Anson was in continuous production in Britain for a remarkable span of years, from 1935 to 1952. During this period, 8,138 were built in Britain, and another 2,882 in Canada. Very few aircraft in history were produced in such numbers.

Designed as a general reconnaisance aircraft with a crew of three, the Anson was a monoplane of steel-tube, fabric-covered construction, with a one-piece cantiliver wing built around two wooden box-spars, with spruce ribs and plywood covering. Extensive use was made of glue in the fabrication of the wings. A gun turret was originally fitted. In the Anson Mark 1, the undercarriage was retracted by means of a hand-crank, very slow and inefficient in comparison with hydraulic mechanism. Later variants introduced hydraulic operation of the undercarriage and flaps, and all-metal wings. The standard engine fitted was the Armstrong Siddeley Cheetah IX of 350 hp, although changes were made in some later marks. These gave the Anson a maximum speed of 303 kph, and a normal ceiling of 4877 metres. The single-engine performance was only fair; height could be maintained with wheels retracted, but an engine-failure on take-off would almost certainly result in a crash.

The Anson was introduced into the RAAF in 1937, and over 1,000 were eventually employed. Many of the older machines were sold off through Disposals during 1946. Several individuals and companies specialised in the purchase — for as little as £750 — and conversion of Ansons for civil use. Conversion could be basic, sufficient only to pass the Certificate of Airworthiness tests, or elaborate. Some Ansons were lavishly upholstered and fitted, with airline passenger seats, enlarged entrance doors and luggage compartments. The result was always a noisy, fairly roomy, quite slow aeroplane with sufficient performance for safe operation in PNG — provided there was no engine failure on take-off.

In June, 1946, one of the surplus Ansons was purchased by E.H. Loneragan of Mudgee, New South Wales, who with his brother, James, was one of the principal dealers in these aircraft. In the following month, the aeroplane was sold to a Sydney businessman, F.W. Sutton, who had it converted for a Certificate of Airworthiness by de Havilland Aircraft at Bankstown. It was registered as VH- AST. Sutton formed a company called Australian Air Traders Pty Ltd.

The company came under the control of the Jamieson brothers, Sam and John. They were two of the sons of a Newcastle master builder, and were well established in various lines of business. Sam Jamieson told of the formation of Australian Air Traders in an interview in August, 1983: 'Fred Sutton established Australian Air Traders. We got together on the matter, in 1946. Things were not going too well, so I bought the aeroplane from Fred Sutton. Straight away we got a charter to take a group of buyers to the big Disposals sales being held in Darwin. Came back, and had a call to fly buyers to the sales in New Guinea...

'So we flew up to New Guinea with the buyers. I bought a lot of stuff myself at Finschhafen. At Lae, I was approached by Jack Simpson, the field manager of Bulolo Gold Dredging Ltd. He had lots of freight that he wanted flown up to Bulolo, too bulky to put into the MAL Dragons. He asked me, "How much to fly it up for me in the Anson?" I said to him, 'I wouldn't have a clue, but ruling rates.' "Too dear!" said Jack.

'So I said, "Let's have a load, I'll fly it in and then cost it out." And that's what we did. That was the beginning of my operations in New Guinea. I gave the buyers I flew up there their money back, and told them to go home with Qantas! I went back to Sydney then and looked up Les Ross, the old Guinea Airways pilot. He later became my flight superintendent. The name Australian Air Traders didn't fit to my way of thinking. I was flying in New Guinea, not Papua, so I said, 'Guinea Air Traders, that'll do...' '

Guinea Air Traders was formed in October, 1946, as a subsidiary of Australian Air Traders, with Sam Jamieson as the managing director. Les Ross, a famous pilot who began his New Guinea career way back in 1928, when it all began, stayed with Sam Jamieson throughout his association with GAT.

In later years, some unkind things were to be said of GAT and its operations, but in February, 1947, when the company commenced charter services out of Lae, it was the only one of the companies then flying that was covered by a proper licence. In the words of J.H. Harper, who investigated the crash of Anson VH- AKH: 'Guinea Air Traders are the only concern operating with the sanction of the department. Both Qantas Empire Airways and Mandated Airlines are flagrantly contravening Regulation 79 by operating much more frequently than once every twenty-eight days between any two places...'

This was a reference to the 'twenty-eight day' rule governing the operation of charter flights that was to cause so much trouble and concern in PNG over the years. Charter aircraft operating between any two places on two or more occasions in twenty-eight days were deemed to be operating a Regular Public Transport — RPT — service, requiring an airline licence. The canny Sam Jamieson had applied for such a licence — as of course had Qantas and MAL — but as we have seen, the policy aspects of airline licencing were still under consideration. But DCA granted Guinea Air Traders a Certificate of Exemption, No. 1, issued on 21 January, 1947. This exempted GAT for a period of twelve months from the necessity of obtaining an airline licence to operate an RPT service, subject to a number of conditions. GAT was authorised to run RPT services from Lae to Madang, either via the Markham and Ramu valleys, or alternatively along the coast via Finschhafen; from Madang to Wewak, Wewak to Aitape, Lae to Bulolo, and Bulolo to Wau. The Certificate also allowed the operation of a service from Lae to Mount Hagen and intervening airports. Maximum permissable all-up weights for the Ansons for each aerodrome were specified, ranging from 3402 kilograms at Mount Hagen to 3719.5 kilograms at Lae. Operations and maintenance manuals had to be provided by GAT to the satisfaction of DCA. Services could be flown under visual flight rules only, between sunrise and sunset. Similar conditions were to be attached to the Certificates of Exemption subsequently granted to Qantas and MAL.

GAT was thus in a remarkably favourable position when it began New Guinea operations. It was covered by a legal licence; it had extensive routes; it operated bigger aeroplanes than the opposition; it had guaranteed loadings. In short order, GAT had a monopoly over freight services into Bulolo, where the amazing Bulolo Gold Dredging company was rapidly restoring the great gold dredges and the goldmining industry. New Guinea Goldfields Ltd at Wau also made extensive use of GAT aircraft. GAT looked set to become the post-war successor to Guinea Airways.

One of the first pilots to be engaged by GAT was John Spiers, an ex-Lancaster pilot who subsequently flew for MAL and Qantas. Another was Tom Deegan, a man who was to become a legend as one of the world's most experienced DC3 pilots.

Tom Deegan was born in Gulgong, western New South Wales, the son of a farmer. As a young man he ran bus services in various country centres and in Sydney, before taking a job in New Zealand, in 1937. He was still there when war broke out, so he joined the New Zealand Air Force.

'I'd had a bit of flying at Mascot, paying thirty bob for a bit of a drive around, although I never acquired enough hours for a licence. I used to fly Gipsy Moths. I knew Smithy and all the old boys...I had a sit in Smithy's aeroplane, *Lady Southern Cross,* when it was under bond in Mascot. I was always keen on flying, so when the opportunity came in New Zealand, I hopped into it...'

Tom Deegan was posted to No. 4 Squadron in Fiji, flying Hudsons. Later, he flew Lockheed Venturas, heavily-armed developments of the Hudson, on ship convoy and anti-submarine duties in the South West Pacific. He was transferred to England, and posted to Coastal Command with the RAF, and then to Transport Command. He flew C47's in Burma during the fierce campaign there, and then came back to Australia from Ceylon via the Cocos Islands, a very long flight in those aeroplanes.

For the remainder of the war, Deegan flew on the courier service from Parafield north to Finschhafen or Lae, and on to Momote. After his discharge, he went looking for a flying job. He was by this time highly experienced on C47 and DC3 aircraft.

'I came back to Sydney and joined Guinea Air Traders. The headquarters were at 10 O'Connell Street. Sam and John Jamieson, they were GAT...Sam could see the future in the development of Bulolo and Wau, so he came south and bought another Anson. That was when I joined. They had this extra Anson out at Mascot, and engaged Johnny Spiers as the pilot. I moved in the day after Johnny, but I had more experience, so I became chief pilot. That was about the end of March, '47. Johnny was endorsed on Ansons by Sam, which was unorthodox, because Sam didn't have a licence himself.

'Anyway, he was endorsed, and we set off for New Guinea in the Anson, VH-AVQ, Johnny in the cockpit, and me and Tom Watson, GAT's engineer, as passengers. We got just past the Sydney Harbour Bridge when Johnny got airsick, went green, and collapsed. So I took over the Anson, and I'd never flown one in my life...We headed for Oakey, because Brisbane was weather-closed. I pressed on, hoping that Johnny would get better, but he didn't, he got worse. When we got up to the Queensland border there was a lot of dust and smoke about, and I couldn't find Oakey. So I landed in a farmer's field at Dalby, with about ten gallons [about forty-five litres] left in the tanks. I got Johnny out and laid him under a wing to recover. Then through the fields came a big flash car with an old grey-haired farmer, all excited; it was the first time anyone had landed an aeroplane on his property. He took us back to the ranch for a Scotch. That helped Johnny to recover...

'We pressed on, Johnny flying it. Half way between Iron Range and Daru we ran into an inter-tropic front. We were overloaded, and I knew how vulnerable the old Anson was in heavy rain. Johnny decided he'd press on — I decided he wouldn't. I was going to hit him over the head with the engine crank-handle if he didn't turn back. He did, and on the turn we lost a motor. Wet leads. Johnny wanted to put her down on a little island, but I said, "Let's get back to Iron Range"...we did that, and GAT sent down the Anson they had in New Guinea with an engineer and some gear, and we changed the leads and set off again. The pilot of the other Anson was Alan Emmerick, an ex-fighter pilot.

'When we got to Lae, I decided I'd do the landing. I wanted some practice on the thing, and I wasn't endorsed. You had to wind the wheels down by hand. So I landed, and ran off the end of the strip, among the rubbish where those old gliders used to be — didn't hit anything, and didn't wreck the aeroplane.

'We operated from an old tin shed at Lae. Next day Alan and I flew the two Ansons to Bulolo, and then Goroka. I didn't know New Guinea then, so I used to stick behind Alan. I ran off the strip again, at Goroka, out amongst the *kunai*. I said to myself, It's no good, I can't fly this aeroplane, every time I try to land I run off the strip. When I got back to Lae, I asked them to have a look at it. The Anson had cable brakes, and they found that the right-hand cable was connected up to the left-hand wheel. When I put on the right-hand brake the aeroplane went to the left. No wonder I couldn't keep the thing on the strip...

'There were only the two Ansons at Lae then, but Sam decided there was a lot of work to be done in the Highlands and the rest of the country as well as Bulolo, so we got some more Ansons, all war disposals. We got more pilots and engineers from south, and they would fly up in the Ansons. We were doing well, rehabilitating the Highlands and the Goldfields...

'The Ansons lived outside, in the wind and the rain. We lived in this old tin shed, and there was no toilet, so we used to go across to the little wharf that was there then and do our business off it. Used to be funny in the mornings, the local kanakas looking at the blokes sitting on the wharf with their bums over the edge. I built the first toilet at the airstrip between flights — got sick of popping over to the wharf. I built a beautiful toilet, wooden seats and pans, the works. We christened her, too...

'We used to fly seven days a week, then. Some good pilots joined us...Peter Lavender, Bill Burdus, Tom Drury...'

Thomas Peel Drury was a child of the Great Depression, born in 1917, near Newcastle. When, as a youth, he decided he wanted to be a pilot he found the cost of private instruction beyond him. Dual instruction in a Gipsy Moth cost twenty-five shillings per hour, and one pound solo. There was no aerodrome at Newcastle, but the New South Wales Aero Club (it was not 'Royal' then) sent an instructor and a Moth to the city twice a month, operating off District Park, and so the Newcastle Aero Club was born. Drury took flying lessons whenever his finances permitted, but, as with so many aspirants during the Depression years, his ambition to become a pilot could never have been fulfilled, but for the Second World War.

Tom Drury joined the RAAF, and was fortunate enough to be one of twelve in a course of 220 to be chosen as trainee pilots. He converted onto Ansons, and then did a flying instructor's course at Camden, New South Wales, on Airspeed Oxfords. He spent the following two years at Bundaberg, as a flying instructor on Ansons. Later, he flew C47's in New Guinea, and in Morotai, the Philippines and Borneo. Drury was a flight lieutenant in New Guinea when the war ended. He returned to Australia in 1946.

In an interview in 1982, Tom Drury told of his time with Guinea Air Traders: 'I was approached in Newcastle by a fellow called Sam Jamieson, to go with him and start a charter service in Lae. He wanted me, since I'd flown in PNG during the war...

'My father knew the Jamieson boys' father in the building industry in Newcastle, and he said to me, "If they are half as good as their father, they are all right." So I went down to Sydney, set sail out of Mascot in an Anson, VH-AKH I think it was, and headed for New Guinea. Tom Deegan and Johnny Spiers had gone up a bit before me. This was May '47. The company had only been going for a few weeks. Sam's three brothers, Arthur, John and David, were involved in GAT. David was the manager at Lae when I arrived, but he didn't stay long...

'GAT had all the BGD business on contract, and we did well. Sam liked to push things hard and quickly — he was a bit of a bulldozer, and he'd be the first to admit it. But we pilots were mostly RAAF-trained, young, not properly equipped for commercial flying. And we had little experience of New Guinea. I had the most, about 700 or 800 hours during the war.

'There was practically no surveillance from DCA. Sam made me chief pilot, and I got tired of writing over to Port Moresby, trying to get them to do something about the condition of the Lae strip. At the end of the runway there was a bank of earth which had been built up during the war, obviously to prevent anyone overrunning the strip with brake failure and going into the sea. This earth mound was ten or twelve feet [over three metres] high. We were carrying heavy loads in rather antiquated wartime machines. I'd seen our pilots miss that bank by a matter of inches taking off on a cool morning, and as the day got hotter the inches got less. I could visualise losing a crew and aeroplane on that bank...

'There were still a few Air Force people in Lae, and I went to a corporal who seemed to be in charge and asked him to bulldoze the mound away. They had a lot of equipment there, waiting to be sold. "No way," said the corporal, "the bulldozers are all up for Disposals — I dursn't start one of them." I said, 'I'll give you five quid.' "No way," says he, "don't try and bribe me." So I said, 'How about ten quid?' He says, "When do I start!"

'So in he went with his bulldozer and pushed all the earth off the end of the strip, down onto the road and into the sea. Didn't quite complete the job though, because a great shower of water gushed out of the middle of the mound, followed by a shower of blue sparks. The town water and power supply ran through the middle of that mound, which nobody knew about. Next minute the town reservoir is empty and the town blacked out.

'That got DCA off their backsides! That's when I met John Arthur for the first time, the district superintendent I'd been writing to all this time without reply. He roared over from Moresby, and blamed me for all the damage. Possibly he was dead right, but I wasn't going to be abused by some nut who thought he was running DCA, but wouldn't get off his bum until after the event! So I ordered him off the premises.

'GAT never had a proper check and training organisation such as we knew it in later years. Things of that sort hadn't really been invented. Main thing I insisted on when I became chief pilot was checking every pilot at least five times over every route...Ansons and Dragons with *both* engines going couldn't outclimb the hills — no chance at all with one. Treated them as single-engined aeroplanes...'

The combination of inexperienced pilots, unfamiliar terrain and low-powered aeroplanes soon began to hit Guinea Air Traders. As mentioned in the previous chapter, Qantas, Anson VH-AKH crashed into the mountains of the Upper Wampit valley only weeks after GAT began regular operations, killing the pilot, R.G. Peters, and the three passengers. Sam Jamieson remembers the tragedy vividly: 'The Anson was a wonderful aeroplane, very safe. That was our first and only fatal accident with an Anson. People laughed when I said I was going to carry BGD's freight in little Ansons, but look what we did: we broke records!

'Eddie Peters. An ex-RAAF pilot. I thought a lot of that boy. Nobody would give him a job, he was a dead-beat. But I gave him a job although he didn't have an endorsement on Ansons. We were in Sydney. I had come down to bring another Anson back, VH-AJJ. Bought it from Loneragan, at Mudgee. We got just past Williamtown and we ran into thick cloud, and then I found that Eddie wasn't instrument rated. I used to fly Ansons in New Guinea, although I didn't get my licence until 1948. I knew the idiosyncracies of the Anson! Eddie said, "You fly it, you fly it," and I took us clear through to Brisbane.

'On that Sunday when Eddie Peters was killed...I had just pinched George Price as my chief engineer from Arthur Butler: a top engineer. Then I pinched Pat Wheatherstone, another top man...he arrived from Australia this Sunday morning on Qantas, with two men for BGD, an engineer and a dredgemaster. They wanted to get to Bulolo right away.

'Sunday was a rest day with us, unless there was an emergency. Our boys flew hard enough during the week, and they deserved their rest. We were up long before sunup, and as long as you could see 100 yards [over ninety-one metres], we took off. No navigation aids, and we flew safely, although human error takes place sometimes in the air. But Eddie said, "I'll fly them in." 'No,' I said, 'They can wait until Monday.' Said Eddie, "Look, they haven't had any mail at Bulolo for two weeks...I'll go in." Pat Wheatherstone wanted to go along too, for the ride.

'So they went, and that was the last that we saw of them.'

The last man to see Peters' Anson before the crash was the MAL pilot, Dick Burt: 'I was flying out from Bulolo that afternoon. The weather was bad, and I came out very, very low, on the north-west side of the gap. I cleared the village only just above the treetops, but I knew the area very well. As I came out I looked back and up to my right to see how deep the cloud was, just in time to see an Anson poke into the clouds. Well, they were all dead a matter of seconds after that. I gave evidence later at the DCA investigation. I came out under the cloud, but with no clearance at all, the cloud was on the ground, virtually...

'Eddie Peters didn't have the experience, didn't have sufficient New Guinea knowledge to fly in that sort of weather. Arthur Collins used to roast me for flying regardless sort of thing, but I always thought that I was fairly safe...'

Sam Jamieson remembers that the weather was good in Lae when Peters took off: 'But it changes rapidly in those mountains. A pilot can get trapped — go through a gap into a valley, can't get through, turn and try to get back, and the gap has closed behind you. Then you are really in strife. But I'd agreed that Eddie could fly these two men in to Bulolo, and that was that...

'They found the plane a long time after the crash. We brought coffins in to Timne Village and carried them, up and down, up and down, to the crash. Strange — the bodies were in a perfect state of preservation because of the altitude, high in the mountains. We wrapped them and carried them down. When we got them into the warm air and the heat, they became a mass of stinking putrefaction...'

The DCA verdict on this accident was hard to argue with — pilot error. Inexperience. It was surely no coincidence that both Doug Tapsall and Eddie Peters killed themselves flying in bad weather over unfamiliar terrain within weeks of commencing flying duties in PNG. The crash investigator, J.H. Harper, in his report wrote: 'The whole situation with regard to internal air services is most unsatisfactory...the District Superintendent and myself made enquiries outside the scope of the investigation with a view to collecting up-to-date information on requirements and the existing air services...'

The DCA survey recommended by Arthur Affleck and assented to by the

Director- General, Air Marshall Williams, had still not been carried out. Harper went on to describe the existing situation in PNG. The Administration was supplying the important outstation of Goilala, in the mountains 121 kilometres behind Port Moresby, by native carrier and mule train because the only airstrip was but 503 metres long, and not even Arthur Collins and the boys of MAL were prepared to fly Dragons into that. Another strip, and an aerial service, was desperately needed. Goroka was still — in March, 1947 — being supplied largely by fortnightly RAAF C47 freighters, which were on the verge of being withdrawn. Dragons were too small to replace them. Communications to Wewak and Madang were still so bad that the Administration was continuing the illegal contract arrangements made with MAL. There were many such instances.

Bulolo Gold Dredging Ltd was fast restoring the great gold dredges. The general manager, J.D. Simpson, estimated there would be sufficient work to keep the dredges operating for twenty years. To maintain their activities, the company required the movement into Bulolo of at least 5080 tonnes of freight per year. The wartime road to the coast was costing £50,000 per year to maintain, and it was generally agreed that it would prove far too expensive an alternative to an air service. New Guinea Goldfields Ltd were about to commence large scale mining operations at Wau and Edie Creek, and other companies were working in the Upper Watut. The existing aeroplanes of GAT, MAL and Qantas could not hope to fully service these, and many other, enterprises.

Harper pointed out that the rehabilitation of PNG would require heavy air freight movements 'far in excess' of anything in Australia. Yet only Dragons and Ansons were available to do the work. They were expensive to charter, too. The companies charged £12/10/- per hour for Dragons, and £18/10/- for Ansons. The Dragon was fitted with non-supercharged engines, with the result that its performance deteriorated with altitude. It had a ceiling of only 3048 metres or so, and at that height its performance was so poor that it was vulnerable to the down-currents of air common to the high mountains of PNG, where peaks were half as high again as the Dragon's ceiling. The Anson had poor single engine performance and it had to be borne in mind that the risk of engine failure was considerable in PNG, since on nearly all flights it was necessary to stay on climbing power for fifty percent of the journey.

Harper and John Arthur interviewed the managers of MAL, Qantas and GAT in Lae, to try and obtain some idea of the intentions of their companies. MAL expected to have a Catalina in service during April, and 'hoped' to have two C47 freighters in service within six months. Harper commented, wryly: 'As these C47's at present consist of components of many aircraft scattered throughout New Guinea, it is very doubtful whether they will fly before twelve months have elapsed.'

GAT had already purchased a C47 freighter, and expected to have it in Lae by mid-April. It would be used on the Lae-Goroka, and Lae-Bulolo-Wau routes.

Qantas had already notified DCA Melbourne that as soon as permission was given, a C47 and four DH84's would be based at Lae. Harper commented: 'One point worth mentioning when the operation of C47 aircraft from Lae is considered, is that it will be several months at least before any reasonable maintenance facilities will exist here. There are only two hangars, and these are in a poor state of repair, and lack any workshop equipment whatsoever.'

The urgency of the problem, Harper concluded, demanded that steps be taken to 'provide whatever services are possible with the aircraft available, operated in the safest manner. To this end, it is considered that companies should be either forced to operate in accordance with conditions laid down in an airline licence, or else positively stopped from operating under present conditions.' He recommended an immediate survey of air services standards and requirements, by a panel of representatives of both DCA and the Provisional Administration.

This time, the Director-General acted swiftly and positively. On 3 April, 1947, he wrote to the Administrator, Colonel J.K.Murray: 'It has been decided that an airways survey party should investigate the requirements in aviation in the Territory. It is proposed that the party, comprising experts in the various elements of airways operation undertake the investigation in three phases...it is also proposed to send an Investigator to New Guinea to determine the economic aspects involved in the present and future airways requirements...

'It is expected that the party will depart for New Guinea within the next fourteen days...these surveys will determine the elements required to ensure the provision of air services with the utmost possible safety and efficiency with due regard to special limitations applicable to the Territory.'

Even as the DCA survey party was beginning preparations for the journey, more accidents in PNG confirmed the need for rapid action to control flying operations, and again exposed the dangers that lurked for pilots inexperienced in flying in that unforgiving country.

Qantas had begun internal operations in PNG with a single pilot, Len McNeil, and a single DH84, at Lae. McNeil flew so hard, and obtained so much business, that the management in Sydney quickly decided to send a second pilot and Dragon to Lae. The pilot chosen was F.S. Furniss.

An examination of the careers of notable pilots usually discloses an ardent interest in flying from boyhood days. So it was with Frank Furniss. In a letter to the author in June, 1983, Furniss wrote: 'My schoolwork suffered because of the time I spent building model aeroplanes, all rubber-powered, because my pocket money went nowhere near procuring a motor...flying lessons were out of financial reach, but I had a trial instruction flight in a DH60 in 1934. Next was a Radio Operator's course with Marconi School of Wireless, getting my first class W/T licence in early 1938.

'I joined Qantas in October, 1938, as Radio Operator on Short C-Class flying boats...with somewhat healthier finances, I started flying lessons with Airflite at Mascot, and also at Singapore, where flying was much cheaper. Got my B licence early 1943, and transferred to pilot strength on DH86's on the Brisbane-Darwin run. Soon transferred to Flying Doctor service based at Cloncurry and after a few months there, transferred again to a Perth basing for the Indian Ocean run to Ceylon, on Catalinas. Later, the run was extended to Karachi.

'Early 1945 saw another transfer — to Liberators, still flying between Perth and Ceylon. I was a First Officer during the Indian Ocean operations. After the war, I was variously on Hythes, Sandringhams, Lockheed 749's, DC4's and DC3's.....'

Missing from the above account is the fact that Frank Furniss was one of the volunteer members of the rescue party commanded by Orm Denny that made the dramatic evacuation flights with the seventy-eight refugees from Mount Hagen to Horn Island in May, 1942, in a pair of antiquated DH86 biplanes.

Furniss applied for the position of second Qantas pilot at Lae when it was announced by the company, and flew to Lae in Dragon VH-AON, arriving on 1 April, 1947. Just ten days after his arrival, Furniss was flying AON in the Highlands, in wet and windy weather. He was bound for Kainantu, with a load of freight and no passengers, and made his approach from the eastern end of that tricky airstrip. In the belief that the end of the strip was boggy from the recent heavy rains, Furniss decided to touch down further up. Wild gusts of wind rocked the Dragon on the approach, and Furniss overshot. He began a normal baulked approach procedure, and slammed open the throttles. The port engine failed to respond, and Furniss was forced to land, touching down two-thirds of the way down the strip. The surface was wet, the brakes failed to stop the Dragon's headlong progress, and the aeroplane ran off the end of the strip, into a creek. Furniss was not injured, but AON suffered damage to the undercarriage, lower wings, engine nacelles, fuselage nose and port propeller.

This was the sort of accident that absolutely mortifies the professional pilot, particularly one as experienced as Frank Furniss. AON took three months to repair.

Basically, the accident occurred because the port engine of AON failed when the throttle was suddenly opened. The crash was investigated by Rube Murray and John Arthur, and their conclusion was that the operation of the DH84 at high altitudes 'could not be considered safe'. The Gipsy Major engines of the Dragon were not supercharged and there was consequently up to a fifteen per cent power loss at 1524 metres above sea level. The altitude control on the Gipsy Major was simply a manual mixture control. The investigation report continued: 'The DH84...has an interlocked mixture control and throttle, arranged to restore the mixture control to the full rich position when the throttle is closed. Assuming the pilot had closed his throttles preparatory to landing, he would in the case of a baulked landing have to partially open his throttles, then adjust his mixtures to the predetermined setting and then fully open the throttles if the danger of the engine failing to respond was to be avoided.'

The Gipsy Major altitude control was never intended to be used under the conditions applying that day at Kainantu, Rube Murray noted, so Furniss was not responsible for the engine failure. Just weeks before, John Rose, with Arthur Collins, the most experienced Dragon pilot in PNG, was taking off from Kainantu and suddenly opened his throttles to full power. Both engines failed to immediately respond.

As a result of Furniss' accident, and John Rose's close escape, it became standard practice for all pilots operating off high altitude strips in Dragons to hold the aircraft on the brakes while opening the throttles to full power, and then release them.

Frank Furniss' accident graphically demonstrated the traps that lay in store for even the most capable and experienced pilot if he lacked New Guinea experience, and had only limited time on Dragons. Furniss never had another accident in Dragons, but the combination of inexperience of PNG flying conditions, and of Dragons, was to prove fatal to other pilots during the Qantas decade.

Len McNeil and Frank Furniss were to train many new Qantas pilots in New Guinea flying techniques, checking them into the various airstrips, showing them the routes, the gaps, and the avenues for escape. Len McNeil recalls: 'They were endorsed on Dragons in Australia. They came up here, and you gave them a briefing. The best way I found to train them was to say, "All right, I'll go up here, you stay on my wingtip." That was just fine in good weather, but some of these fellows were a bit worried if the weather was really bad. They thought that if they stayed right on your wingtip in that sort of weather, and you had to do a very steep turn to come back, you'd hit them...You'd go back and roast them for not staying with you... "Don't turn back," I'd say, "stay with me." In heavy rain up the Markham you'd be down low with the trees on either side of you, as long as they stayed with you they were right, but once they turned back they were on their own...

'Pilot error, they often said when a bloke went in...probably an error to ever get into the bloody thing at all. I wouldn't say that a lot of those inexperienced pilots who had prangs ignored visual flight rules. They did their best, but sometimes you got caught. You didn't have much performance in those normally aspirated engines at altitude, and if you tried to do steep turns in the confines of a very narrow valley, if you stalled the bloody thing you didn't have the air under you to make your recovery. These blokes were very proud of their records in getting their freight through...obviously, if you had a turn-back record you wouldn't attract business like a fellow who always got through would...Don't forget, though, that in the air there are just as many clowns as there are on the ground. The real professionals quickly learned...'

Two weeks after the Kainantu accident, John Spiers made a forced landing in the GAT Avro Anson, VH-AVQ.

At the time of this incident, Spiers' total flying experience in PNG consisted of one flight from Port Moresby to Lae, one from Lae to Finschhafen and return, and some ten return trips from Lae to Bulolo, a total of about twenty-five hours. He certainly did not lack experience as a pilot, having flown Lancaster and Wellington bombers during the war.

Guinea Air Traders had only been operating for a matter of weeks. No proper procedures for training and familiarising new pilots with PNG conditions had yet been developed. When John Arthur requested the GAT manager, David Jamieson, to send two Ansons on charter to Port Moresby for the DCA Survey Party, which was beginning the PNG work, John Spiers was told to fly one, although he had never flown from Lae to Port Moresby.

Alan Emmerick flew the first Anson across, on 20 April. Spiers was instructed to fly to Moresby via the Bulolo valley and through the Kudjeru Gap. Alan Emmerick had familiarised Spiers on the Lae-Bulolo route over three return flights, and had instructed him on local weather conditions and general navigational methods in that area, including the region of the Kudjuru Gap. Emmerick later told the DCA investigators that he was quite satisfied with Spiers' standard of airmanship, although it was apparent that he was experiencing some difficulty in handling the Anson, which was a much lighter aircraft than the heavy bombers he had hitherto flown.

John Spiers took off from Lae in VH-AVQ early on the morning of 21 April, and headed for the Bulolo valley and the Kudjuru Gap. The weather deteriorated, but Spiers kept flying over heavy cloud which at times blanketed the ground — a breach of the visual flight rules. He lost his way, totally confused by the tangle of mountains and valleys over which he was flying, for the first time. His fuel was fast vanishing, and still he could not find Port Moresby. He reached the coast, and saw below what looked to be an area of open flat grass, near a village. Here he landed, damaging the leading edge of the Anson's starboard wing in the process.

The clearing was an abandoned garden, sticky black soil covered by grass, and the Anson stuck fast. Luckily, the village, Gaile, was only some forty-eight kilometres from Port Moresby and the villagers had a battered old truck, in which they transported the pilot to Port Moresby. Spiers reported the incident to John Arthur.

Next morning, Spiers returned to Gaile with Arthur, Rube Murray and a party of RAAF mechanics and fitters. The damage to the Anson was slight, and was soon repaired.

John Spiers did not lack nerve. A rough strip was cut through the grass and Spiers attempted a take-off. The ground was too soft, and although he gave the engines full boost, the Anson refused to leave the ground. Later, a taxiway was cleared to the nearby beach, and Spiers was able to take-off, at low tide.

Arthur and Murray concluded that there was neglect on the part of GAT for having failed to familiarise Spiers on the safer routes to Port Moresby: via the Kokoda Gap, or via the Markham valley, Tsili Tsili, Bulldog and Yule Island. The magnetic compass installed in AVQ was also found to be faulty. Spiers himself was at fault: he had breached the vital visual flight rules.

It was recommended that the GAT Certificate of Exemption be amended, to force the company to effectively familiarise their new pilots by flying a minimum of six trips on each route as second pilot to an experienced captain, and in addition to be accompanied by an experienced pilot on at least the first three trips on each route as captain. A Handbook for Pilots should be prepared, containing all relevant information on routes, aerodromes, weather, navigation methods and forced landing procedures. A general recommendation was: 'That aircraft operating in this Territory be equipped with suitable radio without delay. To effect this, every effort should be made to suitably modify obtainable war surplus equipment for installation in Avro Anson and DH84 aircraft.'

Scarcely had the dust settled from this incident than another GAT Anson crashed, near Bulolo. This time, the aeroplane — VH-AYD, purchased from the Mudgee dealer, E.H. Loneragan, in March — was a write-off. There were elements of high comedy in this crash, which occurred on 24 May, 1947. Two pilots, and two Ansons, were involved.

In August, 1946, an ex-RAAF pilot, D.W. Gluyas, purchased an Anson, VH-AXQ, from Disposals, and applied for permission to operate it in PNG. Gluyas proposed to call his operation 'New Guinea Air Freighters' and planned to fly, from Lae and Madang into the Highlands, at a charter rate of £15 per hour. A second pilot, P.B. Lavender, would be employed, and two engineers, T. Watson and B. McHugh. However, nothing came of this idea, although in October the aircraft was registered in the names of Gluyas and Lavender.

VH-AXQ was sold to GAT in April, 1947. Both Peter Lavender and Tom Watson joined the company. Another ex-RAAF pilot, Bill Burdus, also joined GAT. Lavender and Burdus were young and high-spirited, and a friendly rivalry developed between them.

As mentioned previously, the bulk of GAT's flying was between Lae and Bulolo. The Bulolo airstrip had been constructed across the vast wasteland of rock tailings from the pre-war operations of the gold dredges, and it was dominated at the northern end by a high and rocky hill, known as 'The Pimple', which blocked the approach Most Dragon pilots turned around The Pimple on their approach, then straightened for the landing. The Pimple was an uncomfortable reality at Bulolo, not to be trifled with.

GAT men are in no doubt about what happened on that morning in May. Says Sam Jamieson: 'Those two boys had an argument about which was the quickest way to Bulolo: via the Snake, or through Zenag. So Burdus went one way in VH-ALS, and Lavender the other in AYD. This aeroplane had a modified wingspar, it could operate at 3992 kilograms ...

'Peter Lavender got there first. He went way down the far end of the valley, away from The Pimple, then turned and came in with wheels and flaps down. Set to land, then round The Pimple comes Bill Burdus, wheels and flaps out, skidding around that Pimple, coming in straight at Lavender. Who'll win? Burdus was the lucky one. Lavender had to dodge The Pimple, and he took emergency action, went round again, but his Anson was very heavy, full up with freight, and he hit the eastern side of the ridge...'

Peter Lavender had no chance of avoiding the crash, under the circumstances. The two Ansons were rushing headlong towards each other, and disaster was imminent. He was approximately fifteen metres above the strip when he commenced emergency action, and travelling at an indicated speed of only fifty-five knots. The all-up weight of the Anson was almost 3946 kilograms. The aeroplane simply did not have the performance to climb fast enough to avoid the ridges on the sides of the Bulolo valley, although Lavender applied full emergency power. The wingspar of the Anson had been modified, but this merely strengthened the airframe structure, and did not provide increased performance under PNG conditions. Moreover, it was the conclusion of John Arthur and Rube Murray that the cargo in VH-AYD had not been properly secured, which would have made recovery all the more difficult.

The pilots at Lae thought the whole affair a huge joke, although Sam Jamieson was not amused. AYD could not possibly be economically recovered, and the aeroplane was stripped for spares, and abandoned where she lay. DCA, too, did not see humour in the incident. The verdict was pilot error. Available evidence did not support the theory that Burdus and Lavender had been engaged in a race, and Arthur and Murray were careful to avoid any direct accusations. In the accident report they said, with considerable circumspection: 'It is conclusive that both pilots knew the destination of each other's aircraft and that on the evidence submitted it seems unlikely that neither of the pilots could not have expected to find the other aircraft in the vicinity of their own destination at or about their individual times of arrival.'

It was considered that the all-up weight at which VH-AYD was being operated contributed markedly to Lavender's inability to make another circuit, and it was recommended that in the Territory, the weight from sea level of all Ansons, whether equipped with a modified spar or not, should be 3719 kilograms. It was also recommended that cargo securing facilities must henceforth be provided in all aircraft. These recommendations were subsequently accepted in Melbourne.

At the time of the crash of VH-AYD, the DCA Survey Party had begun their investigations into PNG aviation requirements and standards, which were to profoundly influence the conduct of air operations for years to come.

27. Auster VH-KSM of Guinea Air Traders.
Photo: Jim Perry

27

28

30

29

31

32

28. Guinea Air Traders Anson VH-AKH which crashed en route to Bulolo in March, 1947. All four on board were killed. *Photo: Sam Jamieson*

29. Guinea Air Traders mess at Lae, 1946. *Photo: Sam Jamieson*

30. Guinea Air Traders Anson VH-AYD after crashing near Bulolo, May, 1947. *Photo: Sam Jamieson*

31. Qantas Fox Moth VH-UZD at Lae, 1949. *Photo: Lionel Purkiss*

32. Mt Hagen evacuees, Fathers Glover and Ross, Dr E. Chambers and Frank Furniss. *Photo: Qantas*

The Department Of Civil Aviation Survey

It was fitting that Arthur Affleck should be selected to lead the DCA party to PNG on this vital survey, which began in April and was concluded in June, 1947. In his book, *The Wandering Years** Affleck modestly claims that he was appointed leader 'because I was the only operational bod in DCA with practical New Guinea civil aviation experience'. The members of the team were Icko Tenenbaum, 'Johnno' Johnson, Jack Mowbray and Jim Glindemann, representing the Airports, Aero Engineering, Air Traffic Control and Operations departments of DCA. Airways engineer Ted Booker from DCA Port Moresby also joined the party.

Affleck and his team were required to conduct:

> (a) A survey of existing operations with the aim of increasing the reliability of such operations by improvement of landing grounds, definition of routes to be followed and provision of airway facilities, and
>
> (b) A survey of new routes, the provision of which was an urgent need in the control and development of PNG.

Affleck describes how the job was tackled.

'We arrived in Port Moresby in early April, 1947, and spent several days making out large lists of foodstuffs on which we hoped to subsist during our movements around the Territory. There were no hotels where we were going, and it would be quite impossible to arrange for the party to be accommodated. The only solution was to be self-supporting. We purchased tents, cooking utensils, camp stretchers, mosquito nets, bedclothes, and the hundred and one other items necessary to the support of a party of six in a savage and war-stricken land.

'For the following two-and-a-half months we used almost every known form of transport, and definitely every type of aircraft in the Territory to move us and our equipment from base to base. As we established our base camps at various places we found the natives obviously pleased to have Australians back in their country again...the fact that I spoke pidgin had its advantages...

'We made a survey of every airstrip on the mainland of PNG, from Daru in the south-west and Aitape in the north-west to Milne Bay in the east. We also flew to each strip on all the surrounding islands from Manus in the north to as far south as Bougainville.'

It was indeed an enormous task, as each airstrip survey was extremely thorough. Jim Glindemann — a qualified civil engineer who also held a commercial pilot's licence — still remembers the pace at which the team worked: 'Jack Mowbray and I measured up nearly 100 aerodromes during that survey. We would sometimes do five a day, and completed the reports and sketches that night, ready to start with a clean sheet next day.'

After completing the fieldwork, the team spent three weeks in Port Moresby processing the mass of data that had been collected. In Affleck's words: 'We argued and discussed, and evolved an operational criterion which would enable DC3's to make use of airstrips which would be viewed with horror in Australia, and wrote a voluminous report which recommended services which I considered should be instituted and airstrips which should be retained and maintained. I handed the completed report to the Director-General on our return to Melbourne, and, thereafter, heard no more of it. However, for probably the first time on record most of the recommendations I had made were adopted.

'Madang was accepted as the obvious airport from which to serve the whole of the Highlands, in place of Lae; the recommended airstrips were retained and arrangements made to operate most of the services on a regular basis. The opera-

*Longmans, Melbourne, 1964

tional criterion we had evolved as an interim measure was accepted on an interim basis and remained in force until 1961, fourteen years later. I made one very strong recommendation in regard to the airstrip at Lae — to swing it by fifteen degrees so that it pointed straight up the Markham valley, with perfect approaches at each end, and the possibility of unlimited future extensions. This must have found its way into somebody's 'too hard' tray, or been ignored because it was considered unnecessary...'

The adoption of a special operating criterion for DC3's which allowed this type of aircraft to fly to airstrips that would have been forbidden to them in Australia was perhaps the most important policy decision DCA made in those early post-war years, for the DC3 was to become the workhorse of the aerial transport industry in PNG. Without the DC3, the development of the Highlands districts would have been severely limited. A great many aircraft served PNG well in those years, but none so magnificently as this wonderful Douglas aeroplane.

Following the Affleck survey, DCA issued a policy document which detailed the action to be taken on the recommendations in the report.* To provide the Provisional Administration with stable air transport, Qantas was instructed to provide a number of regular airline services 'as soon as possible'. Catalina flying boat services were required from Port Moresby to Lake Kutubu, Kikori and Daru and intermediate outstations, once per fortnight; from Port Moresby to Rabaul, and from Rabaul to Namatanai, Kieta, Green Island, Talasea and Gasmata once per fortnight; and from Port Moresby to Samarai, Milne Bay, Kiriwina, and Misima and Woodlark islands, also once per fortnight. DC3/C47 services were required from Lae to Rabaul, Kavieng and Manus, and to Madang and Wewak, once weekly. The document continued: 'Because of the rugged terrain, swiftly changing weather and the difficulty of providing aerodromes in the interior to meet ICAO [International Civil Aviation Organisation] standards the Department has had to take a realistic view of safety problems in New Guinea if air transport, which is essential to the area, is to continue...'

A *New Guinea Pilots Handbook* would be prepared, covering all aspects of flying in the Territory. Existing aeradio facilities and radio aids were limited to aeradio, homer and responder at Port Moresby, Lae and Rabaul, and a homer at Finschhafen. Another aeradio station would be opened at Madang with homer facilities to handle the DC3 traffic which it was expected would soon begin into the Highlands. No decision was made regarding other radio aids, but investigations would continue.

A Flight Information Region known as Port Moresby FIR was established, with the regional control centre at Port Moresby and control staff also posted to Lae. Further detachments of control staff would be made as new services were established. Meteorological forecasting stations were already in operation at Port Moresby and Lae, and two more would be established, at Madang and Manus. Full use would be made of the Administration's radio outstation network to collect meteorological information.

For all regular airline operations, full ICAO standards were retained, but for charter operations into the Highlands and other parts of the interior two reduced criteria were adopted. These were:

> '(a) *Take-off*. The distance under criterion density altitude conditions to reach a height of fifty feet [15.24 metres] with all engines developing take-off power, the aircraft in take-off configuration and under the most unfavourable but permissable wind condition, plus twenty per cent of the above distance, with corrections to be applied for the runway surface and the slope of the strip.
>
> '(b) *Landing*. The distance under criterion density altitude conditions to stop from a fifty feet [15.24 metres] height at the aerodrome boundary in seventy per cent of the available strip length under the most unfavourable but permissable wind conditions, the aircraft to have an approach speed of thirty per cent above the stalling speed at the boundary of the aerodrome and corrections to be applied for the slope of the strip. Independent of runway length, a weight restriction is imposed on some aircraft using strips at a high geographical altitude.'

*The massive two-volume report, *Survey of Civil Aviation Requirements in New Guinea* by Arthur Affleck and Icko Tenenbaum, Melbourne, 1947, is too lengthy to attempt to summarise here.

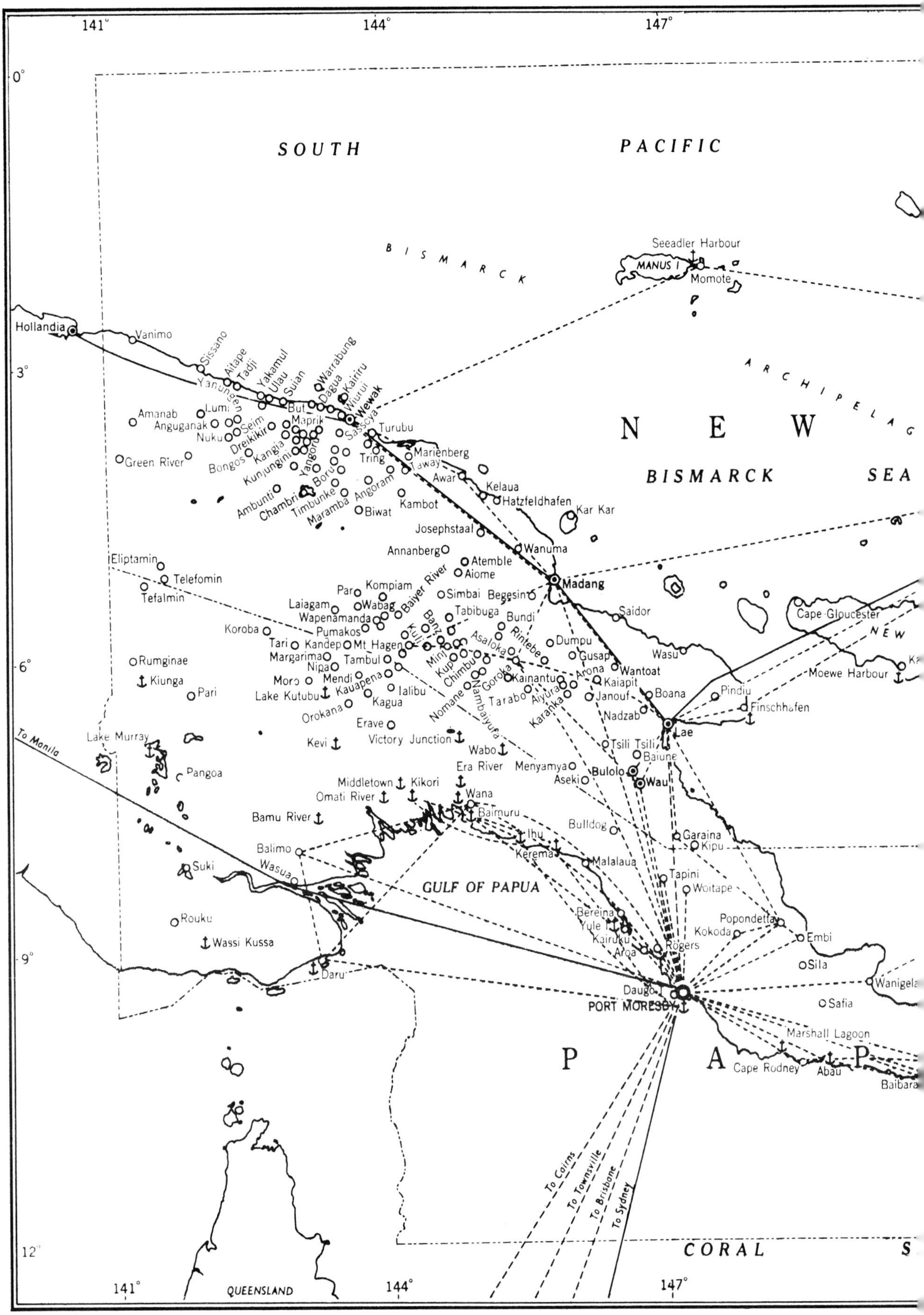
141°
144°
147°
0°
3°
6°
9°
12°
SOUTH
PACIFIC
BISMARCK
ARCHIPELAG
NEW
BISMARCK
SEA
Seeadler Harbour
MANUS I
Momote
Hollandia
Vanimo
Sissano
Aitape
Tadji
Yakamul
Ulau
Suian
Warrabung
Dagua
Kairiru
Wiurui
Wewak
Yanungen
Lumi
But
Amanab
Anguganak
Nuku
Seim
Maprik
Dreikikir
Sassoya
Turubu
Green River
Bongos
Kangia
Kunjingini
Yangoru
Tring
Marienberg
Taway
Awar
Kelaua
Hatzfeldhafen
Kar Kar
Ambunti
Chambri
Boru
Timbunke
Maramba
Angoram
Biwat
Kambot
Josephstaal
Wanuma
Annanberg
Atemble
Aiome
Madang
Eliptamin
Telefomin
Tefalmin
Paro
Kompiam
Baiyer River
Simbai
Begesin
Laiagam
Wabag
Wapenamanda
Tabibuga
Bundi
Saidor
Cape Gloucester
NEW
Koroba
Pumakos
Banz
Asaloka
Rintebe
Dumpu
Gusap
Tari
Kandep
Mt Hagen
Minj
Kup
Chimbu
Goroka
Kainantu
Wasu
Wantoat
Kaiapit
Arona
Moewe Harbour
Margarima
Tambul
Nipa
Mendi
Kauapena
Rumginae
Kiunga
Pari
Moro
Lake Kutubu
Ialibu
Kagua
Nomane
Nambaiyufa
Tarabo
Aiyura
Karanka
Janouf
Boana
Pindiu
Finschhafen
Nadzab
Orokana
Erave
Lae
Lake Murray
To Manila
Kevi
Victory Junction
Wabo
Tsili Tsili
Baiune
Pangoa
Era River
Menyamya
Aseki
Bulolo
Wau
Middletown
Kikori
Omati River
Wana
Bamu River
Baimuru
Bulldog
Balimo
Ihu
Garaina
Kipu
Kerema
Malalaua
Wasua
Suki
GULF OF PAPUA
Tapini
Woitape
Rouku
Bereina
Yule
Kairuku
Popondetta
Kokoda
Embi
Wassi Kussa
Aroa
Rogers
Sila
Daru
Daugo
PORT MORESBY
Wanigela
Safia
Marshall Lagoon
P
A
Cape Rodney
Abau
Baibara
To Cairns
To Townsville
To Brisbane
To Sydney
CORAL
S
QUEENSLAND

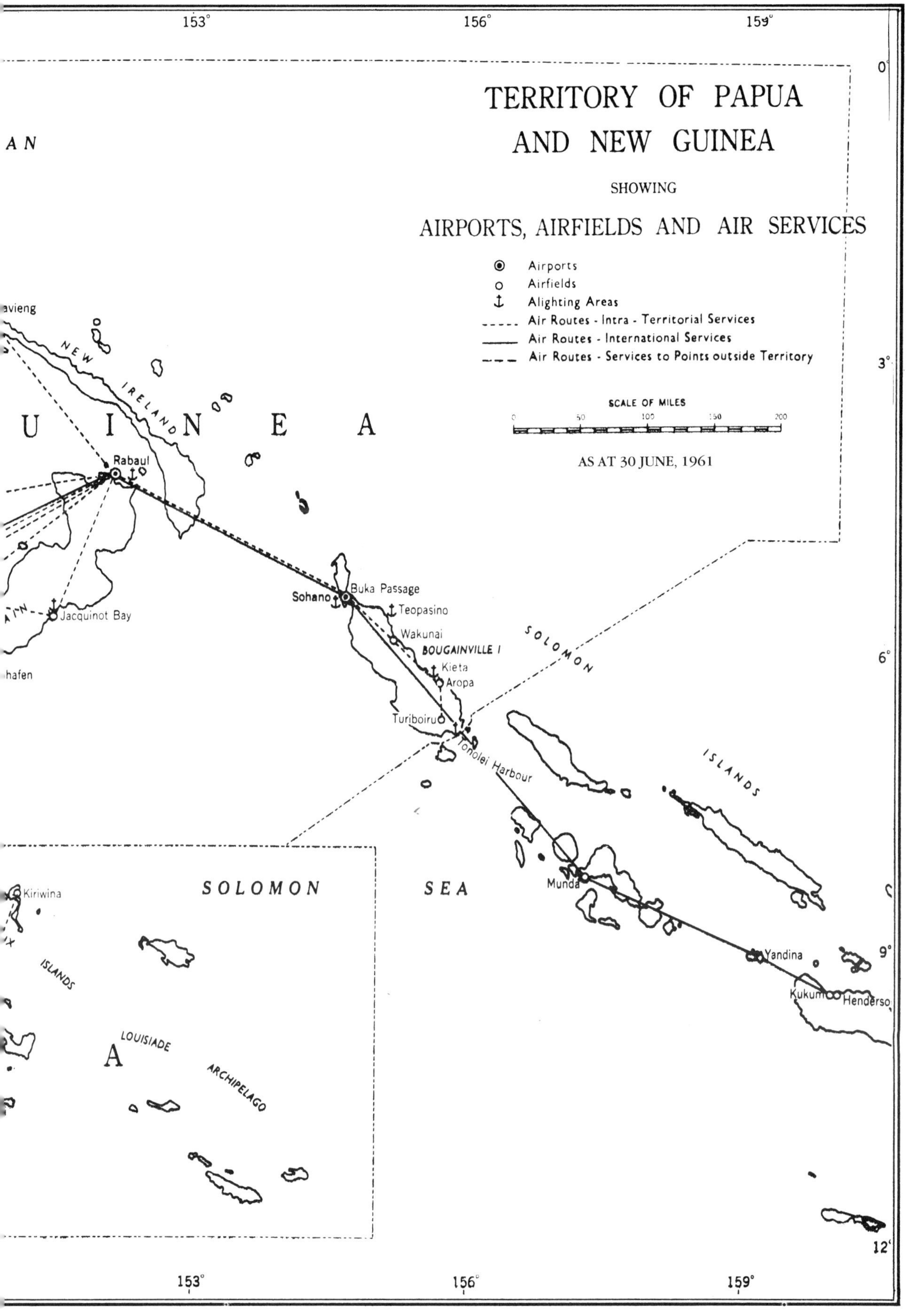
TERRITORY OF PAPUA
AND NEW GUINEA
SHOWING
AIRPORTS, AIRFIELDS AND AIR SERVICES
Airports
Airfields
Alighting Areas
Air Routes - Intra - Territorial Services
Air Routes - International Services
Air Routes - Services to Points outside Territory
SCALE OF MILES
0
50
100
150
200
AS AT 30 JUNE, 1961
153°
156°
159°
0°
3°
6°
9°
12°
AN
U I N E A
NEW IRELAND
Rabaul
Jacquinot Bay
Sohano
Buka Passage
Teopasino
Wakunai
BOUGAINVILLE I
Kieta
Aropa
Turiboiru
Tonolei Harbour
SOLOMON
ISLANDS
Munda
Yandina
Kukum
SOLOMON
SEA
Kiriwina
ISLANDS
A
LOUISIADE
ARCHIPELAGO

The practical effect of this decision was to allow the operation of the DC3/C47 at all-up weights ranging from 12,203 kilograms on take-off from Lae to 9,979 kilograms at Highlands strips like Mount Hagen and Kerowagi, and on landing from 12,203 kilograms at Lae to 10,750 kilograms at Hagen and 10,432 at Kerowagi. Lockheed Hudsons could operate from Lae at all-up weights of 8,391 kilograms, and 6,123 kilograms at Hagen and Kerowagi. Ansons were restricted to 3,719 kilograms at Lae and 2,993 kilograms at Mount Hagen. Dragons could land anywhere at an all-up weight of 2,041 kilograms and at generally lesser weights on take-off at other than coastal strips. The higher the altitude, the greater the weight penalty.

In due course, take-off and landing weights were promulgated for all types of aircraft in use covering all aerodromes. As the years went by, these requirements were often modified for invididual aerodromes and aircraft, as necessity dictated.

The directive again spelled out the fundamental rule of air operations in PNG: 'All operations are confined to Visual Flight Rules, with the one exception of the Qantas Sydney-Rabaul service. In this case certain restrictions on IFR flights are also imposed in the Port Moresby FIR.'

This DCA document made it clear that Qantas was to be the official flagbearer in PNG, responsible for the conduct of developmental services. The operations of other companies would be complementary to those of Qantas, but theirs was still a vital role. Routes would be granted to them on a balanced basis, having regard to revenue-earning potential and viability. Companies operating regular services within PNG would be required to obtain Certificates of Exemption.

Jim Glindemann remained in PNG after the Affleck survey, as District Airport Inspector. Over the following three years he established and developed the new Airports organisation set up after the survey.

The compilation of the *New Guinea Pilots Handbook* was undertaken by DCA, but, wisely, the views of pilots experienced in local flying were sought. Arthur Collins, the MAL manager, agreed to assist, as did Les Ross, Flight Superintendent of GAT; Arthur Jacobsen of Qantas, on behalf of the New Guinea Branch of the Australian Airline Pilots Association, and Hugh Birch, who was chief pilot of Island Airways, Madang.*

Arthur Collins proferred advice distilled from his long years of flying biplanes in the Territory: 'Any pilot using short one-way 'dromes should be competent to skid his aircraft sideways with the landing wheels about two feet [.610 metres] from the ground, and by using this method, quick and effective, of washing off speed, a good pilot should never overshoot an aerodrome when caught with too much speed or having misjudged wind direction and speed. Careful attention should be given to indicated air speed and correction made for altitude at 5,000 feet [1524 metres], allowing approximately one-and-a- half knots per 1,000 feet [3048 metres]. This extra speed not shown on the air speed indicator adds considerably to the landing run.

'The approach is to be made with just sufficient airspeed to maintain control, about ten miles per hour [16.09 kilometres per hour] above the stall, and height just sufficient to clear the end of the 'drome, aircraft nose-up, and in semi-stalled position...excessive speed and height is disastrous, and must bear in mind never to attempt to open up throttles and to round again, no matter how bad the landing may be...'

The submission by Arthur Jacobsen was a mini-handbook in its own right. It was prefaced by a pithy observation: 'A straight line is not the shortest distance between two points in New Guinea. It is usually the quickest way to Press headlines.'

Jacobsen followed this with a succinct list of DONT'S. '*Don't* fly on your nerves, do tomorrow's flying today, take off before you do a cockpit check and clear your motors, fly overloaded, fly in shorts, fly without a raincoat, fly in the dark, fly in cloud, fly without a safety belt, be too proud to turn back, fly without a box of matches, take off before checking for incoming aircraft, fly into any position where you cannot turn back, land on an aerodrome after its time limit, fly loose cargo, push through because you know the other pilot got through...'

Jacobsen continued: 'New Guinea flying conditions are unique in modern aviation. Nowhere, outside New Guinea, do we find men flying aircraft unaided by ground allies. In this country, the safety of flight is not dependent on weather departments, radio stations or modern aircraft. It is through the efficiency of the

*See Chapter 2, The Flying-Boats

ground engineers and pilots that a safe and efficient air service is maintained...

'The New Guinea weather is primitive and mostly unfriendly. The weather is a constant menace and danger to flying if not understood and respected. Most of the aerodromes are short, one way and narrow, with poor approaches and, in many cases, strong cross or down winds blowing over them. Many are more than 5,000 feet [1524 metres] above sea level. The maps of New Guinea are incomplete and inaccurate...compass deviation excessive and inconsistent...

'It is to be remembered that with flying in New Guinea, experience in New Guinea flying is the all important thing. Experience in other parts of the world will not keep one alive in New Guinea... '

Jacobsen recommended that all pilots should carry a light rifle or a revolver, a first aid kit, a sheath knife, two boxes of matches, a full length raincoat for protection if forced down in the mountains, and a small supply of emergency rations. It must be remembered that in 1947 most of the Highlands and many other districts in the interior of PNG were classified as 'Uncontrolled', the warlike tribes under slight — if any — government influence. Under the circumstances, the recommendations were not melodramatic.

The advice Arthur Jacobsen gave was based on the experiences of many New Guinea pilots, and it was all sound. Pilots new to PNG conditions should gradually work up to flying in command, until they had the 'feel' of the aeroplane when handling it at low speeds on landings at high aerodromes. An essential part of New Guinea flying was the ability to do steep turns in a fully loaded aircraft at all heights without stalling. He added: 'I cannot overstress the danger to the inexperienced pilot in "gap" flying. All gaps should be approached on the side, thus leaving room to turn should down- draughts make it necessary...the gap must not be crossed unless the way ahead is clear. If the way ahead is doubtful but there is room for closer inspection, the pilot must turn and look behind to make sure that he can withdraw from the area if necessary. This must be done from time to time as cloud in New Guinea will "curtain" a gap in a matter of one or two minutes, leaving a pilot "locked in" by mountains and clouds. The new pilot if in doubt should turn back, and "live to fly another day". New Guinea demands that a pilot fly in an aggressive and determined manner but it must be controlled by common sense, and when doubts creep into a pilot's mind, it is time to turn back.

'Commonsense is still the best aid we have for safe flight. It costs nothing, weighs nothing, it is unlimited and, above all, it is reliable. It requires exercise, but no maintenance.'

Rather surprisingly, DCA finally decided against the issue of the Pilots Handbook. It was concluded that: 'It is not the function of this department to issue a handbook on "do" or "don't" flying in a particular region. This is a job for each company. The department should issue an Aerodrome Index on New Guinea strips, giving more than the usual amount of information to the pilot.'

Departmental strength in PNG rapidly increased following the adoption of the recommendations contained in the Affleck-Tenenbaum Report. In January, 1948, John Arthur advised Melbourne that with the increasing development of aviation activities in PNG, a sectionalised organisation had been established at Port Moresby: Administration, Finance, and Supply and Stores. At that time, the total DCA establishment in PNG was 151, of whom 108 were local labourers. There were twenty-six meteorological officers. Headquarters staff consisted of John Arthur; clerical officers George Woodward, D.K. Garton, F. Roberts, J.A. Clarke and L. Steen; stores officers R.E. Hannan and R. Hooper. Jim Glindemann had two groundsmen working with him in his section. Ted Booker and Rube Murray spent a lot of time away, in the field.

In January, 1949, John Arthur wrote: 'Civil aviation in PNG is developing so rapidly...strengthening of the administrative staff is very necessary'. The decision was taken to remove PNG from the North Queensland region, and establish a separate PNG region, as from 1 July, 1949. John Arthur was appointed regional director.

It was by this time apparent to all observers that Lae, not Port Moresby, was the centre of PNG aviation, but a headquarters complex, once established, is hard to uproot. John Arthur and his organisation stayed in Port Moresby.

Finschhafen, at an earlier stage seriously considered as the possible centre of New Guinea air operations, rapidly declined. There was no local infrastructure to support it, as was the case in Lae. At the war's end, Finschhafen possessed facilities probably superior to any other centre in the Territory, but after the Americans pulled out, most of the magnificent equipment they left behind was stripped out, and sold cheaply through Disposals. An item in the DCA Annual Report for 1945/46 tells the story: 'Equipment acquired from the USA Signal Corps Depot, Finschhafen, included radio compasses, localiser receivers, VHF transmitters and receivers, test equipment, electron tubes, and spares...'

Geoff Masters, the radio technician who was one of the first DCA men in Port Moresby post-war, was at Finschhafen in 1947, when the Americans were still in the process of withdrawing: 'I was pretty fed up with Lae, and when Ted Booker asked me if I'd like to go to Finschhafen, I said, "Anything to get out of this bloody hole".

'Ted came over to Finschhafen with me, because there were problems, insofar as the Americans were still installed there, and the RAAF, too. We were trying to ease both lots out, and it needed a bit of diplomacy, which at that time I didn't have, anyway.

'Nobody there to meet us. Nobody interested in DCA. Ted made himself known to the American bloke in power at the time, and he said, "Wal, man, we are moving out in two or three weeks, it's all yours..." They had some tremendous gear in that place, gear we had never seen...the tower was something to be seen. Way ahead of the old wooden towers in Lae and Moresby. My eyes bulged when I walked into that Finschhafen tower...

'So I was in the middle of the American and Australian Air Forces. I wasn't liked, I could see that, and I was given really crook quarters...they were all getting ready to move out. LST transports came in one day, and the Yanks just went, overnight. Left it all behind. You have no idea how much stuff there was...

'We were to take over the very large RAAF stores depot at Finschhafen. The time came when we had to make a move, and up until this time they wouldn't allow us to have a look inside...headquarters sent over a bloke to help me, and we started checking the inventories...they were way out! Obviously stuff had been knocked off right and left...they wanted the department to sign for something like two million pounds worth of equipment, but it was below half a million when we added it all up. Jeep parts we were particularly interested in. 10,000-plus spark plugs on paper, you'd go and count them and there'd be 250. Astronomical amount of stuff gone...

'When the RAAF left they ripped out most of the equipment in the tower...we were left with virtually nothing, had to start from scratch...we took over their transmitter station, out beyond the Dregerhafen naval base — the Navy boys were still there. The RAAF boys at the transmitter station were very good. Receiver station was the same. We did very well on that side of it. Blokes there left everything. Banks and banks of AR7 receivers, and they had acquired a lot of stuff from the Yanks before they left, beautiful American receivers built into racks...

'So we started from scratch there, and built it up into a reasonable little station.'

While Finschhafen remained an important aerodrome and DCA station, it was never again a rival to Lae, which continued to expand as the tempo of aviation development in PNG quickened.

In 1947, a fourth air transport company commenced operating out of Lae. This was Taylor's Air Transport — TAT — a small organisation of the pattern of so many that had operated out of Lae pre-war: under-capitalised, poorly equipped, and overshadowed by the larger concerns. It was started by that 'battler' of the early days, the veteran engineer, M.A. 'Joe' Taylor, with the backing of old Goldfields friends, Tom Zoffman and Col Bayliss. Unluckily for Joe, TAT went the inevitable way of its forerunners. The company was taken over by GAT in March, 1949.

In the short time of its existence, though, Taylor's Air Transport tried hard. It will be recalled that Joe Taylor had had dreams of starting a 'New Guinea Co-operative Air Transport Company' in 1945, but this had proved to be impossible. He did the next best thing, left his position with Qantas and went into the air transport business on his own account. He sought, and was granted, permission from DCA to commence charter operations at Lae. He was informed that neither an airline licence nor a Certificate of Exemption could be granted.

Joe Taylor acquired a pair of Dragons, registered as VH-AFK and VH-AHY. He looked around for pilots.

As it happened, Doug Elphinstone, the ex-Qantas pilot, was looking for a flying job. After leaving Qantas, he had gone into business at Wollongong, but this had not been to his liking and he was glad to accept Joe Taylor's offer of the position of chief pilot of TAT. A second pilot, Ross Kerrick, was also engaged. The two Dragons left Sydney on the flight to Lae on 8 June, 1947. Doug Elphinstone remembers: 'There we were in the two Dragons, Joe Taylor and myself in one, Ross Kerrick in the other, two mechanics, and a traffic clerk, all our spares, suitcases, everything to run an airline. Just arrived up in Lae unannounced one day, and started flying the next. Joe had decided he was going to make himself into an airline tycoon.....

'Seems funny now to think it could have been done that way — it couldn't be today. There was no DCA at the time, did have a radio station but there was nobody on the tower. You just steamed off and went wherever you liked. No flight plans or anything...at that stage, MAL was well entrenched, and Guinea Air Traders had some Ansons working, doing all BGD's work. That left more or less the Highlands for poor old Joe Taylor. So we started a regular service every Friday, up to Wabag via Kainantu, Goroka, Chimbu and Hagen, and back again.'

In March, 1948, Doug Elphinstone left TAT, and joined the Strickland Inc. Syndicate on a search for gold in the wild Porgera River country. W.E. Passlow became chief pilot of TAT.

The syndicate members were the famous prospector and explorer, Mick Leahy, and his brother Jim; Administration officers Nep Blood, Joe Searson and John Black, and Doug Elphinstone. It was Elphinstone's task to keep the syndicate supplied from the air, using a DH83 Fox Moth, owned by Norman Wilde of Wau. The whole adventure was a flop. The Administration tried to stop the party from entering the uncontrolled Porgera country by sending Patrol Officer Peter Moloney after them, from Wabag. Moloney failed to turn the prospectors back — Mick Leahy fired a shot over his head — but when they reached the Porgera, they found no bonanza. (It is interesting to note that today the Porgera gold deposits are the subject of world-wide attention).

Pilot John Spiers had already demonstrated a talent for getting into strife. While the Strickland Syndicate members were at Wabag on the outward journey to the Porgera, Spiers made a routine landing in a Dragon with supplies for the government station. He was at this time flying for MAL — many pilots then restlessly changed companies, for a variety of reasons that seemed good at the time. He cut the engines, and walked over to the radio shack, where Peter Moloney was trying to get his battery-charger started. Both Moloney and Spiers were gun addicts, and there were plenty at Wabag then — Doug Elphinstone reckons there were probably sixty or seventy rifles and shotguns with that syndicate expedition. Moloney had a long-barrelled Luger navy pistol, of which he was very proud. Elphinstone and the others were in the nearby office, and heard the two men talking. Says Elphinstone: 'We heard Moloney say, "You should see my Luger." So we heard, Click! Click! Click! and then a roar! A Luger bullet went down past Spiers' knee and through his foot. Johnny jumped three feet [about a metre] into the air on the other foot, and down he came...'

After the Porgera expedition, Doug Elphinstone joined MAL as chief pilot. Arthur Collins had left the company, and Brian Carpenter — one of the W.R. Carpenter clan — was manager/pilot. It was customary then for MAL to fly its Dragons to Sydney for the annual C of A inspection. On 11 May, John Spiers — his foot nicely healed — took off from Lae in VH-AKX, bound for Sydney via Daru, Horn Island and the Australian coastline.

Spiers left Daru early on the morning of 12 May, 1948, on the over-water leg to Horn Island. Half way across Torres Strait, he encountered a terrible storm, and in seconds the Dragon was enveloped in black rain-cloud. He lost sight of the sea, and he turned back for Daru. He was down low, right on the sea, when he saw ahead through the torn curtain of squall-cloud a small, low island, hardly more than a mud-flat: Turnagain Island.

Spiers landed, thoroughly wrecking the Dragon in the process. It was lucky for Johnny Spiers that the tide was out, for Turnagain was mostly covered with water at high tide. But there he was, trapped, with no food, water or weapon — confirmation that the advice Arthur Jacobsen had given in his notes for New Guinea aviators was indeed sound. And the Dragon had no radio, so Spiers could not call for aid. A considerable predicament!

Tom Deegan well remembers the sequel: 'So there we had Johnny missing. Brian Carpenter and I got a DC3 out of the hangar — we had put a couple of C47's

33

into service in March — and off we went to search for him, all the way to Daru, across the Australian mainland, down into the Gulf country. Searched for three or four days, but couldn't find Johnny. We'd just about given him up, and were on our way back from Horn Island to Daru when Brian Carpenter said, "Look over there!" And there he was, sitting on the roof of the Dragon on that mud- flat. He'd been there six days. The island used to flood at high tide, and he'd climb up onto the wing then. When the tide was out, Johnny would walk around looking for something to eat, and there wasn't anything to eat. One day he met a crocodile, just as hungry as he was. So he shot back to the aeroplane and left the island to the big croc.

'The reason he survived was, it rained. He had a lucky beer mug that he'd pinched from some pub in England during the war, and he always carried it with him. When it rained he used it to catch water off the wings....

'Well, we found him. Had to get a RAAF Catalina from Moresby to rescue him. They flew up to Turnagain and landed, and sent an engineer across in a rubber dinghy to get Johnny. Then a big storm blew up, and the Cat had to take off again and get back to safety. So there were the two of them, Johnny and the engineer, marooned on Turnagain. They eventually had to send a pearling lugger from Thursday Island to pick them up. Johnny was flown back to New Gúinea and into hospital. He was very distressed, what with six days without food, and that crocodile.

'Poor old Johnny. He was certainly accident-prone...'

In May, 1946, the Accident Investigation Branch of DCA was established. A DCA search-and-rescue organisation was later set up in PNG, to plan and coordinate search and rescue procedures. In the meantime, search and rescue facilities were provided by the RAAF.

34

35

36

33. Tom Deegan. *Photo: Tom Deegan*
34. Tiger Moth VH-ASW, withdrawn from service, December, 1961. *Photo: Sten-Erik Molker*
35. DH84 of Taylor's Air Transport. *Photo: Ray Harris*
36. Captain Jack Stammer and patrol officer Peter Moloney, Menyamya, 1953. *Photo: Jack Stammer*

The Flying-Boats

Qantas began the regular scheduled operation of Catalina flying-boats in PNG, as required by DCA and the Administration, in May, 1949. The use of the Catalina in civil transport operations in PNG was, however, pioneered by W.R. Carpenter & Co.

In December, 1946, the famous aviator, Captain P.G. (later Sir Gordon) Taylor invited a highly experienced ex-RAAF flying-boat captain, Hugh Birch, to join his newly-formed air company, Marinair.

Hugh Birch graduated as a pilot from No. 1 Flying Training School, Point Cook, in December, 1937, and flew DH60A seaplanes, Seagull amphibians and Supermarine Southampton flying-boats. In April, 1938, he was transferred to No. 5 Fleet Co- operation Squadron which provided the pilots then carried in each cruiser of the Royal Australian Navy, to fly the Walrus reconnaisance amphibians. His experience of PNG flying began in May, 1939, when he was sent to Port Moresby to commence the first aerial survey of that district as a preliminary to the later RAAF build-up there.

In July, 1939, Hugh Birch was one of six pilots chosen to go to England to commence the delivery of Short Sunderland flying-boats for the RAAF. Initially, the six pilots were converted onto the big four-motor biplane Short Singapore boats that had been in use in the Royal Air Force since 1926. When the Second World War broke out, Birch and his companions joined the famous No. 10 Squadron, RAAF, and he spent nearly three years on operations out of England. Birch was at age twenty-six the youngest Group Captain in the RAAF. He was awarded the DFC, and was twice mentioned in despatches. This background well qualified Hugh Birch to join P.G. Taylor's new company, Marinair. Birch remembers: 'He had bought five Catalinas from the RAAF at Rathmines for about a thousand pounds each. The idea was to operate a Sydney-Lord Howe Island service. We made one flight to Lord Howe in A24-350 on 28 December, 1946, and then flew her from Rose Bay to the Cooks River, Mascot, where the aircraft was beached and a conversion began to turn her into a passenger aircraft. I also flew two more "Cats" into Cooks River — A24-358 and A24-384. But P.G. Taylor had to abandon the project when they refused him a licence. With the exception of A24-354, the aircraft were later sold to the Dutch Air Force. On 15 May, A24-354 was registered as VH-ALN and christened *Island Chieftain*. On that day, the aeroplane was purchased by W.R. Carpenter & Co, and I joined the company. A subsidiary of Mandated Airlines was formed, called Island Airways Ltd.

'I was to fly *Island Chieftain* out of Madang. I was the only pilot/navigator. We had a radio operator, two flight engineers — Tom Mitchell and Ken Smith — and a ground maintenance engineer, Charlie Brush. On 19 May, 1947, we set off for Madang via Bowen and Port Moresby, and arrived at our future base on 22 May. There was an old RAAF seaplane slipway there, and on the following day we beached the "Cat" and carried out minor maintenance before taking off on our first charter, to Wewak.

'Whilst we were operating from Madang, I received my instructions from Jack Sedgers at Rabaul, who was the WRC manager in New Guinea, and from Bert Perriman, a director of WRC in Sydney.

'That flying-boat, *Island Chieftain* was a most remarkable and reliable aircraft. It was a PB2B-2, a straight flying-boat, not an amphibian. Most of our work was transporting native labourers from all parts of PNG to plantations in New Britain, New Ireland and Bougainville. We carried fifty or sixty at a time. We made landings in places like Woodlark Island and Sissano Lagoon, and I believe were the first to land at Ninigo Lagoon in the Western Islands. Carpenters had a plantation there. We also flew to Mushu Island, where we discovered an enormous pile of caskets containing the remains of Japanese soldiers who had died in the Wewak-Aitape area...

'About that time John Rowe became my co-pilot. Up to that time I had been flying the Catalina on my own, with the aid of the radio operator and engineers. In September, 1947, we flew to Port Moresby under charter to Oil Search, and after landing in the fast-flowing Kikori River we took off and made supply drops to Oil Search geologists in the Samberigi valley, using storepedoes. Also made a survey of the Fly and Strickland rivers for them. We made flights all over PNG, quite a number on medical emergencies...

'In November, we bought *Island Chieftain* to Sydney for overhaul, and returned to Madang early in December. January '48 saw us once more chartered by an oil exploration crowd, this time the Australasian Petroleum Company. We were again based at Kikori, and we made a number of drops over a period of a week to a geological expedition at the very remote Carrington Junction, on the Strickland River. We dropped 414 bags and seventy-three storepedoes of supplies during that week, almost 32,000 pounds' [14,515 kilograms] weight. Again in March we flew back to Kikori for APC, and dropped to geological survey parties on the Cecilia River, the Strickland-MacGregor junction, and Mount Bosavi. This was exciting flying. The country was extremely inhospitable-looking, and we dropped from a height of 100 feet [thirty metres]. Dropped in the Black River area, too. Those geologists were always very glad to see us...

'In April, we did a drop in the Porgera River country, to the Leahy Brothers and the Strickland Syndicate party. Quite exciting — we had to fly very high to get into this vast valley. That month, too, we dropped to Ivan Champion's camp on the shores of Lake Iviva, at 8,500 feet [2590 metres].

'Finally, on 7 May, 1948, we returned to Rose Bay. The Island Airways operation finished because DCA brought in stricter regulations governing the use of Catalinas that meant we could no longer carry fifty plantation labourers at a time. Seating was the limiting factor, and the Catalina did not have proper passenger seating, of course. This made the recruiting flights an uneconomic proposition. Fortunately for me, *Island Chieftain* was purchased by Qantas, and I joined them, too...'

Island Airways had another Catalina flying-boat ready for PNG operations, VH-BDQ, christened *Island Warrior*. It, too, was eventually sold to Qantas.

Hugh Birch became a Qantas Catalina captain on the routes to Lord Howe Island, Noumea, Suva, Vila and Espiritu Santo. In June, 1948, he did a conversion course onto one of the most unusual aircraft operated by Qantas, a Grumman G44A Widgeon amphibian, VH-AZO. Although operated by Qantas, the aircraft was the property of the Australasian Petroleum Company, and it was being flown by a Qantas pilot in Western Papua, P.L. Oakley. While Oakley was on leave, Hugh Birch flew the little Widgeon, finding it quite a change from the heavy Catalina.

Phil Oakley was another RAAF-trained pilot. His interest in flying was captured after his brother joined the RAAF in 1937, when Oakley was still a schoolboy. Pilot of a Bristol Blenheim bomber, Oakley's brother was shot down and killed while attacking the Maas River bridges during the swift and bloody Low Countries campaign of May, 1940. Phil Oakley turned eighteen in February, 1942, and joined the RAAF. He got his wings, and in due course joined 10 Squadron at Plymouth, England. For two years, he flew Sunderlands out of Plymouth. Oakley remembers: 'Great grounding as a flying-boat captain. Plenty of flying hours — I got up 1,300 hours, which was a lot of operational flying. The Sunderland was an excellent aeroplane, good to fly, easy to handle despite its size, and it was well armed. Tough, too. But we lost a lot, mainly from massed JU88 attacks. We were attacked when I was a third pilot by six 88's, down off Spain, and we only got away with it by dodging into cloud. Couple of the blokes wounded. Aeroplane was badly damaged, and we couldn't control the engines, throttles all shot away, pitch control shot away, still don't know why the engines kept running, but they did. Cannon shells did most of the damage. We ran out of fuel just outside Plymouth and ditched. We hitched a tow from a trawler passing through the boom, madly bailing and pumping. That aeroplane flew again. Plymouth was a wonderful town — the people adopted No. 10 Squadron, and we could do no wrong...'

The war ended. Phil Oakley would have liked to stay on in the peacetime Air Force, but it was not to be. He was discharged, and was fortunate enough to obtain a position with Qantas in November, 1945. The Qantas fleet was then small, the aircraft mainly Lancastrians and Liberators, but it also included a Lockheed 14, VH-ADT, used on the Sydney-Townsville route. Oakley was endorsed onto the Lockheed, and flew the route during 1946, with captains like Phil Miller and Mack Twemblow, both of whom later flew for Qantas in PNG. The Lockheed was written off during an accident in January, 1947 — Oakley was not involved — and Phil Oakley had no aeroplane to fly. He became third pilot on Liberators, and remembers them as 'beasts of things, seating about twenty-two passengers in a configuration not far removed from war'. Then he converted onto Hythe flying-boats, operating between Sydney and Singapore.

'I was still only twenty-three, and I got a bit restless. Still feeling the restlessness of the war. Then I noticed a Qantas advertisement for a pilot with my sort of background to be seconded to APC for one year, to fly a Widgeon in Papua; to get an endorsement, fly it up, and operate it as they required. So I applied. I think I was the only one — of course, there were not too many young flying-boat captains...

'The APC bought the Widgeon new. VH-AZO. They were ill-advised. It only weighed 4,500 pounds [2041 kilograms] all-up, about the same as the Dragon. Had a couple of 200 hp Ranger inline engines. Underpowered, couldn't carry anything. Range was OK, if it had had a bit more payload. They'd have done better with a Goose. But they had it, and got Qantas to operate it. Qantas also recruited a young engineer from their ranks to go with me: Ted Wickham.

'So we set off in May, 1947, from the Qantas servicing base at Rose Bay. Plan was to follow the usual light aircraft route — Sydney, Brisbane, Rockhampton, Townsville, Cairns, Horn Island, Daru. All went well. We were held up at Townsville for three days, bad weather, and waiting there too was Bobby Gibbes, the Middle East fighter pilot. He had just joined Mandated Airlines, and was taking up another Dragon. There was another MAL aeroplane there too, waiting for the weather to lift — a Lockheed Hudson.

'Eventually we got away. I had problems. This aeroplane used 91 octane fuel, and couldn't take the 73 or 100 octane that the Dragon and the Hudson used. There was 91 available at Iron Range, but nothing further north, although there were stocks in New Guinea. The Widgeon had a range of only 400 miles [643 kilometres] or so...I tracked across the Coral Sea from Iron Range in the general direction of Port Moresby, and made landfall at Yule Island. Then flew down the coast to Moresby.

'I was supposed to land at Jacksons — the Widgeon was an amphibian — and there was an APC welcoming party there. But I couldn't tell which was Jacksons or Ward's Strip, so I landed on the harbour, taxied up to the slipway, and asked for directions. Created quite a stir...so we took off, and landed at Jacksons.'

The Australasian Petroleum Company was formed in October, 1938, by the huge Anglo-Iranian Oil Company, Vacuum Oil Company, and the small Oil Search Limited of Sydney. Oil Search held extensive exploration permits in Papua and the Mandated Territory of New Guinea, and Anglo-Iranian had a fifty per cent shareholding in a Vacuum subsidiary, Island Exploration Company. APC was formed to exploit the Oil Search and Island Exploration leases. APC had a nominal capital of one million pounds, and immediately commenced survey and drilling programmes. The company made extensive use of aircraft in its operations in Papua — a Fokker FV11b-3m, and a pair of Douglas Dolphin amphibians — and in March, 1941, commenced to drill a major test hole at Kariava, 122 kilometres up the Vailala River, and 144 kilometres west of Port Moresby. A depth of 1646 metres had been reached by 27 January, 1942, when work was suspended after the Japanese invasion.

Drilling machinery on the Kariava site was dismantled, but not destroyed. In conjunction with Army authorities, APC was able to maintain a skeleton staff of whites and natives at Kariava — oil, after all, is the life-blood of the modern world — and skilled members of the prewar staff were re-engaged by APC as they were discharged from the armed forces. The result was that when operations were recommenced in March, 1946, rehabilitation was not as difficult as it might have been. By May, 1947, APC already employed a staff of 120 whites and 1,100 Papua New Guineans in Papua. The nominal capital of APC was increased to £10 million pounds, and by early 1950 the total of employees in Papua had jumped to 420 whites and 2,300 natives, plus another fifty whites at the Melbourne and Sydney headquarters of the company. APC was then the largest private employer in PNG, with Bulolo Gold Dredging Ltd a distant second.

The chief difficulty confronting APC when it resumed operations was the familiar one of the transport of personnel, supplies and equipment. Shipping between Australia and Port Moresby was sporadic and unreliable, and vital materials from England and USA could literally take years to arrive. And APC operations were not concentrated at Kariava. What eventually grew into a massive headquarters complex was established at Port Moresby, and there were two other sites on the Vailala River; Hohoro, ten kilometres from the river mouth, and Upoia, thirty-two kilometres upstream from Hohoro. The three camps were linked by radio communications to headquarters. The Kariava test hole reached a depth of 5181 metres by May, 1947, and APC announced that a much more extensive programme of work would be undertaken using the very latest methods of geophysical surveying. Two further drilling rigs would be imported, and more exploratory wells would be sunk in other parts of the 72,520 square kilometres of concessions held by the company. It was expected that 'many millions' of pounds would be spent over the coming years.

It was primarily to move technical experts around in the field, and to maintain a courier service between Port Moresby and the camps, that APC purchased the Grumman Widgeon. The massive, heavy equipment required for drilling test wells was transported by sea, in chartered and APC-owned vessels, Airdropping of supplies was done by Hugh Birch in the chartered Catalina, *Island Chieftain* — to the Fly-Strickland Survey, led by the great G.A.V. Stanley in January, 1948, the Samberigi Survey, and others and later by Qantas Catalinas. The little Widgeon was the APC aerial taxi, and Phil Oakley quickly learned the unfamiliar techniques of landing a small amphibian on fast-flowing water. He remembers: 'We lived at the old Moresby Hotel. The Qantas establishment consisted of a senior traffic officer, two traffic officers, an engineer, and a Chevrolet ute, a jeep, and a trailer. That was it. So we increased the establishment by fifty per cent. It was all pretty basic, but then, so was everything in Port Moresby.

'Operations with the Widgeon began by flying to the current drilling site, at Kariava. Found what looked a reasonable alighting area somewhat upstream from the camp. All new to me, operating on rivers and in strong currents. Initially, we just went out whenever there was something to go out. Didn't start off as a planned, regular operation to do a particular job — they had the aeroplane, and didn't really know what to do with it...

'People always ask me whether floating logs weren't a great hazard. I operated on the rivers for ten years, and I never had any particular problems. They were obviously a hazard, but if you took the trouble to fly a circuit and took note of the current flow, you could usually pick a path...you had to be careful taxying in narrow rivers. It was always possible to hit something under the water, and indeed that's what probably happened when I had the accident...'

As was the case with landplanes, flying waterplanes off the rivers of PNG posed special problems. No amount of experience elsewhere fully prepared a pilot for New Guinea. If a pilot new to the country was going to have an accident, it usually happened early on in his career. So it was with Phil Oakley.

On 9 August, 1947, Oakley took off from Port Moresby on a routine flight to Kariava. He had a passenger, APC engineer Frank Edwards. Oakley knew that his landing would be filmed by the veteran Fox Movietone cameraman, Syd Wood, who was in PNG making a newsreel of Qantas operations. Wood got the sequence of his life when the little Widgeon touched down, an extraordinary record of a spectacular 'prang'.

'To be honest, I don't really know precisely what happened. Maybe I had stage-fright, knowing the landing was to be filmed. I'm not conscious of having hit anything, but you can never be sure with those Papuan rivers. The water was particularly glassy, and on touchdown we bounced a bit, two or three times, pretty big bounces. I hadn't realised how serious they were, and when we settled on the water the aeroplane veered off to the left, because the port float had become separated. We glided relatively calmly and slowly into the mud at the side of the river, and stopped. Frank Edwards dived up the front, opened the mooring hatch, and stood up with an amazed expression for some moments, scratching his head.

'There was a fair bit of damage. Port float torn off, trailing edge of port flap and port aileron bent, starboard tailplane fractured...couldn't repair her there. They towed her down to the Kariava river camp, where they had quite a setup, but there was nowhere to beach her. So had to improvise a mooring on the river. Our engineer, Ted Wickham, came out in an APC vessel. They built us a slipway — it took six days — and we winched the aeroplane up as high as we could, and tied her down. Lucky for us, because the river flooded that night...anyway, we repaired her. One of those things. Don't forget, the longer you are at it, the better you get. I certainly learned a lot from those early days on the rivers. The Widgeon was a nice little aeroplane, handled well, stalled at about fifty, no vices.

'Widgeon operations were never large, mainly because it was so restricted in payload. We had to bring her down each year for C of A. Always a hideous problem because inevitably there would be a wait for parts. First time I was away five months. Spares had to come from America. The Widgeon was pretty much of an orphan aeroplane, we had the only one out here. It was the wrong aeroplane for APC, so they sold it to Fred Lamb of New Zealand Tourist Air Travel, the first of the Widgeons that they operated over there.

'Then APC negotiated with Qantas, and I believe that Scotty Allan — he was Controller of Technical Development — recommended the purchase of two second- hand Vickers Supermarine Sea Otters. Almost a Walrus with a puller instead of a pusher engine. They bought them in the U.K. with associated spares, shipped them out, and flew them from Sydney to Port Moresby. They came out early in 1950. VH-AJN and AJO. By this time I was back in the Qantas mainstream. I think it was Hugh Birch who delivered the first Sea Otter...'

In April, 1948, W.L. Milne of DCA, and Orm Denny and M. Baird of Qantas visited Port Moresby and Lae to discuss with Colonel Murray and his officers the introduction of the new developmental services required by the Administration within PNG. Colonel Murray confirmed that the most urgent need was still for the introduction of the Catalina services from Port Moresby to the Papuan coastal stations — almost entirely ignored by the Lae-based operators — and the islands, and the DC3 services from Lae to Madang and Wewak, and to Rabaul, Kavieng and Manus. The Lae companies were fully involved in the Morobe District and the Highlands, and were not interested in developing these new services. The Administration was still paying the £50 per month subsidy to MAL for a fortnightly Lae-Madang-Wewak service, but that was the only subsidised developmental service being flown.

It was acknowledged that the services required by Colonel Murray would be expensive to operate, and would have to be subsidised. The Administrator had made provision for an amount of £15,000 for subsidy payments on his estimates, but it was agreed that it would be difficult to make an accurate estimate of requirements at present. Any eventual subsidy would probably be paid jointly by DCA and the Department of External Territories. It was estimated that the new services would require 700 Catalina hours per annum, and 500 DC3 hours.

Denny and Baird agreed to recommend an expansion programme to Qantas management, in order to implement the developmental services. This called for the establishment of a 'New Guinea Internal' organisation, based at Lae, by 1 June, 1948, with one DC3. Two more DC3's would be provided 'as soon as practicable', as would three additional DH84's. A Catalina would be based at Port Moresby, 'as soon as practicable', to fly the Papuan and Island routes.

At the time of this meeting with the Administrator — 21 April, 1948 — the only airline licence held in PNG was by Qantas, covering the sector of the Bird of Paradise service from Port Moresby to Lae, Finschhafen and Rabaul. Guinea Air Traders held a Certificate of Exemption, and all operators were still unlicenced. It was felt that the issue of airline licences to companies at that stage could involve DCA in compensation claims after Qantas expanded. It was therefore recommended that charter licences should be issued forthwith to all operators covering the whole PNG Region. The GAT Certificate of Exemption would not be renewed when it expired on 30 June, 1948. Qantas would be granted airline licences 'for those routes which they consider warrant same, following experience in the area'.

To enable the level of any future subsidy payments to be accurately determined, Qantas was to submit six-monthly detailed statements to DCA covering passengers, freight and mail carried, and operating revenue and expenses. For the first six months, Qantas would operate on an indemnity basis. The Administrator had already issued instructions that 'where possible', all Administration business was to be offered to Qantas.

The recommendations that flowed from this seminal meeting were basically accepted by Qantas, DCA and the Department of External Territories. The great

Qantas expansion in PNG began. 'New Guinea Internal' was established, a development discussed in the following chapter. DC3 aircraft were by this time relatively easy to obtain, but this was not the case with the Catalina. The Catalina services were later in starting.

A contributing factor was Australia's balance of trade. In July, 1948, John Arthur wrote to DCA headquarters in support of the proposed Catalina services: 'It is pointed out that whereas air services have been and are still being developed in the Territory of New Guinea, Papua is not served by any form of air communication. In the event of emergency, the Civil Administration rely upon the availability of the RAAF Catalina search and rescue aircraft which are based at Port Moresby...

'The Administrative Superintendent of the Australasian Petroleum Company has communicated with me requesting advice as to when the projected Catalina service is likely to commence...Qantas headquarters have advised that the implementation of the service has been deferred for an indefinite period due, mainly, to the restriction in use of aviation gasolene in order that dollar credits may be conserved. APC, an organisation supported by the Australian Government, engaged in work of national importance, wish to add their support to the early implementation of a Catalina service from Port Moresby. The Company has accumulated 60,000 pounds [27,216 kilograms] of cargo to be dropped by air...for this year, to date, the Company has dropped 140,000 pounds [63,504 kilograms] of air cargo. They further anticipate that approximately 50,000 pounds [22,680 kilograms] of air cargo will be accumulated to be dropped over six monthly periods...APC formerly employed Catalina aircraft, VH- ALN, owned and operated by Island Airways...W.R. Carpenter & Co have since withdrawn this aircraft from New Guinea...'

Arthur went on to urge that the implementation of Catalina services be given priority, 'in order that a fundamental service may be established in Papua to Government stations and communities scattered along the coast and at island locations...His Honour the Administrator urges that the circumstances of extreme isolation of the Papuan towns warrants approval of the necessary petrol...Moreover, meeting the air transportation needs of APC may hasten Commonwealth production of petroleum products from our own crudes.'

In November, Arthur wrote again, requesting 'urgent action. It is anticipated that the RAAF at Port Moresby will withdraw their Catalina aircraft and transfer search and rescue facilties to Momote on 20 November. The machine will not be replaced...it is hoped the Qantas Catalina would arrive before the RAAF withdraws...'

In March, 1949, the Director-General of Civil Aviation advised the Department of External Territories that 'Qantas has secured Catalinas, and they are now being overhauled. It is hoped the first would go to Port Moresby in April to make route survey flights. Captain O.J. Denny is at present in Moresby.'

The problems with petrol and dollar balances had been overcome, and Qantas was ready to roll. As previously mentioned, the company purchased the two ex-Island Airways Catalinas from W.R. Carpenter & Co. VH-ALN *Island Chieftain* was the first to be taken over, and was re-registered as VH-EBC. The name was retained. The other, VH-BDG, became VH-EBU.

Hugh Birch returned to PNG operations in May, 1949, once more at the controls of *Island Chieftain*. He joined Orm Denny at Port Moresby, and from 14 May surveyed a number of air routes which would be used as scheduled services: from Port Moresby to the west and east, and to the islands. During these survey flights, Birch and Denny made what was probably the first Catalina landing on the extraordinary Lake Murray.

Hugh Birch retains vivid memories of these flights: 'We landed at Samarai. Very tricky, because of the heavy swell, and I had to do a stall landing for that reason. When scheduled services started we nearly always had to do full- stall landings there, with the tail end of the Catalina touching the water first. Quite spectacular to see, and fairly frightening for the passengers, but I had a lot of experience in landing in rough seas...'

The scheduled Qantas Catalina services began on 30 May, 1949, with a run from Port Moresby to Yule Island, Kerema, Kikori, Lake Murray, Daru and return via the same ports. Hugh Birch was at the controls of *Island Chieftain* with Ross Treadgold as his first officer. On 1 June, the first east-bound flight was made, to Abau, Milne Bay, Samarai, Esa'ala, Losuia, Trobriands, Kitava Island, DeBoyne Lagoon,

Samarai and return via ports.

It is difficult to over-estimate the effect of regular Catalina services on the people who lived in the far-flung outposts of Papua and the Islands. At one stroke, the terrible isolation of the outstations was abolished. Mothers no longer lived in fear of sudden sickness or injury striking down their children, with no hope of emergency evacuation. Regular supplies of fresh and frozen foods were all at once available. Mail was now a weekly or fortnightly event, no longer at the mercy of the vagaries of coastal shipping. The pilots who flew the Catalina services over the years — and, indeed, PNG pilots in general — were ever ready to oblige the outstation people; to shop for women in the big stores in the towns, to see that schoolchildren on holidays were properly looked after; to run personal errands of all kinds. And in return the pilots knew that a cup of tea or a sandwich were waiting at the many stops along the route, and overnight accommodation if the weather turned sour. Lifelong friendships were forged in those days, between people with a mutual dependence upon each other.

The Catalina was perhaps the best loved of all the aeroplanes that flew in PNG. On the water flat and ugly, in the air as graceful as a soaring pelican, this ex-patrol bomber, with its vast parasol-mounted wing, two Pratt and Whitney radial engines of superb reliability, and roomy, blister-adorned hull was noisy and slow, but it had an extraordinary range, and was as tough as an old boot, although the ravages of salt water corrosion caused continual problems for the engineers. The Catalina did a job that no other available aircraft could have done, and did it supremely well. It sometimes appeared as if there was nothing the Catalina could not do.

Hugh Birch recalls one particular run to the western coastal and inland stations carrying a load that was typical for the Catalina: seventeen passengers, three fowls, 200 day-old chickens, a dog, two cats, and 363 kilograms of freezer meat. If it would fit into a Catalina, Qantas would carry it.

Without the Catalina, the Administration would have found it difficult, if not impossible, to re-open the remote outpost at the serenely-beautiful Lake Kutubu, first established by Assistant Resident Magistrate Ivan Champion and Patrol Officer Bill Adamson in 1938. The magnificent Lake Kutubu was the base from which the Southern Highlands district was explored and developed during the 1950s.

The officer placed in charge of the Lake Kutubu work was Assistant District Officer S.S. Smith, assisted by Patrol Officer D.J. Clancy. Hugh Birch was asked to fly Smith and a forward party to Kutubu in *Island Chieftain* so that the work could begin: 'We made several attempts to get in to the lake from Kikori, but each time the weather turned us back. We finally made it, on 22 August, 1949. Took off from Kikori with Syd Smith and his police, and made the long climb over the ranges, levelling out at 7,000 feet [2133 metres], over those wonderful Beaver Falls and the Mubi River...through a break in the clouds we could see Kutubu, deep blue and so calm that the clouds were reflected in its surface. Commenced a gradual descent and landed on the western edge of the lake but I found the aircraft couldn't be anchored there — the water was well over 1,000 feet [305 metres] deep. So we drifted until a canoe came out with about twenty natives on board, led by a rather fierce-looking chief with bird of paradise plumes, and a big bone through his nose.

'One of the native policemen on board the Catalina was a local, and he spoke with the chief who quickly climbed on board through one of the blisters and told us we were at the wrong end of the lake. With the canoemen screaming their heads off we took off and flew at an altitude of about ten feet [about three metres] for eight or nine miles [thirteen to fourteen kilometres] to the far end. Found a perfect place to anchor and Smith and his police set up a camp. We decided to stay the night there in case there was any trouble with the people, but there wasn't. We unloaded the stores, and then we went fishing...

'In that period when we were based in Port Moresby in *Island Chieftain* from May 1949 to May, 1950, I made 331 landings and take-offs, without incidents. Also flew the APC Sea Otter and the Grumman Widgeon when Phil Oakley was on leave. In June, 1950, I completed a conversion course onto the DC3 and was transferred to the Bird of Paradise service. Fred Fox took over the Catalina services...'

Captain Cedric Gordon Fox was known throughout his considerable flying career as 'Fred'. Fred Fox was a Queenslander. When war broke out in 1939, he joined the Army, just before his eighteenth birthday, and next year applied for a transfer to the RAAF. He completed his elementary flying training, on Tiger Moths, in October, 1941, and with ten others was sent to Canada for operational training under the Empire Air Training programme. Fox's group passed through Honolulu the week before Pearl Harbour. They were the last course of Australian pilots to go across in peacetime conditions.

Fred Fox trained in Calgary, on Cessna Cranes, then went by sea to Britain where in due course he was attached to the RAF, and did a tour on flying-boats based at Gibralta. From there he was posted to the Marine Aircraft Experimental Station, where for several months he flew every available type of flying-boat, including scale models of the Short Shetland and Short Solent, and the first of the Mark 5 Sunderlands, fitted with Pratt and Whitney instead of Bristol Pegasus engines. He flew a wooden-hulled Walrus, and the American Martin Mariner and Consolidated Coronado boats.

In June, 1944, Fred Fox was returned to Australia, where he did a test-pilot's course at Laverton. Eventually, he went to the Aircraft Performance Unit at Laverton, where he flew many types of high-performance aircraft: Spitfires, Kittyhawks, Mustangs, Mosquitoes, Beauforts, Beaufighters; and also Catalinas and Mariners. He was discharged in April, 1946.

This unusual breadth of experience gave Fred Fox a solid background as a marine pilot. Qantas was then operating Hythe flying-boats on the Sydney- Singapore leg of the UK-Australia service, and in April, 1947, Fox joined the company, obtaining his Hythe command in August, 1948. In January, 1949, he went onto Catalinas. But the day of the flying-boat on international routes was drawing to a close. The Lockheed 749 Constellation was coming, and this aircraft would spell the deathknell of the flying-boat. There was still a place for the flying-boat in out-of-the-way places like PNG, though, and when Hugh Birch and Ross Treadgold began to run out of hours, Fred Fox volunteered to go to Port Moresby to lend a hand. When Hugh Birch went onto the Bird of Paradise service, Fox took over. He was joined in November, 1949, by First Officer Ron Roberts.

Like many of the Qantas captains and first officers who flew in PNG after the war, Fred Fox went on to command the giant Boeing 747 jets before ending his career, but also like most of them, he remembers the New Guinea years with some nostalgia, and affection: 'It was a lot different flying the old "Cat" in New Guinea than Hythe's on the Sydney-Singapore route. There, all the paperwork — the load sheets, trim sheets etc., — was handled before the aircraft departed. In New Guinea all we had was a load sheet made out at the commencement of the day's flying, and then a form detailing passengers on and off, cargo on and off...at the end of the day the traffic officer would sit down and complete his loadsheets. The "Cat" was fitted with an inclinometer, and after operating on "Cats" for a long time we all knew where the trim should be on the bubble. So no problems in determining whether the aircraft was in trim.

'We kept a close check on the fuel, and we had a pretty fair idea of where the cargo was. We were only allowed to operate the "Cat" at 28,000 pounds [12,700 kilograms] in RPT service, and this was about 6,000 pounds [2,721 kilograms] less than the normal wartime load with mines or depth charges, so we always felt we had a pretty big safety margin.

'Operating in New Guinea required the co-operation of the crew in many ways. The traffic officer was really a purser, he looked after the documentation and cabin side of the business, seeing that the passengers had seatbelts fastened and so on. He put out the drogues when we were taxying before mooring. The two pilots and the flight engineer were responsible for the rest of the handling of the aircraft and the radio, picking up moorings and any maintenance required while we were away from base.

'Navigation was purely by visual means, but there were no great worries about other aircraft on our routes. Just the odd light aircraft. We carried radio, of course, and always reported to Port Moresby and gave them our cruising altitude. Kept pretty close to a straight-line route between the various ports of call.

'We got to know the people at the various ports very well indeed, and they, us. Pretty much like visiting friends. Not so many ports, and at first only one "Cat". So people at all the bases, who handled our moorings, knew us, and we knew them. A very friendly service.

'Passengers came out on canoes or double canoes, with baggage and freight. No cargo hold in the Catalina, so the freight was lashed down the centre of the cabin between the passengers. On some sectors, particularly Samarai-Moresby, the load was such that it was lashed down more by name than effect. The passengers all contributed in making sure it didn't come adrift. It was accepted as part of the developmental flying we were doing. We pilots had seats, but in the back it was well laden with bodies and gear.

'Sometimes we did carry bodies. On one occasion a chap came in from Samarai and during a riotous party in Moresby harbour he fell overboard from a ship, and was drowned. We flew his body back to Samarai in a large box. Didn't want to upset the passengers, so we put him fore-and-aft in a blister, and put a tarpaulin over him, so the passengers would think it was a convenient seat.

'At Talasea, the District Officer would always bring out two canoes with a table on a platform between them, beautifully set with a cloth, cups and saucers, silver spoons, and we would have tea in the shade of the wing. Same thing at Lake Kutubu...

'The outstation people gave us very good weather reports. They had crystals for their transmitters and receivers so they could operate on the recognised frequencies. The weather out in the delta could be atrocious through early morning into early afternoon. Mists and humidity caused fog extending up from the trees, and gradually the sun would lift the base of the mist, above the trees to maybe 1,000 feet [305 metres].

'At Kikori, the particular point of interest was Aird Hills, over 1,000 feet high, and seven or eight miles [eleven to thirteen kilometres] from the station. The weather report would always include the cloud base, and "we can see Aird Hills" or not. Finding Kikori could be very difficult, with low cloud over the trees and visibility only about 100 to 200 yards [91 to 183 metres]. One of the phenomena of these sorts of conditions is that fog forms over the trees, but over the river it tends to be domed, with maybe a one or two hundred feet [thirty to sixty-one metres] cloud base. Generally forms just in from the coastline or over the coast. So to get to Kikori under these conditions we would fly low along the coast, keeping tabs on it, until we reached Cape Blackwood, then turn left on the second or third river and follow it closely until we saw higher ground, just south of Kikori, then land. You could go as far as the fourth river mouth, but if you miscounted the next point was the precipitous Aird Hills...

'We called this the MW service — "Moresby West". Most interesting area, all essentially muddy water, as the coral reef finished not far west of Moresby — Yule Island, Kerema, Kikori, Daru: only Lake Kutubu had clear water. Required a lot of local knowledge to operate safely. Landing at Hohoro was interesting. It required a curved path on take-off, round a bend in the river, quite a familiar technique for flying-boat pilots, but rarely used...

'The service east — ME service — was over much cleaner and more attractive water. Samarai, Milne Bay, up through Esa'ala to Losuia in the Trobriands, down to Misima...all clear, clean water, landing in channels or lagoons. DeBoyne Lagoon at Misima was a lovely place, totally enclosed, crystal water. At Kiriwina we landed in a narrow channel, close to the shore. Inevitably there were crosswinds, at times up to twenty or twenty-five knots, particularly in the late afternoon when the S.E. Trades were fully developed. All required different skills to operate them safely, and all had their fascinations.

'Main attraction of the Trobriands stop was the airdrop to old King Cameron on Kitava. He got his freezer meat, parcels, mail, freight on every flight. It was the first event of the day after leaving Losuia. We'd sweep round low over his front yard, and drop the stuff on his lawn. We all became pretty skilled at free drops. Dropped from the blister on the command of the captain...

'The Samarai landing was pretty much open sea except for a small sheltered area directly in the lee of the island itself. Required quite a bit of skill to touch down on the right spot, so that by the time you were out of the sheltered area and into the waves from the open sea, the aircraft was travelling slowly enough to take it, without being thrown back into the sky. Tricky business during the monsoon season — wet, too, with low visibility.

'The other sector we operated was from Moresby to Rabaul, and then around New Britain and Bougainville. Ran from Rabaul to Talasea, then Kandrian, thence into the good clean water of the Coral Sea to either Lindenhafen or Jacquinot Bay...Kandrian was beautiful. The alighting area was confined by three islands. We called it the "Blue Lagoon", and when conditions were favourable, everyone would go swimming...

'One of our problems was to get the passengers onto the canoes to board the aircraft. They always lingered ashore, saying goodbye to friends. You'd yell and wave, and they'd just turn round and wave back. Trick was to close the blister and turn an engine over — then see them come!

'The other route was Rabaul-Buka Passage-Sohano-Inus Lagoon-Kieta-Buin. Landing at Buin was just like landing off the coast of Bondi, quite a lot of skill in rough-water landing techniques required. Lots of wrecked Japanese ships, and a couple of Jap torpedoes buried in the beach. Not safe for passengers, so we later changed to Tonolei Harbour. The ADO at Buin on one occasion told me a workboat was on its way up from the Choiseuls, in the Solomons, with a passenger for the "Cat". "There it is," he said, pointing away to the south, where there was a tiny speck on the sea. Couldn't wait, it was too far away, would have been full dark before we got to Rabaul, and there were no night landing facilities there.

'So I took off, and there was this boat beneath us. It was only about eighteen feet [five metres] long, which had made it appear miles away. So we landed, and took him aboard — Qantas' shortest ever flying-boat sector, less than five minutes. Didn't log that sector for the engines — would have too much explaining why we did it.

'Inus was another interesting landing. This was Paul Mason's plantation. There were two entrances to a narrow lagoon. Could be quite rough in the S.E. season when waves came right across the reef and through the channels. Had to touch down in just the right spot so that the "Cat" ran across the tops of the swells for a smooth landing. Same thing on take-off. We'd tuck ourselves right back against the reef, and the aircraft was getting onto the step about the time you struck the swells...mostly we carried plantation labourers for Paul Mason — Chimbus. We picked them up at Rabaul.

'We often did things for people, and they reciprocated. From Buka Passage we were generally loaded down with baskets of mud crabs — mangrove oysters, too. On the westerly run we occasionally got bananas and pineapples. On the eastern run, we'd get chickens, roosters, and sometimes a couple of hens. That's how Hugh Birch before me established the poultry run we took over from him...

'We carried some strange cargoes. On one occasion I carried a box of small live crocodiles to Moresby, and one of them escaped. All I could write up in the aircraft log was, "there's a crocodile loose in the bilge". Small, but needle-sharp teeth, and I reckoned the ground crew could find it far easier than we could in flight...'

The last regular airline service flight of a Qantas Catalina in PNG was completed on 24 July, 1958, with Captain Ken Montagu in command. The passing of the Catalina was mourned by a whole generation of outstation people, to whom the roar of the Pratt and Whitney engines and the sight of that flat hull suspended beneath the great sweep of the massive wing were the most welcome things in their world.

37

37. Phil Oakley. *Photo: Phil Oakley*

38

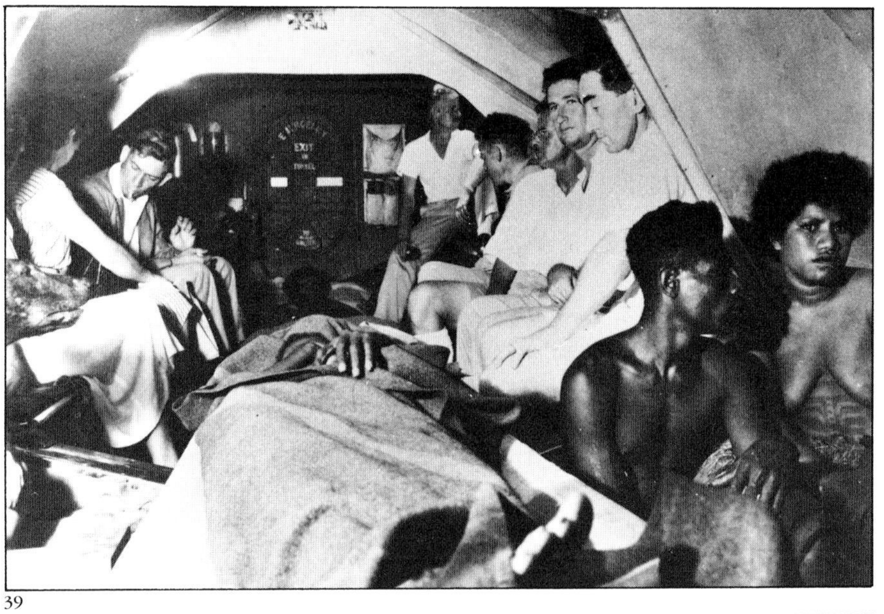
39

40

38. Grumman Widgeon VH-AZO at Sydney, 1947. *Photo: APC*
39. Interior of a Qantas Catalina, 1954. *Photo: Qantas*
40. Sea Otter, VH-AJN. *Photo: APC*
Following page:
41. Captain Hugh Birch in Catalina VH-ALN *Island Chieftain* about to take off from Cooks River, Mascot, May, 1947. *Photo: Hugh Birch*
42. Qantas Catalina VH-EBC at Port Moresby. *Photo: Qantas*
43. APC's establishment at Badili, Port Moresby, c1957. *Photo: APC*

41

4

4

New Guinea Internal

On 29 May, 1947, the Chairman of Guinea Airways Ltd, Sidney Powell, addressed the Annual Meeting of this famous pioneer New Guinea aviation company: 'We are informed that it is the government's policy to use Qantas Airways, in which it has acquired or is acquiring the whole of the shareholding, as its instrument in ex-Australian flying services...in the face of these circumstances we have reluctantly reached the conclusion that the company's prospects of resuming operations in New Guinea on a payable basis are remote...'

Guinea Airways never did resume operations in PNG. The company was refused a licence: Qantas was, indeed, the chosen instrument of the Australian Government.

Guinea Air Traders, at the beginning in a strong position to take over the dominating role in civil aviation that Guinea Airways had enjoyed pre-war, failed to survive. The fortunes of the company were severely depressed when a Lockheed Hudson, leased from Captain Lionel Van Praag, crashed with the loss of thirty-seven lives, to this day the most costly accident in the history of civil aviation in PNG. But this was not the only reason for the decline in GAT's fortunes.

On 15 November, 1947, the Certificates of Airworthiness of Ansons VH-ALS, BGD, AJJ and AST were suspended by DCA. The company relied heavily on Ansons, and had continued to add to the fleet. VH-ALS had been purchased through Australian Air Traders in March, 1947; BGD — registered as a tribute to the company's principal customer, Bulolo Gold Dredging Ltd — in April, and AJJ in February. VH-AST was of course the original GAT Anson. In mid-1947, two more Ansons, VH- BBC and BNS, were purchased from Cairns Flying Service, together with two Tiger Moths and a Dragon.

The grounding of the four Ansons was a serious blow to GAT, and was a factor in the later decision by Bulolo Gold Dredging Ltd to transfer their business to Qantas, a move which spelled the end for the company.

The failure of the Ansons was not due to any lack of care by the GAT engineering and maintenance staff. Sam Jamieson continued to employ well- qualified men wherever he could find them. Two who joined GAT in 1947/48 were to themselves start their own, highly successful, air companies in PNG: L. Crowley and C.R. Jackson.

Laurie Crowley remembers the year 1928, when he was in hospital at Junee, NSW, recovering from a riding accident: 'There was a terrific noise, and I looked up. There was Kingsford-Smith in the *Southern Cross* about twenty feet [about six metres] above the roof. Right there and then my decision was made — one day, I would fly an aeroplane. Just after the England-Australia Air Race a dozen or so aircraft arrived at Junee for a pageant. I had won twenty-two shillings backing Scott and Black in the Race, so my big day had arrived. Even if it cost me the whole lot, I was determined to have a ride. After shopping around, I settled for six shillings for fifteen minutes. The pilot was Bertie Heath...

'We lived in the bush, and I had no other contact with aviation until World War Two broke out, on 3 September, 1939. On 4 September, bright and early, I was down at the recruiting office, and had my application in for pilot in the RAAF...I was not accepted as a trainee pilot, and the officer in charge suggested I join up as a mechanic, and then re-muster as a pilot as soon as there was a vacancy. To cut a long story short, ten or twelve times during the next five years I was accepted for pilot training, but each time my squadron moved on to new areas of war, and I had to start all over again; as a result, the war finished and I was still an engineer. Having maximum priority for discharge after four continuous years of service in battle areas overseas, I was discharged on 4 September, 1945. Next morning I

44. Veteran New Guinea pilot, Bertie Heath. *Photo: Bertie Heath*

44

went to Kingsford Smith Flying School at Mascot to put my name down for flying training, I must have been the "galah" of the century to the hundreds of Air Force pilots recently discharged, and battling to get civilian flying jobs. Trained pilots available: Thousands. Jobs available: None...'

Crowley was by this time a highly experienced ground engineer, and he finally managed to get a job with Kingsford Smith Aviation Services. This entitled him to a reduced rate of twenty-five shillings per hour for flying lessons on Tiger Moths. However, as his wages were only six guineas per week, he could not afford as much flying as he would have liked, but eventually he gained his private licence.

'From then on, it was a continual battle to build up my flying hours to qualify for a commercial licence. Outside being King of England, this was the highest status man could achieve! All my spare time was used up hanging around airports, looking for free flying time. I remember riding a motor bike from Kings Cross to Bankstown to ferry a Tiger Moth to Mascot, and then hitch-hike back to Bankstown for my bike — all this to log another fifteen minutes in my logbook.'

'I left Kingsford Smith and joined New England Airways to get among the big machines — Avro Ansons and DH84's. I volunteered for all the breakdown jobs away from base, so I could fly in the co-pilot's seat; naturally, I got most of the flying, take-offs and landings excepted, while the pilot sat back and read the daily newspaper.

'The company moved me to Coff's Harbour...after some time there, I was called out after hours to refuel a visiting aircraft, an Avro Anson belonging to Guinea Air Traders. Sam Jamieson was the pilot, and after I patched up a few defects to allow Sam to get on his way he asked me how much money I was being paid. I told him,"Six pounds six shillings". Sam said, 'How would you like to double that and free travel to New Guinea and free accommodation?' I didn't hesitate one minute. A few days later, I fronted up at Mascot to board a GAT Douglas C39. The pilot was John Dalrymple, co-pilot Reg Lantzke. Also on board was another engineer, Cliff Jackson...'

Cliff Jackson had joined GAT in May, 1947, soon after the company began operations at Lae. He, too, was a countryman, born in Queanbeyan, New South Wales. He falsified his age to join the RAAF as an engine fitter in July, 1942, and worked on a number of different types of fighter aircraft at various stations around Australia before his discharge in May, 1946. He joined a division of Clyde Engineering, which held a contract at Bankstown to overhaul and convert war surplus C47 and DC3 aircraft for Trans-Australia Airlines. He concentrated on airframe work, and remained in this position until he obtained a position with the newly-formed GAT. He arrived in Lae in May, 1947.

Cliff Jackson well remembers the crudeness of the GAT establishment at Lae: 'We were required to find our own accommodation in abandoned military huts on Lae airport, but prior to my arrival, the party on the first aircraft had set up some form of a camp, and again some form of workshop in one of the abandoned Air Force hangars...personal comforts were completely lacking...GAT operations consisted of charter work, mainly regular and high density flights between Lae and Bulolo for BGD Ltd...

'During my time with GAT, which was until late 1949 when they disbanded, I was involved in the repair and maintenance of all aircraft types, but was mainly involved as an engineer flying around in an Avro Anson conducting ad hoc charters and hauling cargo and passengers from point to point wherever the service was needed. This took us all over the New Guinea mainland, with many trips between Lae and the Northern District of Papua, to take supplies and food to isolated ex-wartime aerodromes which were difficult to identify as being the correct drop-off points...the goods were left on the side of the airstrip, and we never did sight any persons to whom they belonged. They had always gone on our next trip, so we presumed they were in the right hands...

'We also did many trips between Lae and Port Moresby where we based ourselves for many days at a time to conduct local charter work, mainly between Moresby and Milne Bay carrying people seeking military salvage rights...at that time there were no aircraft based at Moresby. The only building at Jacksons apart from the wartime control tower was a black iron hut which was the customs building. There was no form of communication between Jacksons and the town seven miles [eleven kilometres] away, and the only communication we had with the tower was by Aldis lamp and Verey pistol. Our means of communication with the town of Port Moresby was a series of signals we adopted with three business houses: the GAT business agent, Noel Maloney, the taxi driver, Tich Corlett, and the Mobil (Vacuum Oil Company) agent, Pappy Duff. If we required one or another of them, we'd fly from one to three circuits over the town, and they knew which one we wanted. The pilot on all of these trips for a long period was Myles Lewis, who later became chief pilot for my company, Papuan Airlines...'

The GAT Ansons were doomed to short lives, because they were seldom hangared, and spent their days outside, hammered by the blinding rain of Lae, and baked by the tropical sun. Over a period of time, water penetrated the structure, glues used in the fabrication of the wooden box-spars of the wing began to separate, and rot set in. The rapid deterioration of the Anson airframes under tropical conditions was not realised until that day in November, 1947, when VH-ALS, BGD, AJJ and AST were grounded. Tom Drury remembers the occasion: 'We knew that some of our aeroplanes weren't in the very best condition. We'd purchased a lot of our fuel from Theo Mason, I think it was, who bought the disposal rights to the abandoned wartime fuel pipelines that ran from the wharf at Lae to Nadzab. Theo paid peanuts for them, and I remember laughing with him the night he bought the rights. "What do you want with twenty-eight miles [forty- five kilometres] of pipes?" I asked him. But when Theo went to drag the first of the five pipelines out he found that twenty-eight times five pipes were still full of 100-octane fuel! Wasn't the best fuel, it had been lying there for several years. But it was cheap, and it cost about ten bob a gallon [4.5 litres] from Mobil. Sam arranged the deal, all we knew was that it arrived in drums, and we didn't query where it had come from. But we blew a few engines to pieces, and a few cylinder-heads started poking through cowls, and Rube Murray came over to have a look, to find out why.

'I came back from a flight one day, and as I came in all I could see was a grey-haired old individual with a tomahawk in his hand, whom I didn't know and had never seen, chopping the tailplane off one of our Ansons!

'I pulled up and raced over, grabbed the tomahawk, and yelled, "A man ought to split you right down to the spine with this! That's our aeroplane!" He said, 'I'm the departmental airworthiness inspector.' I said, "I don't care if you're the Director-General. You ought to be murdered for this..."

'Rube said, "Well, have a look at what I'm finding in the tailplane." It was full of rainwater, was growing an inch and a half [thirty-eight millimetres] of green moss. None of the screws, glues and tacks which normally held the tailplane were in position. And that aeroplane had gone out that very morning. So I took the tomahawk and said, 'Let's go and look at the next one' — and I chopped the next one off. The wings and tail structures were just rotted away. You could push a screwdriver clear through the main wingspars — just like a sponge.

'That just about finished our Ansons. It was a very expensive, major overhaul to fix them up, and we only did a couple of them. We sold some, and just didn't bother to renew the registrations of the others. Trouble was, those wartime Ansons mostly weren't fabric-covered — just thin plywood, and with the constant humidity in New Guinea and heavy rain in the wet they were just super-saturated, and the water would fill up the wings, petrol tank bays, tailplanes...tailplane and mainplane were plywood-covered. Fuselage was steel, longerons and formers fabric-covered. The few pre-war Ansons the Air Force had were fabric-covered over the plywood, which tended to keep the moisture out, provided the drain-holes at the trailing edge of the elevators, and the wings themselves, were kept open. Trouble was, wasps and hornets in New Guinea — they'd build beautiful little mud-nests in all those draining vents and block them up, and you'd be carrying half a ton [.508 tonnes] of water round with you.....

'That was my first introduction to Rube Murray. From then on, both Rube and John Arthur realised that the hub of the flying universe was Lae, not Moresby. DCA headquarters were in the wrong town. I still maintain this, to this day...'

The grounding of four Ansons at one stroke was of course a matter of great concern to GAT, but it was the loss of the Lockheed Hudson, VH-ALA, on 18 April, 1948, that dealt the company a body-blow.

Sam Jamieson had quickly realised that the Anson was not a large enough aeroplane to satisfy the increasing air transport demands of his principal customer, Bulolo Gold Dredging Ltd. The Anson was not capable of lifting items such as a dredge bucket, which weighed 1587 kilograms. A few months after GAT commenced operations, Jamieson went to New Caledonia and purchased a pair of

C47 freighters. One of these he sold to Butler Air Transport Company. The other, VH-GAT, went into service at Lae in September, 1947.

Sam Jamieson was well pleased with his purchase: 'The advent of that DC3 let us fly incredible loads into Bulolo, and we were never without freight, miles of it round the airport and J.D. Simpson kept it coming.

'One of our Papuan workshop boys had been with the Americans, and knew lots of things we didn't know. We wondered how he got the pots, the heads in aircraft engines, so beautifully clean, with no scraping, no nothing. We found that he just put them into a solution of mosquito repellent, wartime stuff. I knew BGD had plenty, and I asked J.D. Simpson for a barrel. He let us have one, and sent me a bill for it!

'I waited my time. I had purchased from Disposals the rights to all aircraft spares at Finschhafen — got them all for fifty pounds — and I found amongst them this incredible bundle of brass wood-screws. You just couldn't buy brass screws then. J.D. Simpson wanted them all. I gave him my price, got it from the Burns Philp store manager. Too dear! But the BP manager was happy to buy them. Later, J.D. came back with his warehouse man — they just had to have those screws, whatever the price. Wasn't he cranky! I said to him, "Well, J.D., you said my price was too much, but remember what you charged me for that lousy forty-four [200-litre drum] of citronella? You got that for nothing from Nadzab, and we carried it down for you!"

'That was the way it was in New Guinea. But J.D. and I, we got along...'

The ponderous dredge buckets and other heavy pieces of equipment carried in the Douglas soon began to gouge into the aluminium floor. Jamieson decided to lay down sheets of heavy plywood to protect the floor, but none was available in PNG, so he set off for Australia in an Anson piloted by Tom Drury, to buy plywood, and also look for some Anson spares. They obtained their requirements in Adelaide, and flew north again for New Guinea.

'We reached our jumping-off point, Jacky Jacky at the tip of Cape York, and had just passed over East Cay when we lost a motor. We lost altitude — we were heavily loaded, maybe a bit overloaded — and we struggled back to Jacky Jacky. For sure we couldn't double that plywood up if we had to land in the water, like we did to get it in — we had to get back. We made it, and found that on the starboard motor, one of the rocker standards had broken. We had all the Anson spares under the sun on board, rocker standards galore, but not one damned nut to fix it on with. Searched all the pranged wartime aircraft around Jacky Jacky, but they were all American, wrong thread. So I had to send a message through Townsville to Moresby and Lae, and one of our pilots flew a nut and washers across to us — most expensive nut ever put on an aircraft...'

The Douglas proved so successful that Jamieson looked around for another heavy aircraft for the Bulolo work.

L.M. Van Praag, a famous pilot and racing-car driver, had purchased a Lockheed Hudson from the Commonwealth Disposals Commission in November, 1946, and registered it as VH-ALA. The Hudson had originally been constructed in 1941 for the US Army Air Corps, but was diverted to the RAF, and in 1942, to the RAAF. During its career, the Hudson was involved in a number of minor accidents, and in January, 1943, was damaged by enemy action at Turnbull Field, Milne Bay. When it was purchased by Lionel Van Praag, it was in sound condition after what had been a very long war, for a bomber.

Van Praag agreed to lease the Hudson to GAT, and to himself fly for the company. The aeroplane was overhauled and modified for civil transport use at Camden, New South Wales, and began operations in PNG late in 1947. At this time — shortly before the four Ansons were grounded — GAT was operating eight Ansons and the C47 in PNG, and employed six pilots, two flight engineers, fourteen engineering personnel, and eight administrative. Tom Drury was flight superintendent, and the chief engineer was Neil O'Neill.

The Hudson flew profitably for some time. In early 1948, a new pilot commenced working for GAT: Stanley Bowern, a West Australian. Tom Deegan, who was thoroughly familiar with the Hudson, began the process of endorsing Bowern onto VH-ALA.

'I hadn't fully endorsed him on the Hudson. He could fly it, all right, but wasn't capable of being let out as a skipper on it. But I'd been to Moresby in the C47, and overnighted there. Stopped at Wau on the way back. I had a bad starter motor. So

I called Lae to bring me up a starter motor to get me home to Lae on the Saturday afternoon. Didn't arrive, and on Sunday morning I called Lae again, and they told me the starter motor had been on the way in the Hudson, and that it had crashed in the Markham River. Couldn't get details — they were too shocked in Lae. I got a piece of rope, wound it around the propeller boss and got a jeep to tow it, you can swing a prop on a DC3 that way by rotating it and priming the motor. So I got the engine started, and flying down the Markham towards Lae I could see a column of black smoke coming up from an island — the remains of the Hudson...'

Tom Deegan had been scheduled to fly the Hudson to Bulolo with a load of newly engaged Sepik labourers.

When he was delayed at Wau, Stanley Bowern made the flight instead. There were thirty-three labourers on the aeroplane when the time to depart arrived. There were no seats, or seat belts. The men sat on the floor.

Engineer Milton Alway boarded the Hudson, to repair the starter motor of Deegan's aeroplane at Wau after the labourers were delivered to Bulolo. At the last moment, two other GAT engineers, George Price and Bill O'Brien, decided to go along, for the ride. Another who wanted to get to Bulolo was the MAL pilot, Dick Burt: 'I had just finished my one-year-nine-month contract with MAL, and I was waiting to go off to Australia on three months' leave. It was a Sunday morning, and I was sitting with Arthur Collins up on the hill where his house was, overlooking the strip. I saw the Hudson taxying out, so I hopped into the jeep and raced down. The engines were running, and I went to the back door and opened it. I saw George Price. I said I wanted to go to Bulolo and asked, "Who's flying the aircraft?" He told me, and I wasn't happy. I said, "If I come, would they let me fly the aircraft?" and George replied, 'I don't think so.' So I said, "Forget it, I won't come..." '

Burt had extensive experience on Hudsons and Venturas, and he could see that the aeroplane was well loaded. He drove back to Arthur Collins.

'I said to Arthur I wanted to fly it, but they didn't want me to. I wanted to be in the cockpit. The aeroplane was fully laden, and I felt they needed my experience. I'm not being egotistical...

'We saw the Hudson taxi down to the seaward end of the strip, and start its run. They were about half way down when we heard this cough, cough, cough from one engine, like backfiring through the carburettor. There were about five coughs. He should have aborted the take-off, irrespective of what damage he did to the aircraft. Had he closed the throttles and applied brakes he would have made it. There comes a time when self-preservation has to be uppermost in your mind, not only to preserve yourself, but those with you. He should have tried to stop on the brakes, and if it didn't stop in time select on the undercart. Let it skate along on its belly. You'd have wrecked it, but aeroplanes are replaceable. Wouldn't have killed anyone. If he had just gone ahead and run off the end of the strip, the undercarriage legs would probably have pierced the fuel tanks, which were very close to the undercart, and they'd have had a size one fire on their hands.

'But he carried on with the take-off. He climbed a few hundred feet, lost a motor, opened the other throttle, that lifted a wing and he spun in, and killed them all. He apparently tried to return to Lae — nobody will ever know. If I had been the pilot, I'd have gone for Nadzab and tried to land there. But he lost speed, and the yawing effect of the dead motor would have been more than he could handle, and he just fell out of the sky.

'Johnny Rose told me, "If you'd been flying it, you'd have gotten away with that..." '

Tom Deegan has no doubts about the cause of this accident: 'The real truth was the lad shouldn't have been flying it. It should never have happened — he wasn't properly endorsed, and he didn't have the experience. Hudsons were vicious aeroplanes if you didn't know how to operate them on one engine. I flew them in New Zealand and other parts, and I was very cautious with them...'

The Hudson burst into flames on impact, and the bodies were horribly burned. The last seconds must have been terrible for the helpless occupants of the doomed aeroplane, but the end was mercifully quick.

The aeroplane crashed on a small island over four kilometres upstream from the mouth of the Markham River. It was a considerable time before Alan Roberts, District Officer of the Morobe District, was able to reach the scene with a rescue party, but the wreckage was still blazing so fiercely that it was not until morning, after night rains had extinguished the fire, that the remains could be recovered — an appalling task.

When word of the disaster reached the Administrator, Colonel Murray, he immediately despatched W.R. Humphries, the Director of Native Labour, to Lae as his representative. Humphries — ironically, himself to die horribly in the eruption of Mount Lamington volcano — submitted a report on his visit: 'Four bodies of Europeans, only one of which was identifiable, had been recovered from the wreck...after hearing Mr Roberts' description of the place and the state of the remains of the thirty-three natives, I agreed with him that they should be buried there, on the spot...there is no doubt in my mind that all passengers were killed instantaneously, and so were spared the horrors of the fire that followed the crash.'

The dead labourers had arrived from Wewak to work for Bulolo Gold Dredging Ltd only a day or two before the crash. All were Catholics. They were buried in one mass grave on the island. The burial service was conducted by Bishop Leo Arkfeld, and Humphries placed a wreath on the grave on behalf of the Administration.

When the facts of the crash became known, there was an immediate outcry in Australia, and questions were asked in Parliament. It was alleged that that Hudson had been over-loaded, and that it had not been licenced to carry passengers. But what caused the greatest howls of moral indignation was the fact that the thirty-three natives were being carried at freight, not passenger, rates, and had not been provided with either seats or seat-belts. There were calls for a public inquiry.

Reporting the accident, the journal, *Pacific Islands Monthly* commented tartly: 'The fact that the natives were not provided with seat-belts has caused a storm of disapproval. As the plane burst into flames immediately on hitting the ground, it would be interesting to know how safety-belts would have helped matters. It is obvious that Mr Ward* and other departmental heads are looking for a scapegoat...it has always been the custom in New Guinea to carry native labourers as airfreight. It is obvious that if they had to be paid for at full rates, few, if any, would have been transported in this way, and the majority would have been required to walk.

'As far as this goes, most Europeans who travelled by air were also virtually regarded as freight, in that they sat in a transport plane on their gear and took what was coming to them. They were glad to do this, lack of seats and seat-belts notwithstanding. The only real difference between the service given them and that given natives was that while they paid several pounds for the privilege of travelling thus, natives were weighed and charged for at the rate of three pence per pound...'

So loud was the hubbub that DCA despatched the chief investigation officer, J.H. Harper, from headquarters, Melbourne, to take charge of investigations. Harper concluded that the crash was caused by the complete failure of the port motor owing to fatigue failure of an accessory drive gear. Additional factors that probably played a part were inadequate pilot training, and poor pilot technique following the engine failure, and bad load distribution, although there was no question of the aircraft being overloaded.

Reporting these findings in a confidential letter to E.J. Ward, Colonel Murray wrote: 'I asked Mr Harper whether he considered that the absence of seats for the native people, and the absence of safety belts would have contributed to the form and effects of the final disaster...Mr Harper replied that he did not consider that the absence of seats or safety belts really contributed materially to the form the accident took, or its consequences.

'Mr Harper stated that instructions had been given that from the time of the accident, no persons in the Territory were to be carried without the regulations concerning the carriage of passengers being strictly adhered to. My personal opinion is that an open Court of Inquiry will reveal nothing more with regard to this accident than has already been ascertained by Mr Harper. Further, I do not think that an inquiry would serve any good purpose, and it might furnish plenty of headlines, and sensational newspaper articles...European and native people have been conveyed in aircraft without normal seating and safety belt arrangements, and this has been the practice for fifteen years or more...'

E.J. Ward was determined to see Guinea Air Traders prosecuted, if there was any evidence of criminal liability. There were incessant exchanges per radio telephone between Port Moresby and Canberra, but it became apparent that the Administration was itself vulnerable, in that official passengers were commonly

*PIM and its owner-editor, R.W. Robson, were vehement opponents of E.J. Ward, Minister for External Territories, and Labor's 'New Deal' for Papua New Guinea.

carried in chartered aircraft, without seats or seat belts.

Colonel Murray, in a letter to External Territories dated 24 April, made the point that the last formal contract between the pre-war Administration and airline operators, in March 1941, provided for the carriage of native passengers at a rate per pound, and passengers other than natives at a set fare. Colonel Murray continued: 'Certain misconceptions in respect of the term "freight rate" have appeared. The use of the term is unfortunate, in so far as it is a colloquialism evolved in slovenly speech down the years. It is observed that a rate based on weight is charged for mail, goods and native passengers, and that because the rates are the same it is easily understood how misinterpretation can arise and the belief becomes current that natives are carried as "freight". It is clear, however, that the contract was for the carriage of natives as passengers, and that there was not a standard fare, but charges were according to the weight of the person. The reason why natives were carried as passengers on charges computed on the basis of weight is because of the size and weight of themselves and gear in comparison with that of Europeans. In avoirdupois, the ratio is about 185:140 pounds [84:63 kilograms]. The method provided the cheapest possible air travel, and this is a point of importance to those natives who travel on their own business.

'Following the re-establishment of Civil Administration on 30 October, 1945, aerial transport companies began to operate...as all of them have executives or personnel who worked in air companies in New Guinea before the war, it is natural that the previous practice would be followed...'

'From time to time, advantage has been taken of planes not under charter to the Administration going from Lae to the Central Highlands District to transport officers, natives and cargo. In those cases, fares are by arrangement and may be a flat rate per passenger, or according to weight. I believe that private employers follow the pre-war practice and that generally the fares of natives are computed according to weight.'

Without question, natives at this time were regarded as items of freight rather than as passengers, the Administrator's protestations notwithstanding. There were only so many seats in an aeroplane: the average native weighed a great deal less than the average European, and airline operators the world over have always assumed an average weight per passenger when making up their load sheets. It made commercial sense to treat the lighter natives as so much weight. There was certainly nothing racial in the practice, as many critics alleged.

Time dragged on, and a decision on whether or not to prosecute GAT was avoided. Then, in October, 1948, the Minister for Civil Aviation, Arthur Drakeford, announced that instructions had been given for the company to be prosecuted 'for a breach of Civil Aviation regulations' in connection with the Hudson disaster. But, he stated, 'it was necessary to go through certain formalities', and he was not in a position to say when or where the case would be heard.

Talking of this affair many years later, Sam Jamieson was still angry about what he saw as an attempt to penalise GAT for a practice followed by all other air transport operators in PNG: 'They were going to prosecute GAT for not putting seat belts on people! I said to certain people in the Administration, "If this takes place, I'm going to take you people with me." They were people in high places...'

In March, 1949, almost a year after the crash, the official accident report was tabled in the Australian Parliament, and Drakeford announced that no legal action would now be taken against Guinea Air Traders. Although there had only been seating for two on the Hudson, and the Certificate of Airworthiness thus limited passengers carried to two, the all-up weight of the aircraft had still been within the specified limits at the time of the crash. The decision not to proceed with legal action had been taken on the advice of the Crown Solicitor. Drakeford added: 'The Department of Civil Aviation, however, was seeing that regulations for the training of pilots and the loading of aircraft were observed in New Guinea as they were in Australia...'

As far as the air transport companies were concerned, the most significant result of the Hudson crash was that henceforth no passenger, of whatever race or colour, could be carried in an aircraft unless approved seating and seat- belts were provided. It was because of this ruling that the Island Airways Catalina operations out of Madang ceased, as previously mentioned.

By this time, Sam Jamieson had left the PNG aviation scene, and his brother, John, was managing director of GAT. Guinea Air Traders was sliding towards its end, for a variety of reasons but principally because the company had lost the lucrative Bulolo Gold Dredging Ltd business to a formidable new opposition, Qantas' New Guinea Internal.

New Guinea Internal began on 1 June, 1948, with the appointment of Captain W. Forgan-Smith as chief pilot. On 12 June, Forgan-Smith landed at Lae at the controls of the first Qantas C47*, VH-EAP, loaded to the limit with company stores and spares. His co-pilot was First Officer L.W. Purkiss.

PNG aviation has spawned many legendary characters, but few as notable as Bill Forgan-Smith — 'Forgie', as he was invariably known. Born in South Australia in December, 1912, Forgan-Smith moved with his family as a child to Newcastle, where his father obtained work at the Broken Hill Proprietory Company foundry there. Later, this company became a patron of the small Newcastle Aero Club, and presented it with a pair of Gipsy Moths. BHP also provided five flying scholarships, and Forgan-Smith was fortunate enough to be awarded one of these, which led to his private, or class A, pilot's licence.

Says Forgan-Smith; 'So I carried on from there. Wanting to find out more about the game, I decided to build a little aeroplane — little high-winged monoplane, weighed 310 pounds [141 kilograms] with a thirty-five horsepower Twin Anzani engine...built according to DCA rules, and fully aerobatic. Burnt 1.8 gallons [5 litres] of fuel per hour, cruise seventy-five...

'After I built this, the aero club asked me to build a couple of jigs and a couple of sides for those old wooden aeroplanes — heavy landings, and they'd get their sides pushed in. So I did. After that, I got an engineer's licence — C licence — and a D engineer's licence and then an A, allowing me to fully overhaul aeroplanes and build them if necessary. By this stage, I had my commercial pilot's licence...

'I remember the old Klemm Swallow. I used to give joyflights. This was the depression time, five bob for five minutes, ten bob for ten minutes. The old Klemm was a low-wing monoplane with an open tandem cockpit. Had a fellow up one day. We hit a few bumps — we only had lapstraps, and you could leave the seat by six or eight inches [152 to 203 millimetres]. When we landed, this fellow said, "It wouldn't be hard to fall out of this thing." 'Right,' I said, 'hit a decent bump and you'd fly out if you weren't strapped in.' He said, "I'll give you fifty quid if you'll take my wife up, and don't strap her in..." '

I used to go to different places. One was down at The Entrance — used to be a wide patch beside the road. Sit down there, stick your sign on a post, and wait for customers. The Klemm was made of plywood, had a little geared Pobjoy, revved at about 3,200 and geared two to one on the prop, you could almost see it going round when it was idling. Had a kick-starter, like a motor bike. You could start it in the air. Took off at about thirty mph [forty-eight kph] in about two fuselage lengths. You could take her up to 10,000 feet [3048 metres], switch off, and glide about for forty-five minutes before you landed...

'I worked for the aero club as an aircraftsman at £4/8/- a week. Not interested in pay, only in flying. I'd work seven days a week. Get out there at four in the morning, check out the aeroplanes for the day, top up the tanks and so on. I did rebuilds, overhauls, everything. Airframes and engines. Newcastle Aero Club was then one of the best and biggest in the world. Large fleet, about twelve Tiger Moths and DH60's, Hornet Moth, Fairchild 24...depots out at Tamworth, Grafton...we'd do pageants all over the country, joyrides, displays...DCA won't let you do them today, taken all the fun out of flying...

'My trick at pageants was to land off a loop — you'd finish up higher than you started if you did a loop properly. Bit of a side-slip off the loop and bang! you'd sit down. Not dangerous if you do it properly. They stopped us doing pylon racing around the aerodrome...a bit dicey.

'I was going to be an instructor with the aero club when this job came up with Ray Parer in New Guinea, so I went up there, in 1939. Flying pay was a pound an hour — big money then, for you could fly 1,000 hours a year. Tremendous.

'I did a few trips, then finished up in the Bitoi. I didn't know enough about the down currents and bumps in New Guinea then, and I turned the wrong side of a cloud to go up the Bitoi valley in a Fox Moth, got pushed into the down current, and couldn't get enough height to get over the end. So I finally went into the trees. Took me five days to walk back to Salamaua.

'Was only in New Guinea about six months total. War broke out, and I was in the reserve — had been since 1935 — so I went back to Australia, but I couldn't get

*The Douglas C47 was the military freighter version of the DC3. Over the following years, Qantas put seven C47's into service into PNG, but they were almost invariably referred to by the public, and many pilots, as DC3's. This usage will generally be followed in this book.

into the RAAF. Got a job with North-Western Airlines as an engineer. They had a Tugan Gannett, Wackett design, and a DH90 Dragonfly, a lovely little aeroplane. Based at Moree. I serviced the aeroplanes, and did a few flights, but they folded up, couldn't get bits and pieces for the aeroplanes. They sold the Gannett to Arthur Butler, and the Dragonfly — VH-UTJ — to Qantas. I joined Qantas, early in 1941.'

Forgan-Smith became a first officer on the Short C-Class Empire flying-boats, on the Sydney-Karachi run, and took part in the evacuations from Singapore when that paper fortress fell to the Japanese. One of the passengers on his flying- boat was Lieutenant-General H. Gordon Bennett. After Batavia fell, the Qantas Empire boats operated out of Broome into South Java until it, too, was overrun. Forgan-Smith remembers the Empire boats with great affection. They were, he says, 'the greatest boats there were — comfortable, plenty of room, reliable, no vices, did everything they were asked to do...

'By this time I had a pilot's licence, engineer's licence, radio operator's licence and navigator's licence, and after the evacuations I was seconded into the RAAF with a Fox Moth to do special duties out of Darwin. This was 1942, and when I got back from this, the Yanks had gotten into the war, and General Yeager wanted a pilot. I was still employed by Qantas, but I was seconded to the US Army for a year as his personal pilot, flying the Dragonfly. He was always on the move...after a year the war had moved on, and Qantas was given C47's to operate for the US Army out of Archerfield, to go "bully beef bombing" up through the Islands, and I went onto them. Then flew DH86's from Brisbane to Darwin. I was made chief pilot for Queensland in 1946. Then in 1948 I went up to New Guinea again, to open Qantas with DC3's...'

Lionel Purkiss, Forgan-Smith's First Officer on that inaugural flight, was an RAAF-trained pilot, who held both navigator's and wireless operator's licences. He joined Qantas in April, 1947, and became a first officer on the Bird of Paradise run. One day in Lae, Len McNeil said to him, 'Why don't you come up here?'. Lionel Purkiss remembers: 'I asked Lennie, "Why?" He said, 'Come and see.' So he threw a case of biscuits off his Dragon, and took me up to Kainantu. I liked what I saw, and I went up with Bill Forgie on a two-year contract. I stayed for seven...

'When I first went up there, I was co-pilot on DC3's with Forgie, and when we got other DC3's, I flew co-pilot on them. I was checked out on Dragons in December, 1948, by Arthur Jacobsen — now dead. Jake made a pretty thorough job of it. Lae-Wau-Bulolo first, to see how I went there. Then took time off, and we went up into the Highlands. Started at Wabag, and worked our way back to Lae. Bumps and grinds all the way. Stayed one night at Nondugl, with Nep and Olga Blood. We always carried our own bottle of rum and tin of fifty cigarettes. I got a bit the worse for wear that night, and next morning I said to old Jake, "I'm not fit for flying." 'I know,' said Jake, 'I helped to make sure you wouldn't be. It's going to happen that you'll be caught out, and you won't be fit to fly but you'll have to.' So he made me fly all day, and at the end said, 'Oh, I think you'll probably make it.' Hard way to learn. I finished having a hurl at the end of the day...

'I had two bends up there. First was in a Dragon, VH-AXL, at Banz. I had just settled on the ground when the timber at the tail end that held the tail wheel broke. It had rotted. Swung off, and damaged the wing and fuselage, but hardly my fault. Then I three-pointed a DC3 into a sheet of water on Kavieng strip, and the flaps were mangled. I pulled the flaps up as far as they could be pulled, and just kept on going...

'Did some silly things though. Lot younger then. Was trying to do the right thing once and got well and truly scared, in a Dragon. Going in to Minj from the Goroka valley. I suspected a down-draught, and flew past the edge of a ridge, around Mount Otto. There was no down-draught, so over the edge I went, and down I went. Fell like a bucket. I had a load of shovels on board, all looking at me, and four-gallon [eighteen-litre] drums of petrol and some loaves of bread. I reckoned I was going to be tenderised, and burned. I kept working though, staggering around the trees, and got out around Chimbu, and landed there. I'd been on the point of stall, but got out of it. Couldn't talk after I landed. You didn't often get a second chance in New Guinea...that was one time when I did.

'But the Dragon was a wonderful little aeroplane, all you had to do was give it a go. Dennis O'Malley took off from Lae once in a Dragon with a double load on it — put in by mistake. He staggered out over the Gulf, turned round and staggered back past Lae, got up as far as Kaiapit, and couldn't fly any more. That's where he landed, and took a whole stack off, took off again and staggered up as far as Kainantu, and took some more off...'

One other Qantas pilot who was given a second chance was R.K. Crabbe, one of the pilots who joined Len McNeil and Frank Furniss at Lae before New Guinea Internal began. In July, 1947, Ross Crabbe took off for Wau in Dragon VH-BAF, with a load of freight and two passengers, Percy Blanden, the popular Mining Warden at Wau, and his wife.

Crabbe, another ex-RAAF pilot, had until recently been flying a Dragon for the Flying Doctor service out of Cloncurry, and he was thoroughly familiar with the type. But he was new to New Guinea. Crabbe had not been properly familiarised on the Lae-Wau route. He had been told: Just follow the road, go out to sea, turn right, go up the Markham, and follow the road...

On 14 July, Ross Crabbe took off and followed instructions. He picked up the Labu-Wau Road, and followed it. He was so intent on sticking to the route that he did not at first notice that the cloud was getting lower, and the floor of the valley he was following was getting higher. The valley got narrower and narrower, and the sides steeper and steeper. Crabbe was in a position where he could not turn out of it. He had to carry on. He could not outclimb the valley and the thick cloud was closing in, so he did the only possible thing: Sat the Dragon into the spreading branches of a high tree. Just before impact, he released his seat belt, mindful of the cases of supplies immediately behind him.

Crabbe was hurled out of the cockpit, through the nose of the Dragon, and down to a ravine, thirty-six metres below. The Dragon remained in the treetop, stuck fast.

Ross Crabbe injured his back in the fall, but miraculously escaped serious damage. He called to the Blandens, trapped in the Dragon, but could get no response. There was no way he could reach his passengers, and he stumbled away to seek help.

The Blandens had both been injured in the forced landing, Perce Blanden seriously. They could not move, even if they had wanted to. They stayed in the treetop, shuddering with fear as the fabric tore, and the Dragon slipped and again settled.

By great good fortune, a party of native hunters had heard the noise of the crash and they saw the Dragon in the trees and hurried off to Mumeng government station, several kilometres away. Medical Assistants Eric Robson and Kit Atherton were quickly on the scene. They somehow climbed the great tree in the dark and administered morphine to the Blandens, and waited there with them until a rescue party headed by Dr Carl Gunther arrived next morning, twenty-one hours after the forced landing.

The Blandens were gently lowered from the Dragon in large baskets. Both eventually recovered from their injuries, although Perce Blanden had to go to Australia for plastic surgery. Every bone in his face had been broken in the impact. The day after they were rescued, the tree and the still-entangled Dragon crashed to the ground.

Ross Crabbe did another tour in PNG years later, and had his share of thrills, but nothing to equal his treetop landing in VH-BAF.

After New Guinea Internal was established, Qantas made a determined effort to ensure that accidents such as that which had happened to Ross Crabbe, which basically occurred because he had not been properly familiarised over the route he flew, could no longer occur. The freedom of action that McNeil, Furniss and the early pilots had enjoyed obviously had to end when the big Qantas buildup in PNG began. Bill Forgan-Smith was a stickler for proper flying standards, and a formal system of route familiarisation and training was evolved.

Qantas pilots new to PNG began by flying as a copilot in DC3's, over the coastal and island, and then over the more difficult inland routes. This endorsement flying continued for an indefinite period, depending on the individual pilot. After this initial period of training, the pilot began route qualification flights. These consisted of five trips over any new route, two flown as an observer and the final three under the supervision of an experienced pilot. Only as his proficiency increased was the pilot introduced to each new route, until he had achieved full PNG proficiency, able to operate into all strips.

After long hours at the controls of light aircraft — Dragons, Fox Moths, Drovers, Beavers and Otters were at one time or another used by Qantas — a pilot might be considered ready for a DC3 command, the goal of all ambitious first officers. The selected pilot would be given 100 hours in command under the supervision of a check captain, following which he was sent to Sydney for further flight checking and technical examinations. On return to PNG, he was given a final check, and then was sent out in command. The first 300 command hours were flown on coastal and island routes, and then the whole business of flight checks over the critical internal routes was begun. Only after the check captain was fully satisfied with the proficiency of the new DC3 captain was he finally turned loose on his own.

It was a rigorous system, and it generally produced results, although Qantas pilots — and others! — continued to have accidents. Inevitably, aircraft accidents in PNG, as in all countries, were invariably given full and generally sensational prominence in the Press. Many more people are killed and injured on the roads than were ever killed in the air, even in the hostile flying environment of PNG, but few members of the public seem to appreciate this. When the first post-war newspaper, *South Pacific Post* commenced publication in September, 1950, flying accidents were always closely covered.

As Qantas, MAL and GAT introduced more DC3 aircraft into New Guinea operations, the number of accident-free flying hours rapidly increased. There never was a DC3 fatality in PNG operations. Says Bill Forgan-Smith: 'It was the best aeroplane of the lot. Would never hurt you. Easy to fly, reliable, strong, designed for grass strips, could land anywhere. In its element in PNG. Plenty of power to get off. The engines were 1,200 horsepower, and you normally only operated on about 450 hp. During the war, we operated them at 32,000 pounds [14,515 kilograms] all-up, only 26,200 [11,884 kilograms] in New Guinea, so you had plenty in reserve. My motto was, if you could get it in a Doug, and close the door, she'd fly — and I've flown them with things sticking out of the door...'

The DC3 had not been Qantas' original choice for PNG. As mentioned in Chapter 2, Qantas, the company applied in 1946 for permission to operate the Bristol 170 Freighter on the Lae-Bulolo run, and elsewhere, but had been refused on the grounds that aircraft performance standards were still being determined. On that occasion, the request for approval to operate Bristols had been made at the behest of Bulolo Gold Dredging Ltd, who wanted Qantas to transport their freight. This was before Guinea Air Traders was formed.

At the time Qantas was considering the Bristol, it was the most modern short-haul freighter aircraft available. A rugged, all-metal aeroplane of simple construction, with fixed undercarriage and powerful Bristol Pegasus engines of 1,980 horsepower, the Bristol was designed for minimum operating costs. The Series 1 Freighter could carry four-and-a-half tonnes of cargo, and was loaded through big clamshell nose doors and a fuselage door.

High hopes were held for the Bristol, the first new civil transport to fly post-war, but the aircraft, though successful, never lived up to its promise. It was offered at a realistic enough price for a new aeroplane, but the flood of ex-military DC3's and Lockheeds that soon became available, cheaply, to the world's air transport operators killed its chances.

Although Qantas had been refused permission to operate the Bristol in PNG, the company did not abandon its intentions. The Bristol Aeroplane Company had faith in its aeroplane, and during 1946 sent early production Freighters on sales demonstration tours throughout much of the world. In 1947, tours were made in Africa and Australasia. Bristol Freighter G-AIMC was shipped out to Australia with a team of pilots and salesmen, and in October, the aeroplane was leased to Qantas, for evaluation flights in PNG.

Accompanying the Bristol team in C-AIMC was an ex-RAAF engineer-pilot, William Burns. Bill Burns joined the Air Force in 1926 as an engineer-aerofitter, and in due course obtained his A pilot's licence. He left the RAAF in 1938, and joined the fledgling air transport company operated by R.M. Ansett, as foreman of the engine overhaul department. Ansett at this time had just introduced the advanced Lockheed L-10 Electra, and Burns today recalls that the then- struggling company could afford only one spare engine for two Electras. When the Ansett hangar at Essendon was rather mysteriously burned to the ground in February, 1939 — an Electra, a Fokker Universal, several smaller aircraft, spares and equipment were lost — Ansett was forced to suspend his flying operations, and Burns joined the Commonwealth Aircraft Corporation, where he was in charge of the final assembly of the Wirraway general-purpose and training monoplane.

War came, and Burns re-joined the RAAF. He completed an instructor's course at Camden and spent two years there, instructing. He was too old for operational flying, and for most of the remainder of the war Burns was an engineering and maintenance officer. He was finally transferred to Williamtown Operational Training Unit as a pilot, and was converted onto Beaufighters, Beauforts, and Mosquitoes. He was at Williamtown when the war ended and was discharged with the rank of Wing Commander.

Burns returned to CAC, and was involved in the flight testing of the Australian developments of the Mustang fighter. He then joined the Aviation Division of Overseas Corporation (Australia) Ltd under Wing Commander G.D. Nicholl DFC, which represented the Bristol Aeroplane Company in Australia. When Bristol Freighter G-AIMC arrived, Bill Burns was a member of the crew on the demonstration flights in Australia and New Zealand, which were remarkably successful: The aircraft moved twice the volume of freight over short sectors than the DC3 was capable of lifting. Following this tough series of flights, G- AIMC was thoroughly overhauled before the demonstration tour of PNG for Qantas. Bill Burns remembers: 'I was the only engineer licenced on the Bristol in Australia, but another company had the maintenance contract for the Overseas Corporation, which represented other companies besides Bristol, so they did the work. During the overhaul, they replaced the single bowden cable which was the main brake line on the Bristol — a weakness in the whole braking system for an aircraft of that size to have only one cable. The cable went to the single, great big brake lever and was secured by a nipple on the end of it.

'Well, we flew up to New Guinea, and made several flights in to Wau. As you know, it has a slope of one in twelve, and it was standard practice for pilots to park aircraft sideways at the top. Our pilots, the Bristol man Captain Elliott, and Geoff Nicholl, found that the fuselage distorted if the aeroplane was parked sideways because of the slope, and it made it hard to open or close the nose doors. So on 23 October, when we landed there with a load of native labourers, Elliott stopped straight ahead instead of turning sideways, and parked the aeroplane on chocks and the brakes. But that nipple hadn't been properly secured on the end of the bowden cable, and it came adrift...

'The Bristol started to roll backwards down that steep Wau strip, fast. I leapt out and rolled out of the way of the oncoming undercarriage wheels, but Elliott and the passengers rode her down to the bottom, where she broke in two.

'Rube Murray investigated, and the chief engineer of the maintenance company refused to issue a certificate saying that he had replaced the brake cable, so I got the blame, as I was the Bristol engineer. It has always rankled...it was most unfair. The maintenance company did not have the knowledge to do a proper job on the brakes. I got a letter from DCA, but they didn't take my licence away or anything like that...

'We dismantled the aeroplane, took the engines and controls out, instruments, radio gear, all of value...the fuselage broke apart at the main cabin door. It could have been repaired, but only at great cost.

'After getting back to Melbourne, Geoff Nicholl sent me to Bankstown to take over an aircraft overhaul unit called Airwork, a subsidiary of the Overseas Corporation. I spent a couple of years there, then I joined Qantas as their Accident Investigator. I took over from Dicky Mant — Scotty Allan was my boss. I was with Qantas twelve years...'

The wreckage of G-AIMC remained for many years at the bottom of the Wau strip, mute warning to pilots of the danger of relying on brakes on that notorious runway. The fuselage became living quarters for labourers employed on a local coffee plantation.

Qantas never did operate Bristol Freighters in PNG, although Ansett and Trans-Australia Airlines each employed a pair of Freighters there after Qantas left the scene.

In common with all major airlines, Qantas investigated accidents involving company aircraft, generally in cooperation with the DCA investigating teams. From 1950 to the end of the Qantas era in PNG, Bill Burns investigated all major accidents to Qantas aircraft in the Territory.

45

46

47

45. Omkolai airstrip in the Eastern Highlands. *Photo: Qantas*
46. Qantas Dragons VH-AOR and AXL at Minj, 1950. *Photo: Talair*
47. Qantas Fox Moth, VH-URI. *Photo: Qantas*

58

3

Aviation Grows

The Demise Of Guinea Air Traders

The decline of GAT, which began with the grounding of the Ansons and accelerated after the crash of the Lockheed Hudson, was virtually complete by the end of 1949.

During its short life in PNG, the company put into service at different times twenty-four aircraft, and employed among its pilots and engineers men who were to become pillars of the aviation community. The aircraft included three Dragons, thirteen Ansons, an Auster J5, a Fox Moth, two Lockheed Hudsons, two Douglas C47's, and two C39's, the latter a curious hybrid with the fuselage of the DC2 mated to the tail of the DC3, and fitted with twin Wright engines of 930 h.p. As far as can be determined, it was the C39 that first became known as the 'Gooney Bird', a term which became synonymous with the DC3 and its many military variants.

The aircraft met with varying fates. Most of the Ansons either crashed, or were withdrawn from service when their Certificates of Airworthiness were suspended or lapsed, or were not renewed. Some were cannibalised to keep the others flying. Two, VH-AXQ and VH-NGG — the letters were a tribute to good customer New Guinea Goldfields Ltd — were returned to Australia, where both were soon struck off the register. The Auster J5 was lost in a crash at Dreikikir in January, 1949. The Fox Moth, VH-GAV, was destroyed in a crash at Bulolo in April. The pilot was Dick Burt.

Burt joined GAT following an altercation with W.R. Carpenter & Co over the question of his trading and business activities. As well as operating trade stores, Burt also ran a taxi service. On leave in Sydney following the loss of the Hudson, Burt called in to the WRC office and was given an ultimatum: Wind up his private activities, or leave MAL. Burt, ever jealous of his personal rights, resigned on the spot. He did not consider that his business activities were affecting his flying duties, and in any case he was not the man to accept an ultimatum.

'I went out of Carpenters office and up the road to GAT, and said, "Do you want a pilot?" 'What's your name?' said the bloke. I told him, and he said, 'When can you start? Pay's a thousand a year.' I asked him, "Who is the chief pilot?" 'Tom Deegan.' I said, "How is he going to take it when you tell him he's not chief pilot any more?" The bloke said, 'Oh, we're not going to tell him that.' I said, "I'm not going to have Deegan as my boss, because I have more experience than him." It worked out that we became joint flight superintendents...'

The previous chief pilot, Tom Drury, had left GAT by this time, following what he describes as a 'row' with Sam Jamieson, although they always remained personal friends. Drury was becoming concerned about the odd mix of aircraft the company was acquiring, from the tiny Auster J5 to the relatively heavy DC3 and Lockheed Hudson. This created a mighty spares and maintenance problem. Drury wanted to standardise on two or three types only. He was worried, too, about pilot training, which was considerably complicated by the many different aeroplane types in the GAT fleet. As Drury tells it: 'Sam eventually said, "If you're not happy about what we're doing as the owners of the firm, you can go if you wish." I replied, 'All right, if that's what you want, Sam. I'll go and join MAL and fly in opposition to you. You can't run an airline this way — you need more money, more spares, more everything, and better trained pilots...'

'I had left GAT by the time of the Hudson prang, and was in Newcastle, teaching John Jamieson to fly, at the aero club. When we heard the news, John said to me, "Tom, I want you to go back tonight to Lae. DCA will be investigating this crash (and they sure did, it was headquarters that made the investigation, not DCA Port Moresby). I'll give you any price you like to take over our whole operation in New Guinea..."

58. Sam Jamieson and the first Vespa in Australia, in front of Guinea Air Traders C47, 1948. *Photo: Sam Jamieson*

'I didn't accept the offer. I had already joined MAL...'

Dick Burt was engaged by GAT principally to fly another Lockheed Hudson, VH-BLA, which GAT had leased from the owner, Lionel Van Praag. However, he flew other aircraft in the GAT fleet from time to time. On 25 April, 1949, he was piloting the Fox Moth into Bulolo when the engine failed. Sam Jamieson tells of this incident: 'Peter Lavender wrote a bit of poetry about Dick: "He thought she could take it, but it just didn't make it, and he pranged her straight into The Pimple". Wrote the Fox off. Overloaded. As usual!...

'Boy, could Dick fly that Hudson! It was a sight to watch him take off. But he had been an instructor on those things, knew what he was doing...amazing thing to see him take off. Roaring down the strip towards the sea, all of a sudden you'd see the aeroplane starting to lift, not more than a foot off the ground, the wheels tucked up, skipping along — it was a sight to see.'

It was, indeed*. The Hudson bore the appropriate name, *Silver Bullet*. Dick Burt had a take-off technique in the *Bullet* that was dramatic in the extreme to the onlooker. The Hudson would race along the strip and the wheels would suddenly retract, leaving the polished silver aeroplane still 'on the deck', with a rooster-tail of dust leaping up behind, before climbing up into the sky. Most New Guinea pilots thought that this technique involved unnecessary risks, should an engine falter, or fail. Dick Burt denies this: 'My take-off technique? I used to just build up lots and lots of speed before I started to climb, and by the time I reached the end of the strip I had bags and bags of speed, in a very safe position, rather than take off and climb at the steepest possible angle, when you'd stall very quickly if anything happened. It was the procedure we adopted in the Air Force. I used to reach a speed of about eighty knots and then just select up on the undercart, just lift off, take the weight off the undercart, and just leave it there until it built up speed. As soon as you took the weight off the undercart, the wheels would come up in a very few seconds...

'I used to carry forty natives regularly in a Lockheed. In order to get speed I used to put ten of them down in the nose to get the aircraft balanced. I rather liked the Lockheed. They were redlined at seventy-eight knots, I think, under that you were supposed to be flying on borrowed time. Going in to Bulolo I used to make my final approach around The Pimple and straighten up, and I'd be down to about sixty knots when I went over the end of the strip. I didn't use very much flap...in spite of all the things that people say about Lockheeds being dangerous, you could take a lot of liberties with them...'

The *Silver Bullet* was returned to Australia in October, 1949, by which time GAT operations out of Lae were running down. It was purchased by Adastra Airways, and modified at Mascot for aerial survey work. Re-registered as VH- AGG, the Lockheed returned to Lae years later, and met a violent end, described in Chapter 5, Territory Airlines Grows.

The two C39's used by GAT were acquired from Macair, a company controlled by Colonel Charles M. MacDonald, in May and June, 1948. The first, VH-ARB, was finally broken up in July, 1950. The second, VH-ARC, was written off in a heavy landing at Kerowagi in March, 1949 — there were no injuries. Tom Deegan remembers the pair: 'ARC had been all shot up with holes. Hundreds of little holes in the fuselage, patched and rivetted. It was an antique — should have been put in a museum. Those C39's had small engines in them, and they'd ice up if you flew in cloud. Carburettor would ice up. You'd be flying along with hot air underneath the carburettor, and lose about half your power.

'I started off the Lae-Goroka-Madang-Wewak-Manus-Kavieng-Rabaul "round the traps" service in a C39. The people were very grateful. It was a weekly service, and they got mail instead of waiting for ships. It became quite a big run in the end.'

In July, 1948, a pilot who was to have a long and distinguished career in PNG aviation began his association with GAT. J.K. Perry remembers his first ride in an aeroplane at the age of nine, sitting on his sister's lap in the open cockpit of a battered, barn-storming Gipsy Moth, in 1932. Such niceties as a seat belt or helmet were not included in the fare.

Jim Perry volunteered to join the RAAF in 1941, on his eighteenth birthday, and was called up the following year. He trained in Australia on Tiger Moths and Airspeed Oxfords, and was posted to England. He was discharged in June, 1945, and went back to school to finish his interrupted education, aided by the subsidy

*The author was stationed in the Morobe District at the time, and frequently watched the awesome Burt take off in the *Silver Bullet*.

of three pounds per week then paid by the Commonwealth Government to eligible ex-servicemen. Jim Perry went on to university, but his money ran out and he was obliged to abandon his studies. He turned to civil aviation, obtained his commercial licence and made his first flight in May, 1948, with New Holland Airways, flying as co-pilot to Captain Keith Robey in a DC2, on the immigrant run.

New Holland Airways was one of a number of charter companies engaged in a highly lucrative new field, flying out migrants from Europe — Greeks and Italians mainly — to Australia. The Department of Civil Aviation described this development as a 'mushroom growth', that resulted from the ready availability of low-priced war surplus aircraft, the large numbers of ex- service pilots looking for jobs, and the inability of regular airlines and shipping to handle the enormous number of migrants attracted to Australia by the new post-war immigration policy.

In the 1947/48 Annual Report the DCA attitude to these operators was spelled out: 'The policy of DCA has been that operators of these non-scheduled flights are permitted to embark or disembark traffic in Australia only when the regular operator is unable to handle the traffic within a reasonable time. A further consideration is that the fares and freight rates charged shall not be less than comparable rates of the regular operator.'

Sam and John Jamieson were among the first to recognise the potential goldmine of the migrant runs, and they went looking for suitable aeroplanes. They found them, a pair of C47's, in England. Says Sam Jamieson: 'My brother John, he joined me and we bought these two DC3's from Shackleton and Scottish Aviation. G-AGHN and G-AKNB. They were British-registered, and straight away we had 1,000 pounds [454 kilograms] each more payload than if we'd registered them in Australia. We could fly at 28,800 pounds [13,064 kilograms] all-up. I travelled extensively back and forwards with the aircraft — over empty, back loaded with Italian migrants....'

It was the Jamiesons' involvement in the international migrant runs that proved to be the final straw for Guinea Air Traders. G-AKNB was placed on the migrant run between Sydney and Rome, and G-AGHN was sent to Lae, where it was immediately employed on the Bulolo work, GAT's bread-and-butter. With its heavy payload, the Douglas quickly became the key aircraft in the GAT fleet.

Jim Perry first flew for GAT as first officer in a C39 on a ferry flight from Sydney to Darwin. He was offered permanent employment in Lae, but initially declined, having, as he says, 'no inclination at this stage to proceed to such a malarial-infested-rain-sodden-jungle-of-a-place'. He was offered yet another job as first officer on G-AKNB on the Rome immigrant run, but finally took up the Lae offer, arriving in September, 1948. He remembers his first PNG flight, in a C39 piloted by John Dalrymple. He received his Anson command in October, 1948, on one of the few survivors of the GAT Anson fleet.

In mid-1948, Sam Jamieson pulled out of GAT, and John became managing director. The months following the Hudson disaster were difficult ones for the company, faced with the threat of prosecution, and suffering a loss of public confidence in its operations. However, in March, 1949, GAT took over Taylor's Air Transport, with its surviving aircraft and personnel.

Joe Taylor had tried hard, but his resources were insufficient to enable him to survive. Competition in the airline industry was intense, and Taylor was squeezed out. He lost both his original Dragons in crashes, VH-AFK at Bena Bena, and VH-AHY at the very difficult Slate Creek 'drome. He obtained two more, VH-AAC and VH-BJH, and also an Avro Anson, VH-AHG.

This latter aeroplane was acquired in somewhat mysterious circumstances. It was an ex-RAAF machine, which passed into the hands of Baker Motors of Albury, New South Wales. A.E. Baker was a shareholder in the firm Milne Bay Merchants, which operated another Anson, VH-AKB, in Papua.

A.E. Baker engaged a pilot, and in February or March, 1948, the Anson was flown to Milne Bay. In May, it passed to Joe Taylor.

In 1978, before his death, Joe Taylor told G. Banfield how he came to obtain this Anson: 'There was a chap in Lae who had the salvage rights in Milne Bay and he bought an Anson from Disposals. However, the pilots he employed wouldn't stay with him, so he came to me in desperation one day and said that if I would give him forty hours flying when he wanted it, I could have the Anson. I agreed, and sent a pilot over to Port Moresby to bring the aeroplane back — there was no paperwork handed over with it at all; things were so casual in those days. This later caused an argument when I came to sell the company...'

In August, 1948, a pilot who was to spend the next ten years flying in PNG joined Taylor's Air Transport. This was Jack Stammer, a laconic Queenslander who was to become one of the best-known pilots in the Territory. Stammer's connection with aviation began in 1939, when he joined the engineering staff of the Queensland Aero Club. He received a little free flying tuition in return for aircraft maintenance work, and had already decided on a career in aviation when the Second World War came.

Stammer joined the RAAF and trained as a pilot at Sandgate, Narromine and Deniliquin. After completing a course on fighters at Mildura, he became a Boomerang pilot with 84 Squadron, based at Horn Island. He recalls those days: 'Was ten months at Horn Island, a fairly inactive period really. We used to do patrols over Merauke, and nothing ever happened. Chased a few Dinahs, but they flew too high for the Boomerang and we were too slow, anyway...never got near them. Then back to the mainland and spent about eighteen months on ferry flights. All sorts of aeroplanes, mostly single-engined, plus a bit of time on Dragons, Ansons, Oxfords and Beauforts...did another course, on Kittyhawks, and joined 77 Squadron at Labuan, just after the war finished. With 77 to Japan for two years, and back to Australia in 1948. Posted to Canberra flying Mustangs and Austers with 3 Squadron. I could see there was no future for me in the Air Force, so I took my discharge in August, 1948.

'Was roaming around Brisbane not knowing what I was going to do, when a friend, Bob Dalgleish, who had an aircraft servicing organisation at Archerfield, said he had a Dragon to be ferried up to New Guinea to Joe Taylor. So I accepted the ferry job, at two pounds ten a day plus return air fare and expenses, which I considered pretty good...

'After I arrived in Lae, Joe Taylor decided that he'd keep me on as a pilot, and I started flying around the lower Highlands and Wau-Bulolo. After doing this for about a month and getting a bit used to New Guinea and its conditions, Bill Passlow, Joe Taylor's chief pilot, decided I was ready to be checked into some of the other strips — the Watut strips, Rising Sun, Slate Creek, Mumeng...poor old Joe didn't last long, he went broke about March, 1949, and TAT was taken over by Guinea Air Traders, lock, stock and barrel, the two Dragons, the Anson, engineers, pilots...we all went to GAT.'

At the time of its collapse, Taylor's Air Transport employed four pilots: Bill Passlow, Jack Stammer, John Keene, and R.H. Harris.

Ray Harris gained his private pilot's licence at age eighteen, in 1936. He was one of the first members of the Aero Club of Broken Hill, and had to obtain his parents' permission before he was allowed to take flying lessons in the DH60 Gipsy Moth. He remembers: 'The trademark of the old Gipsy Moth pilot was a burn across the arms. The exhaust pipe went along the top of the cockpit, and when you put your arms up to get yourself out of the cockpit, you'd burn yourself...I learned to fly and went solo in the Gipsy Moth after four hours twenty minutes in the air. You couldn't do that today — not allowed to. That's why they brought in regulations to stop sending people off so early. Not much money around in those days, and the instructors used to send you off in the shortest time possible, so that you got onto the cheaper rate, which was solo, of course.

'I stayed in the aero club until war broke out, flying as a hobby. I wanted to get my commercial licence, and I had the number of hours and had passed several ground exams.

'I wanted to join up, but the RAAF would only take me as an instructor. I didn't want that — they hung on and I hung on until it became too much for me, and I said, "Yes, I'll become an instructor..." '

Ray Harris was an instructor at Camden, Cunerdin (Western Australia) and Deniliquin for much of the war, before completing an operational training course, on Kittyhawk fighters. Early in 1944, he was posted to 82 Squadron, and ended up at Morotai and Labuan.

'The airstrip at Labuan was really too small for Kittyhawks, but we were experienced, all twenty-two of us, and we were able to operate from it, although it was very marginal. We all went to Labuan together, and our time was up together, so after a bit of a blue they shipped us out together. So the squadron had an entirely new lot of pilots to operate off this small strip. They pranged a lot of aeroplanes, and practically wiped the squadron out. Just couldn't handle that strip...while I was at Labuan, I turned over my 2,000th hour. Lot of flying in those days..'

Harris finished the war in charge of OTU training at Mildura, and was discharged at the end of 1945. He returned to his civilian occupation, learning to be an underground surveyor in the mines at Broken Hill, but like so many young men newly returned from the wars, he was restless. The prospect of a life underground after years of flying was not appealing, and when he was offered a job as flying instructor at the aero club, he grabbed it. Then one day a short, nuggetty young man with an infectious grin and a tall reputation arrived at Broken Hill. Says Harris: 'It was Bobby Gibbes, who came through to buy Norseman aircraft which were being operated by Silver City Airways. He told me about his airline in New Guinea* and said, "Come on up!" He didn't have a job for me, but he gave me some names, and I wrote some letters. Joe Taylor answered, and said that Bill Passlow, his chief pilot, was coming down to pick up an aeroplane and would come through Broken Hill to interview me. The first accident I investigated after I joined DCA later was when Bill Passlow was killed...

'Well, Joe accepted me so we sold up house and home, and headed for New Guinea, to Lae. We arrived November '48. TAT didn't have long to go. We operated all over. Used to look after the Morobe District — Rising Sun. Slate Creek, Bulolo, Mumeng and so on — and up into the Eastern and Western Highlands as far as Wabag. Did a milk run up there every week, all stations to Wabag and back again, in the Anson.

'Then Guinea Air Traders bought Joe out. I hadn't been with TAT very long, but long enough to know the reputation of GAT. After the Hudson crash, it was very bad. They had another Hudson, being flown by Dick Burt, and they were going to teach me to fly this thing. I went riding with him and took the controls a few times. At Wau once we couldn't get it started, and we put a rope and jeep on the prop. This horrified me. Dick was a wild man. I lasted with GAT a month, then I got out, went over to MAL and asked Doug Elphinstone for a job...'

Guinea Air Traders redeemed something of its original reputation late in 1948, when five plane-loads of live sheep were flown from Cairns direct to Kerowagi at weekly intervals for the Hallstrom Trust experiment. Not a beast was lost. Although the Hallstrom Trust sheep venture ultimately failed, it was at the time regarded as exceedingly important, and the Administrator himself took a close personal interest in the joint attempts of the refrigerator philanthropist, Sir Edward Hallstrom, and the Department of Agriculture, Stock and Fisheries to establish sheep in the Highlands.

The aeroplane used for the sheep flights was the English-registered Douglas C47, G-AGHN. The original GAT C47, VH-GAT, was sold to Ansett in July, 1948.

VH-GAT had been responsible for much of the earlier success of Guinea Air Traders, and Sam Jamieson retains a fondness for it: 'It was a great aeroplane. Les Ross used to fly that DC3 like a feather can land. He could three-point it, but he couldn't do a down-wheeler. By keeping the tail up, he would have saved wear-and-tear on the tailplane. George Price stuck damned patches all over her tailplane. When George died in the Hudson crash, Tom Watson became my chief engineer. He was a rare man.

'Jack Leslie of BGD helped us tremendously, where freight was concerned. I remember one incident with our freight manager, Johnny Rowbotham, a big, tall guy. He was loading water pipes into GAT for Bulolo. We had weighed and averaged out each pipe, so we knew how many had to go in to make up a load. Johnny had a very smart Papuan, who supervised the loading. When the aeroplane was loaded to the limit, this Papuan told the boys, "*Inap!*" '

'Rowbotham came along then, had a look inside, and said, "There's nothing in there yet. Put more in!" Trouble was, those pipes were heavy — the aeroplane looked almost empty, but the weight was there. Well, the Papuan didn't argue, he put more in. When Les Ross took off, he wondered why GAT didn't want to get off the strip. He just missed the *Tenyo Maru* — went way, way out into the Gulf, gradually climbed up and up, got over Zenag and made it to Bulolo. Jack Leslie came down to meet him, and Les said, "I want every pipe on this aircraft counted, one by one, and on paper. And I want the load weighed." When they had counted and weighed it all, they found that Les had carried a double load. Twice the weight! Jack Leslie got a message to us by an Anson pilot. "Get Rowbotham out of the way, because Les Ross will kill him! He'll shoot him!" Johnny Rowbotham couldn't be

*See Chapter 3, Gibbes Sepik Airways.

found for a month.

'I'll tell you a story. After I sold out to my brother John and another party, my wife and I returned to Sydney. My business life was in Europe now, and my wife flew over with Qantas, and on board the aeroplane was Bert Ritchie, the Qantas flight superintendent. She heard from the mouth of the only man who could really say it — he said to her, "Why did Sam leave New Guinea? He'd have been king of New Guinea if he'd stayed. Qantas would have had to buy him out." '

'I thought to myself when she told me this, "Yes, and maybe a dead king, too." I'd been more than lucky...I'd been in an Anson, we'd had to quietly turn back to Lae, empty the aeroplane, take everything we wanted out of her, then push her over to the side and burn her. Only thing to do. The wingspar, it wasn't tropic-proofed, you could run a skewer right up through it, like a sponge...

'But I let GAT go, and Bert told my wife that Qantas just laughed when they heard, and sat back and said, "We'll get GAT by default." My brother John and the others thought they could run the operation from Sydney. No way in God's earth. You had to live there and nurse it, get up in the morning before daybreak, make sure everything was ready to go. The night before, go round and close the damned cockpit windows, which were supposed to be closed and never were...'

The blow that really finished GAT was the loss of the Bulolo Gold Dredging Ltd business to Qantas. Bill Forgan-Smith had been given specific instructions to do all that he fairly could to capture the BGD contract before he came to Lae to start New Guinea Internal, but the service provided by GAT was then satisfactory, and the BGD manager, J.D. Simpson, saw no need for change.

The grounding of the Ansons caused problems for GAT, but other aeroplanes were acquired and the BGD contract was still effectively serviced. The Hudson crash was a severe setback, but it was the loss of DC3 capacity that seriously affected the reliability of the GAT service.

VH-GAT went to Ansett in July, 1948. That left a single DC3 for the Bulolo work, the English-registered G-AGHN. Then this aeroplane, too, left PNG.

So lucrative was the Sydney-Rome immigrant work that John Jamieson withdrew G- AGHN from Lae, and put it to work on the Rome run. Tom Deegan has no doubts about the effect this decision had on Guinea Air Traders: 'The Hudson crash was a great factor in the failure of the company, but the real factor was taking that English-registered aeroplane — it had thirty-four seats — and putting it on the Rome work. There was Ken Lockyer, George Cordell, Bill Burdus, Lionel Van Praag, they used to fly this aeroplane to Rome and bring back migrants for about three hundred quid. There were very few ships, and very hard to get berths. We had lost our Ansons, and when they took this English DC3 away we became unreliable as far as BGD went, and they got Qantas to do the work.'

GAT ran into serious trouble on the Rome immigrant run, and in October, 1949, the company was charged with breaches of the Air Navigation Regulations.

Immigrant charter flights had been allowed by DCA only because the regular international airlines, Qantas and BOAC, were unable to cope with the traffic offering. It was official policy to protect the regular operators, who were obliged to maintain an extensive and costly route and maintenance organisation, and provide a service in both on and off seasons. The charter operators, on the other hand, picked up passengers at Point A and deposited them at Point B, often purchased maintenance as it was required, were free to abandon routes if they became unprofitable, and could get by on sketchy route organisations. As a further protection for the regular carriers, charter operators were not allowed to undercut their fares.

Until the end of May, 1949, Qantas and BOAC could not cope with the traffic offering in Europe for Australia, and until that date DCA freely granted approvals for charter flights. Over a period of time, GAT was given permits for many flights, and in fact received favourable treatment in that DCA accepted direct applications from the company, although since the two C47's were English-registered and therefore foreign aircraft, the normal requirement was for applications for the flights to be submitted through diplomatic channels.

At the end of May, Qantas advised DCA that space was now available on their aircraft and those of BOAC for substantial migrant traffic from Europe. Charter flights were then suspended. In August, Qantas found that it could not, in fact, handle the traffic, and a very limited number of additional non-scheduled flights were approved. The very last approved flight was made by Guinea Air Traders.

DCA found that certain of the non-scheduled carriers were charging fares in excess of those charged by Qantas and BOAC. GAT received a warning on this score. An application by the company for approval to make a second August flight was refused, but not because of the fares breach. Qantas had once again decided it could take the migrants.

Arrangements had already been made for the second flight, and GAT sought legal opinion on the validity of the Air Navigation Regulations, upon which DCA were relying. Acting on this advice, the flight was made. G-AKNB was piloted by Captain Ken Lockyer and First Officer G. Hawke, and on board was Sam Jamieson: 'We had gotten under the skin of Qantas. The politicians made it a political issue. On that last flight, DCA withdrew our rights to overfly Australia from Darwin with paying passengers. But our legal man instructed us to leave anyway. We were fuelled and could fly empty, but couldn't take anyone who had paid their passage! All the Italians — they were from the mountains — were running around, and we had no alternative but to get these people to Sydney. So I said, "Anybody wanting to go to Sydney, hop in!" We took off, and they were firing red Verey pistols at us from the Darwin control tower. We landed at Cloncurry to refuell. DCA gave instructions that the aircraft couldn't be refuelled, but Shell refuelled us, and we flew on to Sydney. They took us to court — shows the idiocy of these characters who were supposed to encourage private enterprise...'

In October, 1949, GAT, John Jamieson, and the unfortunate Ken Lockyer — who had simply been following company instructions — were charged in the Court of Petty Sessions, Sydney, with a number of offences against Air Navigation Regulations 109, 159, 164, 260 and 315. GAT was represented in court by R.J.M. Newton. The magistrate found that the company had indeed carried passengers within Australia and discharged them at Sydney without approval, but in mitigation took into consideration the fact that GAT had been acting under legal advice, and also undertook not to make any more non-scheduled flights. GAT was convicted and fined a nominal £50. The other charges were dismissed.

During the hearing, Newton informed the Court that he had been advised that Guinea Air Traders was winding up its affairs. At the same time, John Jamieson advised DCA Port Moresby that GAT was 'relinquishing all operational flying forthwith'. It appeared that the end had come for Guinea Air Traders. The international charter work was gone, and the company had also lost the Bulolo Gold Dredging Ltd airfreight contract.

Bill Forgan-Smith speaks of the capture of the BGD contract with quiet satisfaction: 'J.D.Simpson, the managing director of BGD, was a cranky character, but he didn't worry me. He was the boss of that business, and I was the boss of this business, and that was that. He'd ring up one day and say, "I want a Doug. right away." I'd say, 'Can't give you one.' "But I want one right away! This is J.D. Simpson of BGD!" I'd say, 'Yes, and this is Qantas...what's wrong with Guinea Air Traders?...' "Oh, they have broken down or something..."

'So it went. We had five Dougs. at Lae then, with two more on the way. J.D. Simpson would telephone again, and say, "When can you give me a Doug?" So we'd make a time, and I'd make sure we were there, right on the nose. Another day he rang and said, "GAT have let me down on the freezer meat — can you fly it in, quick?" Well, all our Dougs. were out working, and I told him, 'Not before 11 o'clock.' "That's no good to me!" he'd say — then he would be back again on the phone a quarter of an hour later, and he'd get his freezer at Bulolo, exactly at the time I had promised.

'But he didn't like it. BGD were used to getting what they wanted when they wanted it. I told him he couldn't expect us to have an aeroplane waiting around in case he needed it, but if he had a contract with us, a Doug. would always be available. "Oh, I'm happy where I am," he said. 'That's OK,' I said, 'but don't go getting mad at us...'

'The Administration changed the labour contract laws about this time, and boys were signing on for a year instead of two years as before. BGD used to bring their labour down by sea, and they lost a couple of weeks getting them to Bulolo, and another couple getting them back at the end of the contract, so only had them for eleven months. I said to J.D. Simpson one day, "Why don't you fly them down? You lose all that time, and you have to feed them, I can get you fifty-four boys on a DC3." 'Are you trying to teach me my business,' he says. "No, but I know the airline business..."

'We could do this, by weight. We'd rigged up a set of sixty seats for a DC3, but we could only fill fifty-four, because of time out of emergency exits. This was OK by DCA. When VH-ALA went in and killed thirty-three boys, they were well within their weight limit, but didn't have seats and lapstraps. That's what they got lumbered on. DCA said, "Stop! No seats, no passengers." But it was always so much per pound for boys. Sounds crook now, but that's the way it was...all passengers are only cargo, anyhow, white or black, but they won't face up to it.

'It wasn't a week later when Bertie Heath, the BGD beach manager at Lae, came in and said, "Forgie, the boat has broken a tail shaft at Madang, and it'll take a week to get one up from Sydney. There's 100 boys waiting. Can you do anything for us?" 'Yes,' I said, 'it's only two Doug-loads.'

'Bert Hargreaves was flying the other Doug, so I sent him in first. I got past Shaggy Ridge as Bert was coming out of Madang with his half of the load, and after I landed our manager there, Phil Chandler, said, "Christ, Forgie, we're in trouble...bloody Bert tossed eight boys off because he didn't have enough lapstraps...what will we do with them?" 'How many were there?' I asked. "108!" 'Oh, chuck them on,' I said, 'nobody is going to worry about them. So long as we can close the doors, she'll fly.' Phil had weighed them, and they were only about 100 pounds [forty-five kilograms] each, so we had no weight problems — was about 6,000 pounds [2,721 kilograms], and we could have taken 7,000 pounds [3,175 kilograms] out of Madang.

'So I took the fifty-eight boys to Bulolo. As soon as I was in the circuit, J.D. Simpson was down at the strip, as mad as a hornet about the eight boys that Bert had tossed off. He had his hat in his hands...they had told me that when he did that, look out!

' "Dry your eyes," I said to him, "and get away from that back door." Out came the boys in single file. J.D. Simpson said to me, 'how many are there on this aeroplane?' I said, "You've got 108 boys on two DC3's, that's all you have to worry about. Don't ask how many Bert brought out, or I brought out..." 'Thanks, Forgie,' he said, and away he went.

'A few days later he came down to Lae and said to me, "If we give you the contract, will you always have a DC3 ready for us?" 'Yes,' I said, 'we'll have seven "Dougs" here in a week or so, and one will always be on standby. You're not dealing with a Woolloomooloo Yank now! [It should be noted here that Sam Jamieson spoke with a marked American accent, gained from his extensive dealings with Americans during the war]. You're dealing with quality! Qantas are here forever!'

'So we got the BGD business...'

Guinea Air Traders gradually expired, but John Jamieson did not give up easily. Although he had advised John Arthur in October, 1949, that the company was ceasing all flying operations, a week later he requested an application for a charter licence for a possible new route to stand. In December, Jamieson applied for an airline licence over the proposed route Sydney-Brisbane- Rockhampton-Cairns-Port Moresby-Lae-Rabaul, making two round trips each week, and employing two

59. From left: Sam and David Jamieson, Tom Drury and two DCA officers. *Photo: Sam Jamieson*
60. Guinea Air Traders C39, VH-ARB. *Photo: Jim Perry*

59

60

DC3's. Jamieson estimated that the frequency of the service would be increased to four or six times weekly, as demand increased. He based his application on 'the inadequacy of the present service to meet the demand', quoting his own experience of being unable to obtain a Christmas booking with Qantas.

The application was, of course, a direct attempt on the monopoly that Qantas enjoyed on the Australia-PNG route, as an instrument of the Australian Government, and, predictably, was refused.

GAT operations at Lae slowly ground to a halt. In February, 1950, Jamieson advised DCA that the company would be resuming at Lae with DH84 and C39 aircraft, but in April wrote that GAT operations had 'ceased temporarily while a complete re-organisation, financial and otherwise, is being carried out. The Company is at present disposing of all its old and obsolete aircraft, and negotiating for two Bristol Freighters for PNG...'

John Arthur informed the Director-General in May, 1950, that: 'The company is still in existence at Lae, although the C39 aircraft, VH-ARB, has been sold to MAL, but they still have two DH84's, VH-BMX and VH-AAC. They still intend to operate a regular service Lae-Kavieng, with stops at Madang, Wewak, and Rabaul.'

But GAT was finished. The Dragon, VH-BMX, had been severely damaged in a crash and was eventually purchased 'as-is' by the Catholic Mission of the Holy Ghost, Alexishafen. VH-AAC went to Mandated Airlines, as did most of the GAT personnel. The last survivors of the GAT pilots were First Officer Rex Cuthbertson, and Captain Jim Perry. They, too, went to MAL.

61. Guinea Air Traders C47. *Photo: Sam Jamieson*
62. Guinea Air Traders Hudson, *Silver Bullet. Photo: Geoff Goodall*
63. Guinea Air Traders C47, VH-GAT at Goroka, 1947. *Photo: Sam Jamieson*
64. From left: Tom Drury, Sam Jamieson, Mrs J.D. Simpson, Les Ross, 'Hank' Henry and Jack Mitchell at Bulolo, 1946. *Photo: Sam Jamieson*

61

62

63

64

At this time, Qantas and MAL were struggling to service the Territory with but a handful of pilots and light aircraft, but outstation people were in no position to appreciate the difficulties under which the companies were labouring. All they knew was that their mail was not getting through, and their goods were not moving. However, Jack Sedgers, in charge of the many W.R. Carpenter interests in PNG, was sensitive to this sort of criticism. MAL was beginning to suffer under the label of 'Mean And Lousy', and Arthur Collins was directed to base a Dragon and pilot at Wewak, to handle the growing Sepik traffic.

Soon after his arrival in PNG, Gibbes was transferred to Wewak, where Bruce Brearley was resident MAL engineer. Brearley had finished his apprenticeship, and held C and D tickets, on airframes and engines. He liked Wewak, a tiny township of seventeen permanent white residents — the total white population of the huge Sepik District was but 150, mostly Administration officers, missionaries, prospectors and traders, living on small outstations like Angoram, Aitape, Dreikikir, Ambunti, Lumi, Yangoru, But and Vanimo.

Before Gibbes arrived as the first resident pilot, MAL had from time to time based pilots at Wewak for days on end to clear accumulated cargo for the outstations. The Sepik had suffered heavily during the war, and the litter of war was everywhere. Many of the Sepik strips were constructed during the war. The light Dragons often suffered damage in landing on these all too often neglected landing grounds. Bruce Brearley grew to respect the Dragon, and its Gipsy Major engines: 'The Gipsy Major was fantastic. It would chug along, chug along, hour after hour...bit heavy on its top gear, though, forever changing rings and that sort of thing...The Dragon was a great little aeroplane. I remember one time one of our pilots flew in to Hayfield to pick up a load of boys. Taxying down Hayfield, he hit a forty-four gallon [200-litre] drum, overrode the top of it, split the front spar, shattered it, you could slip your whole hand down into the crack, four or five feet [one to one-and-a-half metres] long. Pilot jumped out, had a look, jumped back in, and flew it back to Wewak with the boys on board. Must have been a hair-raising ride...'

When Gibbes arrived, Wewak was still a tiny settlement, and Jean, when she appeared, was not impressed by the old ex-Army black-iron and sisalkraft hut that was to be her home. But this was the common New Guinea experience then, and like other young wives fresh from city comforts, she adapted, and survived. She was less happy about the Dragon, and the flying that occupied Bobby from dawn to dusk, on most days. There was no control tower at Wewak then, and of course the Dragon carried no radio. Once Bobby took off, there was no way that he could contact base should he have an accident, or be delayed somewhere because of the weather. Nor could he be contacted.

66. Janice Pugh. *Photo: Sten-Erik Molker*

66

Soon after starting work at Wewak, Gibbes landed at Yangoru in a heavily-laden Dragon, and on his landing run the wheels hit a drain concealed by long grass. The undercarriage was damaged. He was given shelter by the patrol officer in charge of the nearby Yangoru Base Camp, and during an attempt by a well-meaning police constable to retrieve the Dragon with a line of prisoners, several wingribs were cracked. There was now no way that Gibbes could take off.

There was no radio at the base camp, and the patrol officer arranged with local villagers for news of Gibbes' plight to be broadcast by *garamut* signal-drums. Early next morning, there was the sound of Gipsy Major engines, and a Dragon appeared, low in the sky. Says Bobby Gibbes: 'This aeroplane made a circuit, and then came in, but much too fast. Yangoru strip was short and nasty, and it could be tricky. As it flashed past me, I could see that it was the Catholic mission Dragon, and the pilot was Father John Glover, a wonderful bloke, but a very inexperienced pilot. What really horrified me was the sight of Jean's face at the window, behind Glover. I was dry-mouthed, watching that Dragon miss the approach and come round again for another go. He made it the second time, and he was dry-mouthed, too, after he landed — it was the first time he had put down at Yangoru. He asked me to fly his Dragon back to Wewak...'

Father John Glover was a priest of the Catholic Mission of the Holy Ghost, Alexishafen. The Catholics and the Lutherans had been the pioneers in missionary aviation in New Guinea in the 1930s, and John Glover was famed for his amazing flight to Australia in March, 1942, with engineer Karl Nagy in an ancient DH60 Moth, to seek aid from the authorities for the people trapped at Mount Hagen by the speed of the Japanese invasion. It was this flight of John Glover's that led to the wonderful evacuation flights by the two ex-Qantas DH86's between Hagen and Horn Island referred to in Chapter 1, The Pacific War, of this book.

Hero or not, Glover had but limited experience as a pilot. The Catholic mission was quick to re-enter the aviation field after the war. Their inland stations were scattered, and aerial transport was the obvious way to supply them. In June, 1946, the mission acquired an ex-RAAF Tiger Moth, VH-BXA, in the name of Father Glover. He flew the Tiger in the service of the mission until he wrecked it in a crash near Kegsugl, in July, 1947. BXA was a tough little aeroplane, and was successfully rebuilt.

In January, 1947, the mission purchased a DH84 Dragon, VH-AMO. It was in AMO that Glover and Jean went looking for Bobby Gibbes at Yangoru. A second Dragon, VH-AOE, was purchased in December, 1947, and was ferried to PNG by John Glover. This was initially piloted on mission service by the veteran W.C. Gatenby, mainly into the stations in the Highlands.

In August, 1948, the newly-appointed Bishop of Central New Guinea, Reverend Leo Arkfeld, created a stir at Madang when he landed an hour after dark in a tiny Auster J5, VH-KSR. While Arkfeld circled the town, residents turned out in force in their jeeps and trucks to light the airstrip for his landing. Arkfeld was not at all concerned, and was immediately dubbed 'The Flying Bishop'. Arkfeld was a better priest than he was a pilot, but he put his trust in God. In 1949, he inaugurated the flying service of the mission in the Sepik District, and during an overhaul of the Auster, it was found that the Bishop had replaced the standard control column with an unauthorised wheel, and had substituted a St Christopher medallion for the vital compass correction card. DCA was not amused.

Tragedy struck the mission on the last day of December, 1948. Father John Glover crashed on landing at Mingende in the Dragon, VH-AMO, and was killed instantly. He was greatly mourned, by laymen as well as his colleagues. His death was a severe setback for Catholic mission aviation, but two more flying priests had joined the service, Father Henry Hoff and Father Joe Walachy, and both were to become well-respected members of the PNG aviation fraternity. Both were to meet their deaths in the crashes of light aircraft.

Bobby Gibbes, in his flights through the vastness of the Sepik District, quickly realised that there was money to be made in business there, for the people had war-damage money to spend, and little on which to spend it. He opened a trade store at Yamil, four hours' walk from Maprik. To keep it supplied, he chartered his Dragon from MAL on Sundays at commercial rates. Later, he opened a second store, at Burui. Both prospered. The stores were licenced in Jean's name, and as far as Bobby was concerned, his business activities were quite apart from his flying duties.

As already indicated, however, MAL was not in favour of its pilots running private businesses, and Gibbes was told to cease trading, or cease flying. He resigned on the spot. This was in August, 1947.

At first, Gibbes had no intention of operating an air transport company. He wanted to develop his trade store business, but to do this he really required his own aeroplane. He still owned the Butler Bat, which could lift over 180 kilograms, sufficient for his purposes. Gibbes applied to DCA for approval to use the Bat in the Sepik, and was granted permission to use it, but strictly for the carriage of goods only, not passengers, and subject to the issue of a daily Certificate of Safety.

A certificate could be issued only by a licenced engineer, and the only engineer in Wewak was employed by MAL. Arthur Collins refused to allow his engineer to sign-out the Butler Bat.

Gibbes had already decided to dispose of the Bat. It was an 'orphan' aircraft with a very limited performance, and would inevitably be difficult to maintain. He needed an aeroplane with a better performance, for which spare parts could be readily obtained. For, encouraged by promises of support from his many friends in the Sepik District, he had decided to seek a charter licence, and operate in competition with MAL. On 25 September, 1947, he wrote to John Arthur: 'I have purchased a new Auster from Kingsford Smith Aviation Service, and expect delivery in six weeks. I hope to get a Dragon in five or six weeks...I will use these machines for my inland trade stores and general charter etc. I have erected a hangar of welded tubular steel, covered with iron...'

He went on to apply for a Certificate of Exemption to allow him to operate over the routes from Wewak to Maprik, Kiarivu, Yamil, Burui, Angoram, Aitape, Lumi,

Mimi, Madang and Lae, and from Madang to Mount Hagen, Chimbu, Goroka and Josephstaal. In another letter he asked for the twenty-four-hour Certificate of Safety requirement to be relaxed and replaced by a seven-day certificate, until such time as he could engage an engineer with C and D tickets. In the meantime, engineer-pilot W.C. Gatenby would carry out the seven-day servicing at his current base, Madang.

There is no doubt that John Arthur frequently bent the rules to help Bobby Gibbes and his company over the next eight or nine years, but Gibbes often raged at decisions he felt were capricious. The relationship between the two men did not begin happily. While Arthur supported the issue of a charter licence to Gibbes, he opposed the replacement of the normal twenty-four-hour Certificate of Safety with a seven-day certificate.

Problems began for Gibbes immediately. The Director-General approved the use of the Auster J5 only for the carriage of his own trade store goods, and not for passenger or general charter operations. The Dragon could be used on RPT services or charter operations, but if passengers were carried, a daily certificate would be insisted upon. If only freight was being carried, then a seven-day certificate would be acceptable.

Gibbes appealed against the Auster decision. It was plain that he would have to obtain the services of a licenced engineer, fast, if his plans were to come to fruition. There was a further complication. Air Navigation Regulations now provided that after 1 January, 1948, aircraft could not be used in aerial work or charter operations except with an airline licence. Exemptions could be granted in certain cases, but such cases were limited to holders of charter licences who proposed to operate RPT services. Gibbes was therefore asked to apply for the appropriate licences, and prepare and submit operations and maintenance manuals.

Gibbes was in too deep to back out now. He had obtained a bank loan to purchase his Auster, and could not afford to use it solely to service his trade stores. The aeroplane had to pay its way. He took delivery of the Auster in Sydney in late December, 1947, and with Jean and his dog, Paddy Gibbes, flew the little aeroplane, registered as VH-KSD, back to Wewak. Fortunately, the problem of signing-out the Auster had been solved. Bruce Brearley had left MAL after finishing his apprenticeship, and had applied to join the Fleet Air Arm.

Before leaving for Sydney, Brearley walked for three days down the stony beaches of the Rai Coast from Madang to visit his sister, who was married to the owner of Bau Plantation, near Cape Ridney. Brearley recalls: 'I found that my brother-in-law had bad tropical ulcers, and he had to go to Madang, to hospital. So I stayed for three months, running the plantation. One morning I heard an aircraft engine, a good old Gipsy, and there's this Auster, buzzing the plantation. It was Bobby Gibbes. He dropped a note, asking whether I'd come to Wewak to work for him. Then he throttled right back, came in low, and yelled out the window at me. Well, I liked Bobby. He made several passes while we yelled at each other. I said I would come, but that I'd have to go to Sydney first to see the Fleet Air Arm people first. "OK!" Bobby yelled, "I'll pick you up in Madang when you get back..."

'The plantation boys all stopped work, of course, and were all lined up, goggling at this Auster as it buzzed up and down...

'So I went to Australia, and they told me I was too late for the Fleet Air Arm. So I went back to Bobby, and we started Gibbes Sepik Airways...'

In February, 1948, the Auster was approved as a type for charter and minor RPT work in PNG, and Gibbes abandoned his intention of purchasing a Dragon, and instead decided to get a second Auster J5. The first, KSD, could not handle the flood of work offering; instead of using the aeroplane primarily for his trade stores, Gibbes was making mainly general charter flights. A second Auster made financial sense, and Brearley could as well maintain two Austers as one. But obviously Gibbes would need a second pilot.

A pilot was at hand. An old friend of 3 Squadron days, Tom Briggs.

Briggs was another who caught the flying bug early in his life. He remembers meeting the famous aviator, C.W.A. Scott, as a schoolboy, and also Lester Brain and Amy Johnson. Briggs left school in 1937, and joined the Brisbane Gliding Club. The following year, at age eighteen, he got his private pilot's licence, in a DH60 Moth, the wonderful little biplane that introduced so many eager young men to

67. Gibbes Sepik Airways chief engineer, Eric Cross. *Photo: Sten-Erik Molker*

67

flying. It was inevitable that Briggs would join the Air Force when the Second World War began. Says Tommy Briggs: 'I was called up for my medical late in 1939, but didn't start flying until 1940, when I signed on the dotted line at Mascot. There was this short bloke standing alongside me, signing at the same time — it was Bobby Gibbes, and that was the first time I met him.

'After we had done our elementary training at Mascot, most of our course was posted to Richmond, but I went on to Point Cook. The Richmond people had the honour and glory of flying Wirraways, which were just starting to come off the production line, but we did our training on Hawker Demons and Westland Wapitis. The Demon had a Rolls Royce Kestrel engine, forerunner of the Merlin, but the ones at Point Cook were all clapped out, and as a result there were a lot of forced landings. I had one myself, motor just gave up the ghost. The old Wapiti looked like a relic of World War One, had a gigantic propeller like an enormous clothes prop, and a great big thing at the back end of it like a shovel, the tail-skid. Had to almost get a step-ladder to get up into the cockpit...'

After getting his wings, Briggs was transferred to Laverton, and then to Townsville with 24 Squadron, flying Hudsons and Wirraways. He remembers the Hudson as 'a very violent aircraft, had a very high wing loading, and when it decided to stall and spin, it spun'. Luckily for Briggs, he was again transferred, to 3 Squadron in the Middle East before the Japanese entered the war, for 24 Squadron was sent to Rabaul, and was almost wiped out there.

In February, 1942, Tommy Briggs was shot down in his Kittyhawk just south of Tobruk. He was one of two victims of an ambush. The Germans sent over a Messerschmitt 110, easy meat for the Kittyhawk, as a lure for the 3 Squadron fighters. Briggs and Frank Reid were on standby, and were scrambled to intercept the 110. Unknown to them, a formation of Messerschmitt 109 fighters, far stronger than the Kittyhawk, were waiting, high up in the sun, and as Briggs and Reid took off, they pounced.

'I didn't know what hit me. Anti-aircraft and everything was exploding in all directions, and out of the corner of my eye I saw my No. 2, Frank Reid, go in to the ground like a dart. I saw two 109's looming up. I was just off the deck and the whole thing happened so quickly...Was trying to pick up my No. 2, next thing, Bang! Bang! Bang!...an explosive cannon shell hit my foot, and if I hadn't been wearing big fleecy-lined flying boots it would have blown my foot right off. I went into a vertical climbover onto my back into a spin, had no control. Don't know how many cannon shells hit me, I just dived over the side, didn't have time to disconnect my oxygen mask. I went through the wireless aerial, and I can remember being hit in the back with the tailplane, pulling the ripcord at the same time. I had a death's grip on that ripcord, tore all the fingernails off my right hand...'

Briggs was knocked unconscious by his parachute landing in the desert, and he had lost a great deal of blood when he was picked up. He went into hospital at Tobruk, where his mangled foot was operated on. It was eighteen months before Briggs was pronounced fit for flying duties, but the foot troubles him to this day.

There was no more operational flying for Tommy Briggs. A fighter pilot with half one foot shot away was deemed to have done his bit. He was discharged the day the war ended. His family were in the timber business in Cairns, and there Briggs went.

'I heard that Bobby Gibbes was in New Guinea, so I wrote to him and told him I was interested in starting in the timber business up there. Asked him if I could stay with him, and back came a telegram, saying he desperately needed another pilot, and would I come up and fly for him. I went up to have a look first and did a bit of work for him as a traffic officer, then I went back to Cairns and told the family I'd be leaving for New Guinea. Got my commercial licence in Sydney without much trouble...'

Bruce Brearley went South on leave in March, 1948, and obtained his private pilot's licence. Cliff Haydon left MAL and joined GSA as an engineer, and Gibbes' cousin, Peter Howse, joined the little company as traffic officer. In May, Briggs and Brearley flew the new Auster, VH-KSK, to New Guinea.

Briggs recalls: 'It was a dicey old trip. The DCA bloke at Coff's Harbour must have thought the old Auster was a high-speed pursuit ship or something, because he gave us permission to come through to Archerfield, even though it was very late. When we arrived at Archerfield there were all these DC3's buzzing about and the Auster had no lights at all, nothing in the cockpit, no navigation lights or anything. The bloke in the tower panicked a bit. He didn't know what to do, and ended up waving a red light blindly around the tower. I circled around the tower as it was the safest place for me to be with all these DC3's about. The DC3's were all panicking — the tower had told them there was an unlighted aircraft flying about. We landed on the flare path in pitch darkness...'

The Auster was a capable small aircraft developed from the old Taylorcraft of pre-war days. It could carry three passengers, or about 270 kilograms of cargo over a short haul, and its range was limited. It certainly could not reach Port Moresby without refuelling. Briggs and Brearley decided to carry jerricans of fuel in the cabin and attempt to refuel in flight, by cutting a hole in the floor and feeding the petrol into the slipper tank beneath the belly. It was a radical, and illegal, method.

'Didn't work too good. The petrol kept blowing back in our faces, and we just didn't seem to be able to get the petrol in. Then the tail-trim wire broke — one of the faults of the Auster — so we had to keep full pressure on the stick. We were over the sea, and we were just about out of juice, and the two of us got a bit browned off at all this. We thought we should be near Kerema, but we couldn't find it at all. I said to Bruce, "Well, what do you think?" 'We'll just have to plonk her down on a beach somewhere,' and that's what we did. On a beach somewhere between the Fly River and Port Moresby, God knows where. Indigenes came running from everywhere while we filled the tank. We waited until the tide went out, and we took off in pouring rain, and went on to Port Moresby. Then we went up to the pub, and had a few drinks...'

The two Austers were worked to the limit. Gibbes and Briggs were up before dawn and in the air at daybreak, and often returned to Wewak after dark. The official limit on maximum flying hours was ignored as the two pilots strove to get ahead of the work offering. The little Austers could get into the shortest of the Sepik District strips, and Gibbes and Briggs seldom turned back.

Much of the business handled by GSA was for private customers, but the Administration was also a steady and valued client. In July, 1948, VH-KSK was chartered by DCA for a survey of the aerodrome at Telefomin, located in the most remote region of New Guinea in the mountainous headwaters of the Sepik River. The Administration had decided to establish a station at Telefomin.

Telefomin was discovered by Charles Karius and Ivan Champion in 1927, during their great North-West Patrol. In 1936, members of the gold prospecting party led by the American, J. Ward Williams, built a small aerodrome at Telefomin, which was enlarged during the war for possible emergency use by aircraft flying across New Guinea to Hollandia. In September, 1947, at the request of the Administrator, an aerial examination of Telefomin was made from an RAAF C47, to see whether the strip could safely be used by a light aircraft. With the party in the C47 was District Airport Inspector Jim Glindemann, who reported his opinion that a landing could be made on the long-neglected strip, in the right aircraft. The gliders and mechanical equipment that had been flown in during the war for the strip extensions — dump trailers, a roller, a drawn grader and two miniature rubber-tyred tractors — were still there, obviously much the worse for wear from long years in the open in a bitter climate. They could perhaps be used to repair the strip.

Telefomin was so isolated, close to what was then the Dutch New Guinea border, that an Administration station could only be established there if reliable aerial communications could be established. To achieve this, the strip had first to be repaired. A preliminary examination from VH-KSD by Jim Glindemann, Ivan Champion and Gibbes in April, 1948, disclosed great pools of water over the surface, and as it was the tail end of the wet season in the Highlands it was decided to wait until the weather conditions improved before attempting a landing.

On 1 July, the three men set out from Wewak in the new Auster, KSK. The little aeroplane was loaded to the limit with camping gear and rations, and the fuel tanks were topped up at Burui. It was a two-hour flight from Wewak to Telefomin for the Auster, over incredibly difficult country for much of the way, and with no emergency landing grounds should anything go wrong.

The first attempt to reach Telefomin failed. A build-up of cloud forced Gibbes to turn back to Burui, but on the following morning KSK landed safely at Telefomin.

Gibbes left Glindemann and Champion there, on their own in one of the loneliest places in PNG, among wild tribesmen who in November, 1953, were to kill two patrol officers and two police constables. But the people were friendly on this occasion, and willingly helped the two white men as they made a survey of the strip, pegging a line down the centre for levels, taking rough cross-sections about every ninety metres and digging holes to test the sub- surface. This work occupied three days. Gibbes returned on the fourth, and they flew back to Wewak.

This was a daring survey, and it yielded sufficient data for Glindemann to be able to draw up plans for the reconstruction of Telefomin strip by the Administration officer who was to establish the station, Assistant District Officer D. Clifton-Bassett.

Telefomin was kept alive during the early months by the little Austers. It was a long flight from Wewak, and the treacherous, ever-shifting weather pattern often beat the pilots. It was an expensive station to maintain, but once opened, the Administration could not withdraw from Telefomin. The problem of supplying the station was eased, however, when Gibbes acquired his first Norseman.

The Noorduyn Norseman was a large, rugged, fixed-undercarriage, high-winged, strut-braced monoplane designed for bush operations in Alaska, and which first flew in 1935. It remained in production until 1950, some 900 of different variants being manufactured. The portly fuselage was a welded-steel tube framework, and the wings and tailplane were of wooden two-spar construction. The whole aeroplane was fabric covered, and the normal engine was a strong Pratt and Whitney Wasp nine-cylinder radial of 600 horsepower. The cargo cabin was capacious, and the loading door large. The Norseman could lift over 900 kilograms of cargo off rough strips and carry it at a speed of 209 kph. A total of thirteen saw service with the RAAF during the war, and the aircraft that Gibbes purchased was one of these.

Before committing himself, Gibbes applied to John Arthur for approval to operate Norsemans from Wewak to Aitape, Lae, Bulolo, Wau, Hayfield, Madang and Mount Hagen, and from Madang to Goroka and Hagen. Always impatient, Gibbes did not wait for confirmation, so sure was he that the Norseman would be approved. In April, 1948, he took an option on a Norseman, registered as VH- BNE, from Kingsford Smith Aviation Service, the purchase price to be £5,500 — a considerable sum: good used DC3's could be had for £10,000. Gibbes was stunned — and enraged — when DCA refused approval for the operation of Norsemans in PNG, 'because of high stalling speed, fairly high take-off and landing runs...in the event of a forced landing, likely serious injuries could result'.

This really began the long battle between Bobby Gibbes and the bureaucrats, a battle in which different skirmishes were won by one side and then the other. Bobby never was a man to tamely accept the fiats of officialdom, and he was ever-ready to use the influence that his outstanding war record had earned for him. An impartial observer would conclude, however, that overall he received very fair treatment from DCA.

Gibbes won the first round. He appealed directly to the Minister for Civil Aviation, Arthur Drakeford, and to the Director-General, a staunch ex-RAAF man. He had a strong case. The RAAF had considered the Norseman to be a suitable aeroplane for New Guinea operations, and it had an outstanding reputation in outback Canada. It was at least as safe as the Fox Moth, Dragon and Auster. Approval was given to operate the Norseman at a maximum all-up weight of 3,193 kilograms at Wewak, Aitape, Lae, Bulolo, Madang and Hayfield, and subject to a dry surface at Wau, Angoram and Goroka. The weight limit for Mount Hagen was 3,044 kilograms and Telefomin, 3,030 kilograms.

The decision was given on 28 October, 1948, and Gibbes finalised the purchase of VH-BNE the following day. The Norseman was an instant success. The charter rate of £28 per hour was reasonable, and the aeroplane was very popular with customers, for it could accommodate bulky items of freight, and it had a good turn of speed. Gibbes set about acquiring more. VH-BNL was purchased in late June, 1949, and during 1950 three more: ASS, BLM and BNT. The Auster J5 fleet was increased by the purchase of KSQ, KSS, KST and KSU, and a final Auster, KSX, in February, 1951. A branch office of Gibbes Sepik Airways was opened in Port Moresby in October, 1949, and pilot Ted Marriott was stationed there, with an Auster. In June, 1950, pilot Pat Tainton was stationed at Goroka, with another.

Gibbes was starting to expand from his Wewak base. He was fast becoming a force in PNG aviation, particularly after GAT folded, and he was taking business away from MAL and Qantas. The going was not uniformly smooth, however. On 1 October, 1948, Gibbes lost his first aircraft. The pilot was Tommy Briggs.

Jim Glindemann had borrowed VH-KSD to make an airstrip inspection, and on landing at Tadji in windy conditions the aeroplane bumped heavily, crumpling the sternpost. There were no licenced airframe welders in Wewak, and it was necessary to fly KSD to Lae for repairs. Briggs flew the Auster, with Bruce Brearley as passenger. As Briggs tells it: 'Well, I had the honour and glory of having the first prang for GSA. On the way down to Lae the weather caught us, and we stayed overnight at Kaiapit with the patrol officer and his wife. They were very good to us, so we said we'd bring them some freezer on the way back. 'Well, in those days the only way you dropped anything off an aircraft like the Auster was to open the door, and throw it out. Bruce just opened the door and chucked the meat out, and I think he managed a direct hit on the PO's house. The only unfortunate thing about it all was that we followed the meat in.

'I don't know what happened exactly, we got into a loose bit of air, we couldn't seem to be able to climb, and the only way to climb in the Auster under those conditions was to sort of bend the throttle lever, you didn't have any reserves of power at all. Anyway, there I was bending that throttle lever trying to climb out, and in a few split seconds I realised we weren't going to make it, so I just pulled the throttle off and we sat down in the PO's garden. We were lucky that we really didn't cause too much damage to ourselves, although the Auster wasn't too good.

'Bruce was actually a bit worse off than I was. He had his lapstrap done up. I wasn't wearing mine. He sort of jack-knifed onto the instrument panel and bit right through it. They've still got that dashboard somewhere with a couple of teeth in it. I had a cut over one eye and a busted nose but anyway we were both all right. Bruce broke a couple of ribs on the dual controls but apart from that we were OK.

68

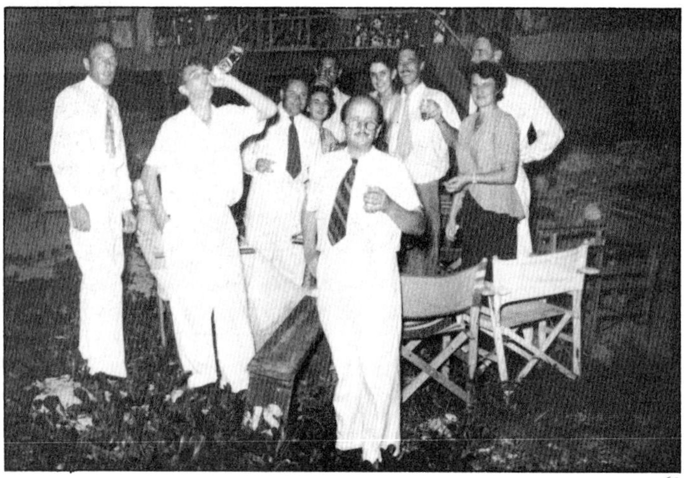

69

7

7

68. Gibbes Sepik Airways terminal and freight shed, Goroka, 1957. *Photo: Sten-Erik Molker*
69. Gibbes Sepik Airways personnel having a session at the RSL Club, Port Moresby; Dennis Buchanan on the bottle. *Photo: Frank Goossens*
70. Sepik River Trading Co. (Hall's Charters) DH84, VH-AOT at Madang. From left: C. Brush (pilot), Morton Johnson (engineer) and Ben Hall (owner and manager). *Photo:* Pacific Islands Monthly
71. Goroka, 1957. *Photo: Sten-Erik Molker*

'They took us out to Lae Hospital, and the DCA investigating committee interviewed us about this prang while we were there. That was the first contact I'd ever had with DCA. Rube Murray, the senior investigating engineer, said, "Well, you blokes have just about broken every law that DCA ever made..." '

KSD was a write-off. The bits and pieces were purchased by Guinea Air Traders and shipped to Kingsford Smith Aviation Services in Sydney to be rebuilt, but KSD never flew again.

Four months after this incident, Tommy Briggs overturned in VH-KSK while landing for the first time at Telefomin. He hit a soft spot, and the little Auster flipped over and was badly damaged. It was at first thought that repairs would be impossible in such a remote locality, but after Briggs had been marooned for ten days, Bobby Gibbes flew Bruce Brearley in to Telefomin in Norseman VH-BNE — which had been damaged in an accident at Angoram even as KSK was flipping onto its back — and the Auster was repaired, and flown out.

Tommy Briggs was never endorsed onto the Norseman. His true interest was in sawmilling; he had only taken on the job with GSA as a favour to Bobby. And as GSA expanded, other pilots, and ground staff, were employed: Alf Clare, Alan Mossman, Pat Tainton, Peter Flanagan, Roy Shaw, Frank Goossens, Peter Manser — an accountant, but a born pilot who soon talked his way out of the GSA office and into the air — and Rinus Zuydam. Says Briggs: 'After Rinus arrived, I more or less gave it away. I wanted to get this sawmill going. I still did occasional trips for Bobby, familiarising new pilots, and I did a few trips — still had my commercial licence. I spent my time building up my timber business, in Angoram, Wewak and Madang. Sold out in 1970, and came South.'

Among the staff that Bobby Gibbes employed in 1949 was a young lad who was to become a leader of the aviation industry in PNG: Ranald Dennis Buchanan.

74

72

75

3

72. Loading Norseman VH-ASN at Tari, 1956. The pilot, Adrian Nisbet, is near the tailplane. *Photo: James Sinclair*
73. Alberto Buzzi (on wing) and Peter Manser. *Photo: Sten-Erik Molker*
74. Ben Hall's Tiger Moth, VH-BDR after crashing at Goroka, October, 1948. *Photo: Alf Lane*
75. Tom Briggs beside his Auster which turned upside down on landing at Telefomin, 1948. *Photo: Tom Briggs*

Buchanan

Dennis Buchanan was born in Sydney in November, 1932. His father, a general storekeeper in a country town, was a Queenslander, forced like so many others in those grim Depression years to seek work in the city when his business failed. Buchanan's earliest memory is of life on a farm near Riverstone, east of Windsor, as the youngest of a family of five brothers and three sisters. A fourth sister was born before Buchanan's mother died, in 1939. An elder sister, Mary, took over the running of the household upon the death of her mother.

War came, and the family began to separate. Two brothers, Joe and John, joined the RAAF. One sister moved to Bondi, and another to Kensington. Soon Dennis Buchanan was the only one of the children still living at home with his father, and at the age of nine, he was sent with his little sister, Lilla, to stay with their uncle on a farm in the Hunter valley. The two Buchanan children attended a little country school over three kilometres away, with their cousins. Buchanan remembers: 'We'd be up in the morning about five or five-thirty, and chase up the cows before going to school. My uncle was milking fifty or sixty cows. Then we'd walk to school, and after walking back to the farm in the afternoon we'd chase up the cows again, and milk them, and go to bed by six or six-thirty. Lilla was only five, and I had to watch her carefully. Very hard on her, I thought, to walk all the way to school. It was a good life, but hard...'

It was too harsh a life for young children, in the opinion of Joe Buchanan, newly returned from the war, and he took them back to Sydney. Lilla went to live with her sister, Joyce, and Dennis went to the nearby home of a family friend, Mrs Una Brown. It was the last school term of 1943, and he remained in Una's care until the end of 1944. Her husband was sick and in hospital, and she was obliged to work for a living. Young Dennis was left pretty much to himself, largely free from supervision, and he admits today that he 'might have been a little unruly.

'I went to Eastwood Public School, a mile or so away, and I walked there, under the railway in the storm-water drains, through the culverts, having a ball. I'd usually get to school late, and sometimes I'd just go to the bakery instead and eat the damaged bread. Una Brown could see I was a little difficult, with no discipline. She never knew whether I was at school or not, so she decided to talk with my father, sisters and brothers. Towards the end of 1944, I was organised into a boarding school at Bathurst, All Saints' College. Sounded good enough to me, and it was back in the bush...'

Buchanan did well at All Saints' at first, but as he got older he became interested in sport, particularly football and cricket, and his studies took second place. He managed to scrape through his exams for the Intermediate Certificate at the end of 1948, with B passes in English, Economics, Geography and History. This enabled him to carry on into his fourth year of secondary schooling at All Saints', but the end was approaching.

'My father was only a working man, and he was finding it difficult to keep me at boarding school, and he had my little sister to support, as well. In May, '49, he told me, "Look, I can't keep you on at All Saints'. You had better prepare yourself for leaving school by August." I said I didn't mind. I went back to All Saints' and talked to the headmaster, a man called 'Sno' Evans who had previously been in Charters Towers. He said to me, "We don't think you should leave school. You're quite a good student, even though you're only just passing." He offered to keep me there, without fees, until Fifth Form, but I was a bit proud and pig-headed, and I said, "No, I can't accept charity."

'Then, in about June, '49, Bobby Gibbes wrote to the school, wanting a couple of kids to come to New Guinea as trainees in his office. He was an All Saints' old boy, and very popular. His letter came just before the time I was going to have to leave, so the Bursar, Albert Emms — he was also the football coach, and a good friend to me — wrote to Bobby and recommended me.

'I got the job, but Bobby couldn't afford to fly me up to New Guinea with Qantas, and he told me to wait in Sydney until he came down to pick up another Norseman. So I stayed with my sister, Mary, at Summer Hill. My brother, David, got me a job while I was waiting. Bob wrote down in September, saying that the aeroplane would be ready in October, so I resigned from my job. But November came, and the Norseman still wasn't ready, and I was broke. The Norseman was being overhauled at Mascot, and I went out a couple of times to have a look at it. Didn't look too safe to me. The only aeroplane I'd flown in was a Butler DC3, and I looked at the Norseman and thought about how I was going all the way to New Guinea in it, and I wasn't too disappointed that the departure was delayed.

'Towards the end of November, Bobby came down to Sydney and said, "Right, we're off." I didn't have any money left, and had to borrow fifty quid from David to buy clothes and things for New Guinea — white shorts, shirts, socks, and a pith helmet Bobby said I'd need, tiger shooters, we called them. Dad came out to Mascot, and he was horrified to see his little son going off to New Guinea in a thing like the Norseman.

'We waited all day. There were four of us in the Norseman — me, Bobby, Alan Mossman and another pilot, Percy Lusk — and it was loaded right up with fuel and gear. Bobby was flying it, and he was mad to get away, but things kept going wrong. But Bobby was mad to go, so we got on board, said goodbye, and headed up the coast, very low. Bad weather, and the engine kept on cutting out, and making funny noises. Bob kept saying, "The engine's icing up, but it'll be all right." We'd lose a few hundred feet, then she'd pick up a bit, then she'd lose a bit more. It kept doing this all the way to Newcastle, so Bob said we'd better turn round and go back. So back we went to Mascot, and stayed that night while they fixed the engine. Didn't tell anyone we were back.

'We got away again early next morning in bad weather, which got worse. It was wet and raining in Brisbane, and we got over the Glasshouse Mountains, following the road. Somehow we found our way to Maryborough and refuelled there, in the rain. Made Mackay that night, and took off in the dark next morning for Daru. I wasn't worrying too much by this time.

'We had 101-gallon [459 litres] belly tanks in the Norseman, and two others of fifty gallons [227 litres] each, with one thirty-seven gallon [168 litres] tank right in the back of the fuselage. Was a long hop for the Norseman. I had to climb all over the cargo to turn the fuel on at the back. We landed at Iron Range to top up, and we were supposed to go on to Horn Island to clear customs. We climbed out of Iron Range to 10,000 feet [3,408 metres], and we were going so smoothly that Bobby said, "It's a bit silly going down there," and we just did a couple of circles over Horn Island to let the customs bloke know it was us, and just kept on to Daru...'

Buchanan settled down happily enough to life in Wewak. He shared at first an old wartime house of black iron on a concrete slab with pilot Alf Clare.

'I liked Wewak, a real frontier town then, maybe 80 or 100 people including Chinese, Mission and Administration. I'd turned seventeen just before leaving Sydney. First week I was there, Jerry Stewart died, Tom Briggs got married and a couple of babies were born, and they seemed to be always having parties. Worked all day and partied all night. They called me "Junior". Bobby and Jean were supposed to be my guardians, and they looked after me. Wasn't allowed to drink — you couldn't legally drink until you were twenty-one.

'I was the office boy-rouseabout. I did paperwork for the accountant Peter Manser, who was becoming more and more involved as a pilot. I had to look after the paperwork for the store. Everything was in the same building, the hangar they had bought from GAT at Wewak airstrip. Jean was in charge of the funds, and she used to help Peter with the accounting. I did a bit of journal work, entries, checked the day's flying, collected fares, did the banking.

'An MAL DC3 used to bring fresh meat to Wewak every Thursday, and we ran an Auster to Lae once a week and one to Madang, for supplies. Got fruit and vegetables from But and Dagua. They taught me to drive the old Willys jeep, and the blitz trucks, then put me onto aircraft loading, then unloading ships. Cargo was taken off ships in the bay in lighters, and unloaded on the beach. No cargo sheds in those days. We'd pick up stores off the beach for people like Bill Royal at Maprik, and Tom Ellis at Yamil, clear the stuff through customs, sign for it, truck it to the airport, scale it up, load it onto an aeroplane, and send it off. Then I had to help with the engineering a bit. Everyone had to help, because we were short of engineers. Even the pilots used to help with the paper work. It was quite a family show, Gibbes Sepik Airways. I was learning a lot about the airline business...'

It was the best of schools for a lad who was to become the operator of one of the world's largest fleets of light aircraft. Bobby Gibbes was 'Junior' Buchanan's mentor, and from Gibbes he learned much. He learned in particular how difficult it was to run an airline on limited capital. For the whole of its existence, Gibbes Sepik Airways was under-capitalised, and Gibbes' swift expansion was achieved at relatively small cost, but he seldom had much cash in hand, and could afford only the most basic facilities. The mess that he built at Wewak for his single staff was constructed mainly of bush timber, and his first hangar at Goroka was thatched with *kunai* grass. But Gibbes employed good pilots and engineers, another lesson learned well by Buchanan. In 1949, when DCA was considering Gibbes' application for a charter licence, GSA operations were examined by John Arthur and Rube Murray. Arthur reported to the Director-General: 'The company organisation, operations control and aircraft maintenance engineering was thoroughly examined...the company is operating on sound lines...safe and effective control...'

The licence was granted. GSA had problems, though, which rapidly multiplied. Gibbes sent his pilots and aeroplanes into rough strips that were generally shunned by other operators, and inevitably the aircraft suffered. And there were accidents.

Alf Clare was flying Auster VH-KSS to Wewak on 27 October, 1949, with Dr John McInerney as passenger. McInerney was District Medical Officer of the Sepik. They ran into foul weather near Yapunda, and Clare, a seasoned flyer, decided to land while he could still select his ground. A violent storm was building, and Clare was too experienced a pilot to risk an unnecessary battle with the elements in a fragile aeroplane like the Auster. He landed safely on a patch of open ground — you could land an Auster almost anywhere — but the undercarriage was written off.

Clare and McInerney started walking. The alarm was quickly raised when the Auster failed to arrive at Wewak, and Gibbes set off in another to search. He found the two, almost through to the coast, and dropped them food and a medicinal bottle of rum. Clare and McInerney reached Aitape the day after the crash, none the worse for the experience.

In February, 1950, KSS crashed shortly after take-off from But, with Alan Mossman at the controls. The Auster was completely wrecked, and Gibbes sold the remains to the insurance company. What remained of KSS was shipped to Sydney, where it was rebuilt. It passed into the hands of the Kingsford Smith Flying School, and crashed, finally, at Mittagong in 1953.

Dr McInerney was killed in an Auster crash, in March, 1953. He was an extraordinary man; brilliant doctor, womaniser, drinker and hell-raiser, who risked his life countless times flying in bad weather to isolated outstations when his services were needed. He was revered in the Sepik. He was taught to fly by W.C. Gatenby and Bobby Gibbes, and obtained his private licence in December, 1948. In March, 1952, he gained his commercial licence.

In August, 1948, John McInerney purchased a Tiger Moth, VH-BLT, and used it to assist him in his medical duties throughout the vast Sepik District. He found the Tiger a little too slow and stark for his needs, so he sold it, and in August, 1950, purchased an Auster J5, registered as VH-KSY.

This was a considerably more expensive aeroplane than the Tiger Moth, and the Department of Public Health agreed to pay McInerney the sum of £7/10/- per hour while he was flying the Auster on official duties — the GSA charter rate was £10 per hour. In 1951, McInerney applied for a charter licence, but the matter had not been concluded at the time of his death.

McInerney was killed when KSY crashed into Vanimo Harbour immediately after take-off. His two passengers, Administration officers Ian Skinner and George Wearne, were strapped in and went down with the Auster into more than fifteen metres of water. Both were injured, but both managed to escape. McInerney's body was not recovered for two days. Wearne and Skinner were rescued by Frank Galliano, a DCA officer, and Patrol Officer Graham Taylor, who happened to be on the scene.

John McInerney was thirty-seven when he died. He had amassed a total of 2,030 flying hours, mostly in the Auster, and should have been capable of controlling his aircraft under all normal conditions. However, when KSY was recovered from the bottom of Vanimo Harbour, it was found that the airspeed indicator and tachometer were unserviceable. DCA investigators concluded that the cause of the accident was pilot error: 'Incorrect technique in effecting a recovery from an inadvertent stall after a steep turn at low altitude without an airspeed indicator and tachometer'.

Bobby Gibbes lost another Auster in June, 1950. Pilot John Dawe, newly engaged, crashed at Wom in VH-KSU, writing off the aeroplane. Dawe was badly injured, and was flown to Adelaide where he was under medical care for months, at Gibbes' expense.

At this time, aircraft operators were obliged to carry passenger insurance, but the insurance of crews and aircraft was optional. Gibbes had no cash to spare for anything but essentials, and prior to the crash of KSU he had carried his own risk. After the crash, GSA pilots and most of the aeroplanes were insured.

It was just as well. Auster VH-KSQ was lost in August, 1950. Pat Tainton was airdropping emergency medical supplies to a ground party near Chuave with a passenger, Dr D. Jamieson, and was caught in a violent down-draft. That was terrible country for an aeroplane in trouble: a tangle of sheer, contorted mountains and deep, twisting valleys. Pat Tainton was forced to fling the tiny Auster into so steep a turn to avoid disaster that he stalled, and crashed into a mountain-side. Both Tainton and Jamieson were injured; KSQ was totally destroyed.

In October, Alan Mossman crashed near the dangerous Tapini strip in Auster VH- KST. The aeroplane was destroyed, but Mossman escaped with slight injuries.

More serious was the loss of the first Norseman. The Austers were useful little aeroplanes, but the real money-spinners were the Norsemans, and Gibbes could ill afford their loss. The aeroplane was VH-BNL, and the pilot was F.J. Goossens, who was to play a significant part in the development of aviation in PNG.

Frank Goossens joined GSA in January, 1950. Born and raised in the Dutch East Indies, he was living with his parents at Bandung, Java, when the Second World War began. The bloody battles in Europe were remote from Java, but as relations between America and Japan steadily deteriorated, the grim spectre of war began to hover over the Pacific.

The Dutch had minimal forces in the East Indies, and a desperate effort was made to strengthen these before it was too late. There was a particular need for pilots for the Army Air Service, and an emergency training scheme was introduced. One of the volunteer trainees was Frank Goossens.

When the Pacific war began, the Japanese advanced with such incredible speed that the Allied forces were almost paralyzed. Singapore fell, and the Japanese swept on to the East Indies. Just before Batavia was taken, Goossens and the other trainees left in the steamer, *Boisevan* for Australia. Also on board were the gold reserves and archives of the Netherlands East Indies Government.

Frank Goossens and his fellow volunteers were trained in Australia and America, and most were then posted back to Australia. Goossens flew Kittyhawk and Mustang fighters against the Japanese out of Merauke and Biak, and in 1948 left the Air Service and flew for a time with KLM in the Netherlands. He had married an Australian girl in 1947, and both found the bitter cold of Europe and the way of life in a small, overpopulated country too much to bear. They returned to Australia, for good, at the end of 1949.

Frank Goossens went looking for a flying job in Sydney. He found it impossible to obtain work with the major companies, whose policy was to give preference to ex-RAAF pilots, of whom there were a great many. There was some prejudice, too, against European migrants in those days. In January, 1950, Goossens heard that Bobby Gibbes was in Sydney, looking for pilots for his rapidly expanding company. He contacted Gibbes, and was given a job.

Goossens flew Austers and Norsemans out of Wewak for a time, and then was transferred to Goroka. On 11 July, 1950, he was returning to Goroka in BNL, late in the afternoon. It was the dry season, and the broad Asaro valley was filled with smoke and haze. Visibility was deteriorating alarmingly, and when within a few kilometres of the airstrip, Goossens was forced to land on a patch of open grass.

76

77

76 & 77. Crashed Norseman at Tapini airstrip. *Photos: Frank Goossens*
78. Goroka as seen from a DH84, c1954. *Photo: Alf Lane*
79. Goroka, 1954. *Photo: Doug Elphinstone*

He spent the night trying to sleep in the cockpit. The morning dawned bright and clear, and Goossens decided to risk a take-off.

It was an unwise decision. The *kunai* grass was tall and dense, concealing a small hillock. Goossens was almost airborne when his port wing struck the hillock, and BNL ploughed in. It was badly smashed up, and was out of action for years before it was rebuilt. Goossens escaped with scratches and bruises.

Following this accident, Frank Goossens was sent to the Port Moresby branch of GSA, and most of his subsequent flying career was spent in Papua.

Other Norsemans had minor accidents during this period — the DCA finding on one of these was that the pilot (who shall be nameless) had been 'under the influence of alcohol' — but there were no GSA fatalities.

Qantas and Mandated Airlines were not so fortunate.

78

79

80

81

82

80. GSA Norseman VH-GSA airdropping supplies to patrol. *Photo: James Sinclair*
81. Pat Tainton and Dr Jamieson being carried on stretchers after leaving Watabung. *Photo: Alf Lane*
82. Dennis Buchanan (left) loading a Norseman at Goroka. *Photo: Sten-Erik Molker*

The Trials Of Mandated Airlines

It will be recalled that when Guinea Air Traders folded, MAL took over most of its staff and aeroplanes. As the decade of the 1950s began, MAL appeared to be in a strong position. The company fleet then consisted of eight DH84 Dragons — VH-AAC, AOP, AOQ, APJ, APL, ARJ, BDB, BDC — one Lockheed Hudson — VH-BDN — and two Douglas C47s — VH-BFV and BFW. (The ex-GAT C39, VH-ARB, was scrapped in July, 1950). Brian Carpenter had departed, and the manager of MAL was now Doug Elphinstone. There was at first no chief pilot, but the chief engineer was the well-respected G.H.(Bill) Humphreys.

Under Elphinstone, MAL expanded its route network, particularly into the Highlands. The opening of the Highlands to labour recruiting in 1950 gave a further impetus to MAL's operations. Doug Elphinstone remembers: 'When I became manager there were practically no regular services throughout New Guinea at all. I had in mind the very successful service that Joe Taylor had run, from Lae to Wabag, so I started one to Madang-Wewak-Momote-Kavieng-Rabaul and back the same way. Another service Lae-Goroka-Moresby-Goroka-Wau-Lae. They were really successful, too. That sort of put us more or less on an airline basis, although we were still operating under a charter licence.

'When they decided to let the Highlanders out to work on the coast, a terrific passenger trade developed, between Goroka and Rabaul in particular, because the Rabaul plantations always seemed to be strapped for labour. And those Highlanders all seemed to want to get out and earn some money. Les Davis and I would often spend a full week in a DC3 just running from Wewak to Rabaul and back the same day. So we got to know that inter-tropic front that lay across the Bismarck Sea very, very well. And didn't it give us some rough frights, too.

'We really got stuck into that recruiting business. I remember going in to Burns Philp's Sydney office, and coming out with a contract to transport 12,000 boys a year. To Rabaul from Goroka. It was a big operation...'

It will be recalled that the 1947 DCA Survey led by Arthur Affleck recommended the development of Madang as the aerial supply point for the Highlands. This recommendation was accepted, but it was a long time before the Madang aerodrome was suitable for unrestricted DC3 operations.

The adoption of reduced standards following the 1947 survey resulted in the division of PNG aerodromes into three categories: A, B and C. Category A 'dromes were maintained, developed and controlled by DCA, and included all the major airline service aerodromes, and strips at key points. Practically all water alighting areas were also classed as Category A. Category B 'dromes were those considered necessary mainly for civil administration purposes, and included aerodromes such as Angoram, Chimbu, Garaina, Kainantu, Telefomin, Kokoda, Maprik, Tapini, Wabag, Mumeng. Category C 'dromes were private aerodromes, licenced to non-Administration organisations (principally missions, but also including companies like Bulolo Gold Dredging Ltd) and maintained to DCA standards. Some of these were Banz, Bena Bena, Kegsugl, Nondugl, Ogelbeng, Yamil.

Madang was, of course, a Cat. A aerodrome, and by early 1950 both Qantas and MAL had developed sizeable branches there. The Catholic Mission of the Holy Ghost were also heavily engaged from their base at Alexishafen, twenty-four kilometres from Madang, in supplying their Highland stations. But Madang 'drome was only 1286.3 metres long. Before it could become the major airport for the Highlands, it would have to be lengthened.

The matter was urgent. The Administrator, Colonel Murray, was becoming concerned at the mounting cost of supplying the Highlands from Lae. The flying time from Madang to most of the Highlands ports was much shorter than from Lae.

In December, 1948, Murray directed the government secretary, S.A. Lonergan, to take action: 'We have now been administering the Highlands for a period approaching three years. It has been generally accepted that the least expensive and the most efficient route for the supply of the area from Goroka to Wabag is from Madang...I wish action taken which will bring about the situation as we know it should be, i.e., use Madang as the main port from which material is flown into the Highlands...

'I am not impressed with the difficulties with regard to supplying the Highlands from Madang, and these must be overcome, and you are to direct that such work be undertaken as is required...'

The buildup of a large, comprehensive government stores organisation at Madang began. Wharves in Madang harbour were improved. For its part, DCA commenced to upgrade the Madang aerodrome. Airport Inspector G. Coughlin prepared plans for the lengthening of the landing strip to 1615 metres, to allow all-up operations of DC3's on Australian standards. Plans were prepared for the erection of a huge Pentad hangar.

The lion's share of the rapidly increasing Highlands business out of Madang fell to MAL and Qantas. Madang-Highlands operations were particularly important to MAL, and pilot David Robertson was sent to Madang to take charge of the growing branch.

David Robertson — known throughout the PNG aviation world as 'Dave Robbie' — was born in Sydney, and remembers having his first flight in 1930, paid for with ten shillings swiped from the till of the garage his father ran at Warwick Farm. Ten shillings was a noticeable amount of cash then, and the theft was quickly discovered. Robbie was belted by his justly-incensed parent, but the elder Robertson was an understanding man, and he could see that his son had been well bitten by the flying bug. He paid for the next flight, with the famous pioneer aviator (and later PNG identity) Goya Henry.

Dave Robbie became an apprentice motor mechanic, but his heart remained in flying. He commenced private flying lessons just before the war, but when he joined the RAAF it was as an aircraft engineer, since he was a qualified mechanic. He was re-mustered as a pilot in 1943, and was posted to Port Pirie as an instructor, where he remained until the war's end.

Improbably for a man still in love with flying, Robertson joined the Singer sewing machine company as an engineer after he was discharged from the RAAF — although as he points out, a lot of World War Two aeroplanes had Singer electric motors in them. But it was inevitable that he would return to flying. In 1947 he obtained his commercial licence, and joined Butler Air Transport company. He obtained engineer's licences on the DC3 and on Pratt and Whitney engines, and in 1948 answered an MAL advertisement for an aircraft engineer, based at Lae. He told the company that his primary interest was in flying, not engineering, and it was agreed that after six months as an engineer he would be transferred to the flying staff. In November, 1950, Dave Robbie was sent to take over the Madang branch.

By this time, MAL had slipped out of favour with DCA. The company had been reprimanded for inadequate maintenance of its DH84 aircraft following Arthur Collins' mishap in February, 1947, but MAL pilots had full confidence in their aircraft, and in the engineering staff. In fact, the MAL Dragons had a performance advantage over those of Qantas through the use of modified heads on the Gipsy Major engines which allowed them to use leaded petrol, and delivered something like ten extra horse-power. Qantas refused to similarly modify its Dragons. John Spiers' forced landing on Turnagain Island hardly reflected on the Dragon, nor did the forced landing of another Dragon two months later, piloted by V.A. Smith.

Vic Smith was an ex-RAAF pilot, who joined MAL in October, 1947. His service career in some respects paralleled that of Dave Robertson. He joined as a flight mechanic, then became a fitter, and was re-mustered to aircrew in 1943 and was posted to Port Pirie as an instructor. In 1945, Smith did a conversion course onto Liberator bombers, and went on posting to Darwin with 99 Squadron as the war ended. He remained in the Air Force until January, 1947.

Smith well remembers his brief flying career with MAL: 'It was interesting flying, and plentiful. The legal limit of 100 hours in thirty days was reached time and again, flying into airstrips all over New Guinea...west of Goroka was the Wahgi valley, then restricted by DCA to experienced pilots. My introduction to the Wahgi came very quickly. I was flown in to Wabag by Dick Burt to bring out Johnny Spiers after he shot himself in the foot...

'The high ranges and rapidly changing weather conditions were a feature of the Wahgi, and it was like looking down a tunnel, many times, coming back from Hagen.

'I remember carrying full loads of forty-four gallon [200-litre] drums into Rising Sun, with the tops and bottoms cut out. They used to weld them together into pipes for the gold sluicing operations. You had to crawl through the drums to get into the cockpit...

'About May, '48, Len McNeil invited me to join Qantas, and after an interview I was told to front up in Sydney. MAL had a Dragon to go down for a C of A, so I flew it down. Took me twelve days!

'I was on my way south when a pushrod in one of the engines broke, and I had to make a forced landing on an old wartime strip near Terapo mission. I was stuck there for three days, waiting for somebody to find me...'

When the the Dragon failed to reach Daru, the RAAF search and rescue organisation swung into action. A thorough search of the Papuan Gulf failed to discover the Dragon. It was finally located by Bobby Gibbes, who flew in to the rescue.

Vic Smith delivered the Dragon to Sydney, joined Qantas, and returned to New Guinea, where he was to spend the next five years, flying for that company. His name will be encountered again later in these pages.

The journal, *Pacific Islands Monthly* was not impressed with this incident, following so closely upon the loss of Spiers' Dragon on Turnagain. 'New Guinea's air record, which was unsurpassed before the war, is now beginning to look rather dubious,' said the report.

That doughty firebrand, Dick Burt, was indirectly the cause of another DCA reprimand for MAL, late in 1949.

Following the demise of Guinea Air Traders, Burt rejoined MAL. On 16 December, his commercial pilot's licence was suspended for six months, for various breaches of regulations during November while pilot in charge of the C47, VH- BFV.

This was Burt's second suspension. The first took place during his first period with MAL. As already indicated in this account, Dick Burt was distainful of officialdom. During the early days of flying in PNG post-war, such niceties as flight plans were unheard-of: a pilot merely loaded up his aeroplane, and took off. Clearly, this was not a practice that could be allowed to continue, and after John Arthur became District Superintendent and as the DCA establishment increased, flight plans became mandatory. Some pilots were slow to adjust to the new requirement, including Dick Burt: 'I was at Port Moresby, flying in trade store goods to Bert Kienzle at Kokoda, and flying rubber out. I went up to the flight checking office, and this bloke applied some restriction I didn't go along with — can't remember now, 'twas a long time ago. I just grabbed the flight plan that I had made out, tore it up and threw it at him, and got into the aeroplane and went. Without a flight plan. That didn't help things all that much, and they suspended my licence...I went down to Sydney and got it back after doing a bit of yacketty. I was going to knuckle that DCA bloke in Moresby, because he riled me somewhat, but it didn't come to that...

'I was in Rabaul in December, 1949, with a Mandated DC3 when a signal came through that my licence was suspended again, and I couldn't fly the aeroplane home to Lae!

'What happened was I'd been flying a DC3 from Kavieng to Lae, and one engine wouldn't start. So I wound a rope around the propeller and hooked it onto a truck, and started the engine that way...had some cargo and an electric generator in the back that wasn't tied down. I had some DCA blokes on board as passengers, and one of them reported me to John Arthur. Arthur was an awful little squirt, knew nothing whatsoever about aviation...'

Dick Burt had been unwise enough to quite openly flout the regulations before an audience of three engineers from DCA headquarters in Melbourne, plus Ted Booker and George Woodward of the Port Moresby office! It was one of the headquarters officers who reported Burt. In a recent letter to the author, Booker commented: 'Personally, I thought that it was mighty enterprising of Burt to start the engine that way, and it got us out of Kavieng and back to Lae...'

Dick Burt's suspension notice listed a number of breaches of regulations during the period from 17 to 25 November, 1949. It was held that he had 'operated an aircraft in a negligent manner so as to endanger the lives of others and contrary to the interests of public safety'. Cargo had not been properly secured and was stowed aft of the passengers, and the loading had not been properly calculated; persons other than licenced aircrew had been allowed into the crew compartment during flight; passengers had been allowed to occupy the co-pilot's seat; passenger seat belts had not been secured; the aircraft had been operated over the sea in excess of sixty-minute time intervals from land although Burt did not have the required instrument rating.

Dick Burt appealed against his suspension, pointing out that DCA officers had been on board, and would surely have protested at the time had he been endangering their lives. It was certainly true that passengers in those relaxed days frequently wandered up to the flight deck to have a yarn with the pilot. Nor was it unknown for passengers to occupy the co-pilot's seat. Other pilots had been known to start a baulky engine on a DC3 by means of rope and motor vehicle — it was safe enough, if properly done. Not many pilots then were instrument-rated, and breaches of the sixty-minute rule were not uncommon. Indisputably, however, these practices were prohibited by the regulations.

The case went to the Appeals Board. Two nominees of the Australian Air Pilots' Association, captains H.N. Humphries of Australian National Airways and the highly experienced R.O. Mant of Qantas, sat with the chairman, W.R. Roberts, Deputy Crown Solicitor, Townsville. Many witnesses were called. The upshot was that the suspension was confirmed, but the period was reduced from six to four months.

After his licence was restored, Burt resumed his flying for MAL, and in 1952 resigned and left New Guinea. He settled with his wife, a pharmacist, in Barradine, New South Wales. Mrs Burt opened a pharmacy there, and Dick Burt built a garage, sold cars, ran school buses, mail runs, an aerial taxi service, and a small flying school. Later, the Burts moved to Williamtown and then to Tasmania, where Mrs Burt died in 1980. Burt returned to the mainland and settled in Newcastle, where he was still living at the time of writing. He retained his love of flying, and during his long life in and out of aviation accumulated over 9,000 flying hours. He lost his commercial licence at the age of seventy-three because of failed hearing, but was granted a private licence. He was one of the great characters of PNG aviation.

Dick Burt's suspension caused problems for MAL. In preparation for the appeal proceedings, DCA carried out a thorough examination of the organisation and operating standards of the company. It was considered that MAL had 'the best nucleus in New Guinea for the development of a sound operating organisation', but that the existing organisation had been allowed to 'deteriorate to the stage where it may have been potentially dangerous'.

This was damaging criticism. J.B. Sedgers, the senior W.R. Carpenter and Co. representative in PNG, was quick to act, and in January, 1950, John Arthur was able to write to the Director-General that 'W.R. Carpenter and Co. and MAL are in the process of completely overhauling and re-organising the company...and taking immediate steps to ensure prescribed safety standards are maintained.'

Two months later, MAL suffered the worst accident in its history when the Lockheed Hudson, VH-BDN, crashed at Lae, killing the pilot and two occupants of a house.

BDN had originally been purchased by W.R. Carpenter and Co. from the Commonwealth Disposals Commission in March, 1947. It was sold to Fawcett Aviation Engineering Service, Bankstown Airport, in 1949, and was leased back to MAL for use on the plantation recruiting runs later that year. It was flown to Lae from Bankstown by Dick Burt.

Harold Gibson-Lee joined MAL from Venezuela, where he had been flying C47's into the interior. He was an ex-RAAF wing-commander, and was awarded the DFC during the Battle of France, and later an Air Force Cross for his work in blind-flying tests. He was an authentic war hero, and a first-class pilot.

On 25 March, 1950, Gibson-Lee returned to Lae from Rabaul in one of the company's DC3's. Tom Deegan will never forget what happened: 'It was a Saturday afternoon. We'd been having trouble with the Hudson. One of the motors was back-firing, and we couldn't get it right. Bill Humphreys thought that he'd corrected the fault, and when the DC3 landed I was busy, and I said to Gibson,

"Will you test-fly the Hudson — just do a quick circuit, and see if it's OK?" He said, 'Right — when I come back, let's go up to the pub for a drink.'

'So he hopped out of the DC3 and into the Hudson, and on climb out of Lae he lost the starboard engine. He feathered it, and was going down-wind and turning in for his final approach, and there was a Qantas DC3 coming back from Wewak or somewhere had landed just before him. Gibson had no radio, wasn't working, so the tower couldn't tell him. He turned away and back in to let the Qantas aeroplane taxi off the strip. On his second turn he brought the wing up against the dead motor, and the Hudson went on its back. He corrected it, but hit the telephone wires at the end of the strip across the road, then went into a house and killed the young Flarey couple, on their honeymoon...

'We got Gibson out. He was badly hurt, and died of third degree burns ten days later. Our two engineers on board, Colin Wren and Jack Brown, survived — when they saw they were going in, they put their backs up against the bulkhead, and they were thrown out the back.

'Poor old Gibson. He was a good pilot. It was just bad luck, Bill Humphreys, our chief engineer for all those years, was an excellent man. He was with GAT, and I enticed him away. Great engineer, very thorough, ran a great engineering shop, had great facilities. When you remember the conditions blokes had to work under, our maintenance was excellent. Spare parts weren't readily available. MAL was a good company. We were left alone by Carpenters in Sydney, but Jack Sedgers, our chief in Rabaul, had to show a return. As soon as we got a bit ahead and had accumulated a bit of a nest-egg, he'd come across from Rabaul and take it away!

'When I joined MAL, there was no chief pilot. Doug Elphinstone ran the show, and he told you what to do and that was it. He left the rest to the boys, so long as you operated the aeroplane OK...

'The day Gibson-Lee died, Jack Sedgers came over from Rabaul and said, "Tom, we'd better have a chief pilot. Will you take it on? Worth a couple of quid a week."

'A couple of quid, to do all the worrying! But MAL didn't pay much, so I said, "All right." Pretty lousy, you know...'

The crash was closely investigated by DCA, and it was concluded that at the time of the accident the Hudson had been incorrectly loaded, with the centre of gravity 317.5 millimetres forward of the forward limit. The cause of the accident was 'loss of control of the aircraft due to incorrect loading', with the failure of the starboard engine after take-off a 'contributary cause'. The failure was due to a fracture of the exhaust pushrod tube of No. 14 cylinder.

The death of Gibson-Lee and the newlyweds, Mr and Mrs Stanley Flarey, was a cruel shock to the MAL manager, Doug Elphinstone: 'It really got to me, more than anything else. Gibson-Lee had a very distinguished war record, and he was the only pilot available licenced on the Hudson that day. It took a long time for him to die, and those young people...one of the things that really knock you...'

MAL's period of troubles was not yet over. The expansion of routes and services continued, and in Madang a fine new hangar and passenger lounge was constructed. Prior to the posting of Dave Robertson to Madang, the senior MAL pilot there was W.E. Passlow, but he was about to start his own small charter operation, and resigned before Robertson's arrival.

Dave Robbie has a vivid recollection of a flight with Gibson-Lee in the ill- fated Hudson, shortly before the Lae crash: 'I didn't like the Hudson, didn't like them from the war days. We got to Manus Island this day, and for no apparent reason the aeroplane started to go quite crazy, in the circuit area. Gibson got control, put on rudder, he's fighting this thing trying to straighten it up, screaming his head off, "Bloody Hudsons!" this that and the other! We were heading straight for the water but came out of it, don't know to this day what he did to it...we landed at Manus, and I said to him, "This is where I stay, Gibby. I'm not getting into this thing again..."

'We both stayed, refused to budge. They brought a DC3 over with engineers, and went over that Hudson with a fine-toothed comb. Couldn't find anything wrong with it. Same aircraft in which he killed himself...'

Robertson remembers two other nightmare flights during his short time with MAL. One was as second pilot to Dick Burt, in a DC3: 'We taxied out of Lae, heading for Wau. As we careered down the strip I could see the airspeed wasn't going to be so good....I said to Dick, "What have we got on?" 'Oh, nothing much,' says Dick. 'She's right.'

'We did get airborne, and staggered out over the sea and made a very wide arc back up the Markham valley. The old Three wasn't climbing like she should have, but we finally got through the Wampit Gap, and we were on top of eight-eighths cloud, like looking down on a table-top. Couldn't see Bulolo, and you could see the gap between the ranges over at Wau, and that was fogged in, too. Then, approaching Bulolo, an engine quit. I dived to the cross-feed, and when I did the other one shut. I switched back pretty smartly. It was a fuel problem, and we only had one engine. We were so heavy that there was no way we were going to get anywhere...couldn't get back to Lae

'Dick said, "Oh well, looks as though we'll have to go to Bulolo." I looked down on the cloud and said, 'Yes, but where is it?' Dick said, "Don't know, but that looks like The Pimple, so I'll take a good punt, it's just about there..."

'Down we went, straight through that cloud, and when we broke out underneath, there wasn't much to spare, but there, right in front of us, was Bulolo strip. Dived for everything — flaps, gear, all the rest of it, and we landed. I said to him, "What the hell have you got on this thing, Dick?" He said, 'Just a few things for my trade store in Wau...'

'I had an engineer's licence on the Three, so I nosed around, got the cowls off, and found that a fuel line had come off. Popped it back on, we offloaded a lot of stuff, and flew on to Wau.'

Talking of this incident with Dick Burt years later, the author found that he had not been quite so nonchalant that day as Dave Robbie had supposed. 'I was between a shit and a shiver,' said Burt.

Les Davis was one of MAL's most experienced captains. On a flight to Rabaul from Kavieng in a DC3 full of plantation labourers, Davis and Robertson — second pilot — encountered one of the gigantic local storms frequently found in those latitudes. Says Robbie: 'We could see in underneath it, could see Rabaul strip, and we were heading straight for it. I said to Les, "You're not going under that, are you?" 'Oh yes,' he says. I thought, "Oh God, it's only a few miles around the storm..."

'Underneath we went, and the next thing up she went, into the storm, into the cloud. We were hurled about, the instruments were all over the place, I'm screaming at Les, "Get your airspeed back," and do this, that and the other...I got my foot on the control column, he's hanging onto it...I shoved it forward and we eventually got the airspeed back, but nothing left on the clock...we came out about 180 degrees from where we entered it, and going down in a complete dive. Broke out underneath the cloud, and I reckon that's where I nearly bought it. In a Three. I really thought it was the end. All the seats had collapsed, and all the boys were sitting on the floor with the straps around them, howling their heads off. Pretty grim. Always felt good with the old Three, it just kept chugging along...'

Soon after Robertson took over as manager of the Madang branch MAL suffered another serious accident. The aircraft, Dragon VH-BDB, crashed on take-off from Kerowagi strip in the Highlands on 2 April, 1951. The pilot, J.M. Jenkins, and two passengers were seriously injured. Other passengers were slightly hurt. BDB never flew again.

The investigation established that the Dragon was overloaded by at least eighteen kilograms at the time of the crash. It was, moreover, carrying eight passengers, two more than the maximum permissable number. None of the passengers were wearing safety belts, and in any case only four belts were provided in the aircraft. Freight, unsecured, was carried behind the passengers, and this shot forward on impact. The pilot failed to compile a load sheet, compute the centre of gravity, or carry a load chart.

John Jenkins had joined MAL as a pilot in March, 1950. He was introduced to the company by Ray Harris, who knew Jenkins when he was chief flying instructor of the Broken Hill Aero Club. At that time, Jenkins held a commercial licence, and he subsequently gained an instructor's rating while flying with the Aero Club. After Jenkins joined MAL, he was trained by Harris on Dragon aircraft over a period of nine months. When he began with MAL, Jenkins had a total of 420 hours in his logbook. He was rated by Harris as a 'careful, competent, confident' pilot. Doug Elphinstone found his work 'beyond reproach'. Jenkins had accumulated 1,435 flying hours at the time of the crash.

83. Lae aerodrome in 1952. *Photo: Isobel Pert*

83

There was no question of any fault with the Dragon. Jack Hammond and Fred Bird, MAL engineers at Madang, rated BDB as 'the best aircraft we have handled'. The cause of the accident was 'that the aircraft failed to become airborne when taking off from an airstrip of high density altitude due to incorrect technique on the part of the pilot in that he failed to adjust the mixture controls to obtain maximum engine power'. A contributing factor was the existence of a tailwind exceeding the maximum allowable for Kerowagi.

Once again, MAL came under criticism from DCA. In common with other operators, the company had been permitted to carry out RPT services under a charter licence 'to assist in evaluating Airline Licence requirements for New Guinea. A decision regarding Airline Licences is approaching finalisation'. But MAL had still not produced an approved Operations Manual, or introduced an approved training and checking organisation for DH84 pilots, although the training of DC3 crews was satisfactory. Rube Murray, who investigated the accident, considered that MAL DH84 pilots received 'very limited operating instructions from the company', particularly in regard to loading. Questioning revealed that Jenkins had not been properly instructed on the use and effect of the DH84 mixture controls, or on the compilation of loadsheets and the computation of the centre of gravity.

Murray added: 'From personal observations, it can be said that the general practice of DH84 pilots in New Guinea is to use full rich mixture on all takeoffs, irrespective of the altitude, apparently due to a lack of appreciation of the extent of the consequent loss of power. As there is no practical method of determining the correct position for the mixture control levers, other than by running the engines to full power before each take-off, it is understandable that pilots use "full

84

84. Remains of MAL Dragon, VH-BDB after crashing at Kerowagi, April, 1951. The pilot was John Jenkins. *Photo: Roger Wilson*
85. Interior of a Douglas-DC3. *Photo: Jasper Maskelyne*
86. Interior of a 'side-saddle' DC3. This photo was used as evidence during Dick Burt's appeal. *Photo: Ted Booker*

rich" mixture. As calculations relating to permissible operating weights for DH84 aircraft are based on correct use of mixture controls, the failure to use these controls is very serious in view of the marginal performance of DH84 aircraft in New Guinea'.

Since there was 'no practical method' of determining the correct position for the mixture control levers, there is little wonder that pilots continued to have accidents in Dragons at high altitudes!

The Jenkins crash had serious consequences for MAL. Upon receipt of Murray's report and John Arthur's recommendations, the Director-General issued orders that the conduct of MAL operations be 'completely examined and reported upon'. All aspects were to be examined — flying operations, the adequacy of pilot checking, flying practices and techniques, loading procedures. The question of the Operations Manual had to be settled, and John Arthur was directed to provide an answer to the question: Did MAL in any way encourage economy at the expense of safety? At the time this direction was issued, John Arthur was on leave, and the MAL examination was left until his return. This gave the company a little breathing space. It was a difficult time for MAL. The Qantas operation was fast-growing, and to the alarm of the other operators, was being given an increasing share of Administration business, the bread-and-butter of the air transport industry in the 1950s.

Qantas, though, was having problems of its own.

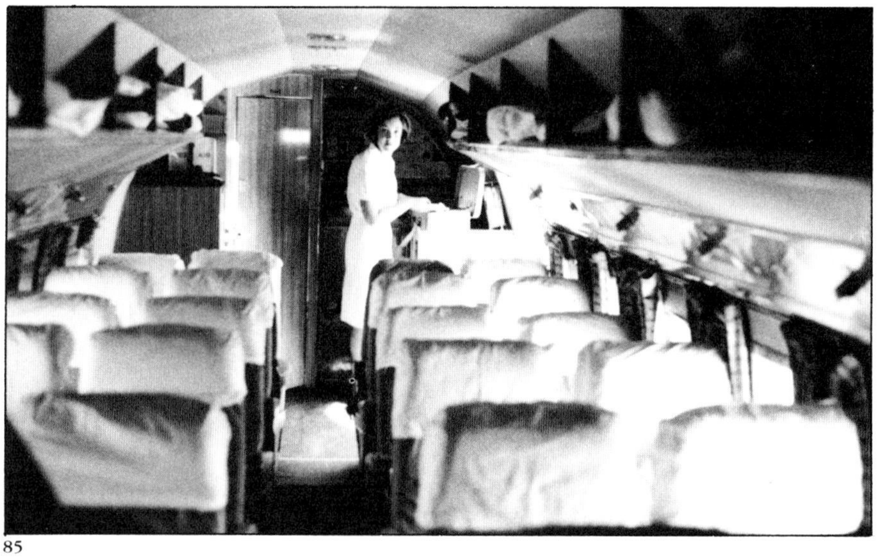

85

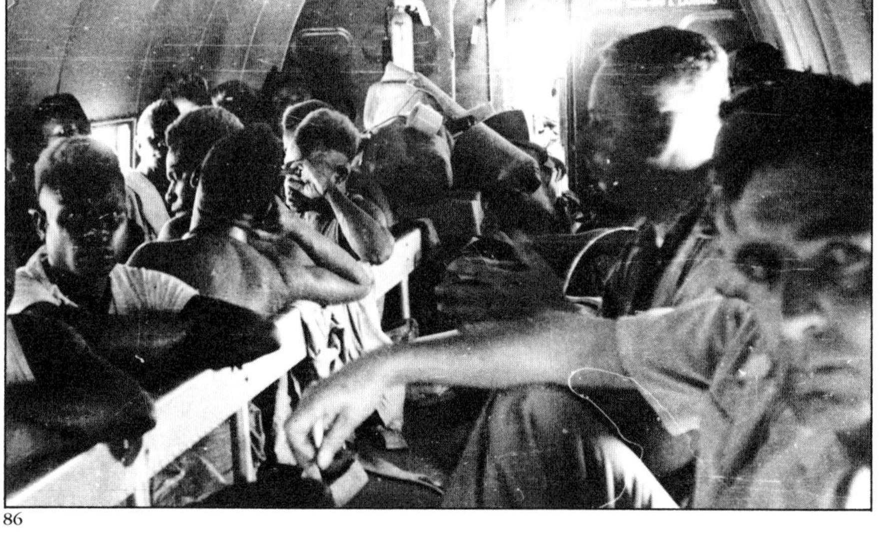

86

The Woes Of Qantas

By 1950, Qantas' New Guinea Internal was indisputably the major air transport operator in PNG. The company had in service six C47's, twelve DH84 Dragons and two Fox Moths, most based at Lae, and two Catalinas at Port Moresby, where two Sea Otters were also being operated on behalf of the Australasian Petroleum Company. Two DC4 Skymasters were in use on the Bird of Paradise route, supplementing the normal DC3 services. During 1949, Qantas aircraft had flown more than 1.5 million kilometres within PNG, double the figure for the preceeding year. Almost 16,000 passengers had been carried, and some 3.6 million kilograms of cargo and mail. These figures continued, rapidly, to increase.

The Lae base had been dramatically developed and improved. In the beginning, living conditions had been extremely primitive. Staff shared a common mess room, toilets and cold showers. Sleeping quarters were provided in long huts with iron roofs and partitions of sisalkraft. Iron ex-military cots served as beds. A Qantas accountant, B.G. Lindsay, who was one of the early arrivals, was intrigued to find that the lampshade in his room — more properly, cubicle — had a neat round hole in it. He was told that the pilot who had previously occupied the room left his light on until late one night. A brother pilot, in the adjoining room, badly wanted his sleep for an early morning take-off, and using his trusty .45, he shot out the light.

Once the decision to begin New Guinea Internal was taken, Qantas poured money into the operation. The new Qantas mess and lodge at Lae, catering for both staff and passengers, was opened in July, 1950. By the standards of the time, the accommodation was excellent. Comfortable married quarters were provided, and single rooms for men and women. Good quality meals were served at the mess, using fresh fruits and vegetables flown in from the Highlands. Hangars, workshops, offices, passenger terminals were built. It was estimated that the total Qantas investment in PNG at this time was around £400,000. New Guinea Internal lost money in the early days — indeed, Qantas was paid a subsidy of £38,200 per annum from 1 October, 1948, in order to provide the developmental services required by the Administration, which were run, generally, at a loss. A formal subsidy agreement was signed with DCA in July, 1950, which allowed a 'reasonable margin' for profit, calculated at 7.5 per cent on the capital value of the assets used to provide the services. Subsidy payments over the next four years ranged from £40,000 to £76,000.

Qantas now had a large muster of pilots and engineers in PNG. It was normal practice for a Qantas pilot to be offered employment in PNG on a two-year contract basis, but many stayed on for years, a few right to the end of the Qantas era. New Guinea flying got into a pilot's blood. It was excellent experience, too — provided one did not stay on too long. Qantas regarded a period of New Guinea flying as first-class training for young first officers on the way up, and headed for the international routes. It was wonderful stick- and-rudder experience, but little use was made of instruments in New Guinea, there was no night flying, and navigation was mostly simple point-to-point. A pilot who stayed on too long ran the risk of falling behind in the development of certain professional skills that would soon become essential as aviation moved out of the age of piston engines, and into the new world of jets.

Pilots who flew in New Guinea for Qantas in the late 1940s and early 1950s included men who would go on to command Boeing 747 jets before their retirements — and Old Father Time has ensured that most are now retired. Among the best known of the early pilots — apart from those already mentioned in this account — were Bert Hargreaves, Bill Hosking, Ron Ballard, Ross Biddulph, Dick

Davis, Eric Sims, Bruce Fawcett, Dennis O'Malley, Ian Taylor, Nicholas Carter, Fred Phillips, Jock McKee, Ron Roberts, Ken Montagu, Tom Howes, Roger Wilson, Jack Bird, Mac Tremlow, Mal Shannon, Gordon McCrae, Clyde Johnston, John Keene and Rex Cuthbertson.

Qantas enjoyed several advantages over its competitors, quite apart from government support. Qantas had been in existence for years, and had built up a reputation for excellence in its operations. The financial resources of Qantas enabled the company to use good aircraft, and few would argue that its maintenance standards were equal, or superior, to any in the world. There was a large pool of pilots and engineers upon which New Guinea Internal could draw. There was an accumulated store of experience of incalculable value. Yet Qantas, too, had its full share of misfortunes in the early days.

Qantas as a rule stuck to classes of aircraft, rather than orphans. In PNG, Dragons formed the light aircraft fleet in the beginning — with the exception of two Fox Moths. Heavy aircraft were all Douglas C47's — with the exception of a single Avro Anson, acquired when Qantas took over Ben Hall's charter company.

The Anson was VH-BBZ, and as described in Chapter 3, Gibbes Sepik Airways, it crashed in May, 1949, and was a total loss. The first Fox Moth, VH-UZD, was being flown by the original pilot at Lae, Len McNeil, when it came to grief near Tapini, on 3 October, 1949. Even a veteran could occasionally have an accident in New Guinea! McNeil recalls: 'I'd been flying DC3's in and out daily, all over the place. Flat out on DC3's, and Harold Hindwood came to me and said, "Len, we want a bloke to go over to Port Moresby to fly stuff into Tapini in a Fox Moth." I said to him, 'You're sending me back to a Fox Moth!' "Yes," he said, "we want to get this cargo in, and we know you'll do it..."

'So away I go. This is going to be a bit dicey, I thought...I hadn't been in a Fox for some time, so I was going to make sure that all my load sheets and everything were all right. I had three police boys with their rifles and gear in the cabin, she was right up to the maximum legal take-off weight...

'Got up there in the morning, and I had a look at Tapini as I came in. You had to fly over a saddle, then downhill over a creek and swamp, then straight up onto the strip. You aimed just across the saddle, pretty close to the ground, and you had no horizon whatsoever. Apparently I struck the katabatic effect of air...I'm coming in there, and I thought, Jeez, I'm going to be flat out making this. I had full bore on the thing, and the nose was up...anyway, I just didn't cross the saddle. Bloody near made it. Wiped the undercarriage completely off, then the Fox slid downhill all the way into the creek...

'I put my face into the instrument panel...had blood all over me. Couldn't see out of one eye, and I thought, Oh Jesus, I've done an eye in. My passengers had all hopped out and left me — they thought that was the way you normally rode — and I groped around trying to find the eye, I thought maybe they could stick it back in. Apparently a flap of skin was hanging down over the eye, and I flicked it back when I was groping around — oh, you beaut, I can see!

'I'm sitting there, cursing my luck — I was just at the end of the strip, missed out by a couple of feet — and along comes the patrol officer from Tapini. Can't remember his name. We've got to sew that eye up, he says. We went up to the station, and he gets a bloody blowlamp going, and starts to heat up some needles so that he could bend them. I'm watching all this, and shock is starting to set in. He said, "You'd better have some anaesthetic." 'What sort have you got,' I asked. "The Black Bitch," he said — Rhum Negrita. So I downed a glassful, and didn't feel too bad. He put the first stitch in, and I said, 'Oh, Christ.' He crossed himself — fairly religious bloke — and gave me some more anaesthetic. I didn't feel the second or third stitches. He did a pretty good job, too. When he finished, I had a face like a Mongolian...

'They got UZD out, had to fly an undercarriage from Karachi or somewhere...hammered out the metal prop, which was all bent...'

Len McNeil left New Guinea Internal at the end of 1950, with some reluctance. He recognised, however, that a move was inevitable if he was to progress in aviation, and Qantas. Over the following years he flew most of the major Qantas types — DC3, DC4, Lockheed Constellation, the various models of the Boeing 707, and finally the 747 — before his retirement from Qantas in November, 1977. Much of his time in the final years was spent in training other pilots. The other Fox Moth, VH-URI, crashed on take-off from Boana in November, 1950. The pilot, Nick

87. Qantas captain Tom Howes with son Johnny. *Photo: Tom Howes*

87

Carter, was unhurt, although his passenger, W. Lamont, received a fractured shoulder. URI was repaired, and in 1952 was sold to the Wau private owner and businessman, Norman Wilde.

Qantas operations continued without serious incident. There were a few minor incidents involving DH84 and DC3 aircraft, but damage in each case was slight, and there were no injuries. The strict flying and maintenance standards of Qantas were paying off, for operations were increasing rapidly, and company aircraft were covering huge mileages.

On 21 January, 1951, one of the worst natural disasters of modern times took place in PNG: the eruption of Mount Lamington volcano. Some 3,500 native people of the Orokaiva tribe perished, and thirty-five whites, including the Director of Labour, W.R. Humphries, and the District Commissioner of the Northern District, C.F. Cowley.

Mount Lamington had been rumbling and belching for days, and a minor eruption had occurred on 19 January, but it had not been considered necessary to evacuate the villagers living on the slopes of the volcano, or the government station at Higaturu, eleven kilometres away. The eruption — a series of colossal explosions between 10.15 and 10.45 a.m. — blew the side out of Lamington. A gigantic black, glowing cloud of ash, pumice dust and fire-balls swept over the surrounding countryside. Most of those who perished died of the effect of the shockwaves, or were overtaken by the onrushing lethal, volcanic cloud as they tried desperately to escape. Their bodies were horribly burned. It was, in the words of the Administrator, 'a holocaust beyond description'.

In the amazing, inspiring and superbly-coordinated work of rescue and evacuation that followed, the air companies of PNG, and in particular Qantas, played a vital part. When Lamington exploded, two Qantas aircraft were in the vicinity. One, a Dragon, was being flown by Ross Biddulph and Fred Barlogie from Lae to Popondetta, carrying three Administration officers. Mount Lamington was within minutes of eruption, and as the Dragon neared Popondetta, passengers and crew took photographs through the Dragon's windows.

Biddulph, in command, turned in on his final leg to the airfield, and then Barlogie shouted, 'Look at the mountain!'. As they watched in awe and horror, the entire mountainside blew out, and Higaturu station was blotted out. The terrible cloud rushed towards them, and Biddulph banked steeply away, and sped back at full speed to Lae. Having no radio in the Dragon, he was unable to send news of the disaster. 'Never has a Dragon been banked so steeply,' he later said. 'The shadow was overtaking me,...I bent the throttles...'

At about the same time, Arthur Jacobsen was on the normal Bird of Paradise flight to Rabaul from Port Moresby in a DC3, flying at a height of 2438 metres above an eight-eighths layer of stratus cloud. At exactly 10.42 a.m. there was a gigantic explosion, and a great black volcanic cloud burst through the stratus, about sixteen kilometres to the starboard side of the DC3. It grew and expanded, like a horrible mushroom, and Jacobsen took two dramatic photographs, later published throughout the world. The DC3 was in danger of being engulfed, and Jacobsen dived away, pushing the sedate Douglas to a speed of 346 kph. Even at that speed, it took the DC3 five minutes to draw away from the cloud, which Jacobsen estimated reached a height of over 15,000 metres and covered an area of 160 kilometres wide. As he raced the lethal cloud, Arthur Jacobsen sent out a report on his radio, although he did not realise at the time what had happened.

Lae heard of the eruption when Biddulph landed, at 12.50 p.m. The Qantas area manager was now H.E. Shannon. Harold Hindwood had recently, and much against his will, been transferred to Darwin to replace Shannon there. When Bill Forgan-Smith, the chief pilot, heard Biddulph's report, he requested permission from Shannon to fly to Popondetta with emergency supplies.

No special request for assistance from the aviation industry had yet been made by the Administration; the scale of the Lamington disaster was still being assessed as reports began to flow in. A DC3 was nevertheless speedily loaded with drums of water, sacks of rice and flour, cases of meat and medical supplies, and when instructions were finally received from Port Moresby, soon after 4 p.m., Forgan-Smith, First Officer Lionel Purkiss and the Area Transport Superintendent, H.R. Goodsell, were ready to go. Forgie recalls: 'As we were departing Lae, the tower told us Popondetta was out, covered with ash. I decided to head there anyway, to see if I could make a landing.

'We got down around Morobe, and you couldn't see anything for dust. We wound our way across the treetops, circled Cape Killerton, then flew in towards Popondetta. The scene of the volcano was just a continuous line of lightning, couldn't hear anything on the radio for static. We found Popondetta, but they told us not to land unless they wanted us. Circled Popondetta, and some Europeans waved to us. Put the wheels down, but they didn't seem to want us to land. Popondetta strip was far from being out. Higaturu was flattened, covered with dust, not a tree standing anywhere, the pumice looked like snow. We circled Popondetta again, and dropped notes we'd written on the backs of loadsheets, telling them that a trawler was on the way...'

They flew back to Lae, and on 22 January, regular Qantas relief flights began. Over the following days, every available DC3 was pressed into service. Aircraft of MAL and GSA and Charter Aviation Service* took part, but it was primarily a Qantas operation. One hundred and seventy-five rescue personnel and over 62,000 kilograms of relief supplies were flown in during this period. One of the principal needs was for clean drinking water, as all of the streams in the area were heavily fouled with pumice dust. In total, the DC3's made 143 flights to the disaster area, carrying over 265,000 kilograms of supplies. Dragons made 117 flights, carrying over 53,000 kilograms. Virtually every Qantas pilot took part. The Royal Australian Air Force sent Dakota freighters to assist in the evacuation of survivors, and over a period these aircraft airdropped some fifty-five tonnes of emergency food to evacuation camps.

Bill Forgan-Smith was back at Popondetta the day after the eruption: 'It was really only a Dragon strip, but we put the Threes in there early in the morning, and started evacuating burned natives out to Lae. Horrible sights. When we filled Lae hospital up, we started flying them across to Port Moresby. I spent four days there before I got back to Lae, no shave or change of clothes or anything. Colonel Murray burst into tears when we flew him around to look at the devastation. Then the press arrived. Administrator said that nobody but the rescue workers would be allowed in, but I flew Syd Wood, the Cinesound newsman, over it, and he took a movie. When I landed there was an uproar from the Administration, but somebody had to know what was going on. Colonel Murray was a bit annoyed with me...

'We sent a Dragon down, and it was based there with the vulcanologist, Tony Taylor...'

The Dragon was under the command of Ron Ballard. The scenes he saw will never leave his memory. Ballard, a wartime pilot, had flown Kittyhawk fighters out of Milne Bay and Merauke during the war, and had survived a ditching into Torres Strait in a C47. He was familiar with the sight and smell of massed dead Japanese bodies.

Ballard joined Qantas in 1947 as a traffic officer — he had only 813 total hours, just about all on single-seaters, and Qantas preferred pilots with at least 2,000 hours on multi-engined types — and was finally accepted as a pilot in 1949. He was a friend of Ross Biddulph's, and it was at his suggestion that Ballard went to New Guinea for further experience, still with less than 1,000 hours in his logbook. Says Ballard: 'I made my first New Guinea trip on 16 June, '49, as second pilot to Captain Les Ross. Famous old guy, been flying up there for years and years...he rather liked the gin bottle. Had a disgusting habit of spitting inside the aircraft, all our C47's had badly stained upholstery on the left-hand side of the flight decks...great old pilot, though.

'I was sent down to Popondetta to help Tony Taylor. I was terribly impressed by the havoc — everything was flattened...

'I flew Taylor around the Lamington volcano area every morning while he observed developments. He wanted to fly right down into the crater — he was gamer than I was — but I was absolutely scared stiff. Great rocks were being hurled up out of the crater to greater heights than us. So Taylor asked for another pilot, and they sent Jock McKee. He was a lot braver then me, and he flew Taylor right in and out of the crater for three weeks. I admired Jock — I wasn't game to do that sort of flying...'

John J. McKee — yet another ex-RAAF pilot — went to Lae to join New Guinea Internal in November, 1949. He spent the first eighteen months of his three years in PNG flying Dragons and DC3's out of Lae before transferring to Madang, where he mainly piloted the new DHA Drover. He admits today that his memories of

*See Chapter 4, Success and Failure

Mount Lamington days remain 'very vivid'. For his part in the work, McKee was awarded the MBE.

Two other Qantas pilots received the MBE for their services during the Mount Lamington emergency: I.C. Taylor and John Rose. John Arthur received the OBE, and the vulcanologist, G.A. Taylor, was awarded the George Cross.*

Ian Taylor was known throughout the PNG aviation world as 'Two Heads'. He served with Qantas at Lae from 1950 to 1957. There are several explanations for his unusual nickname — some not printable — but according to Bill Forgan-Smith, this is the true story... 'When the new Qantas mess was built at Lae, up on Namanula Hill, four single pilots lived together in House 14. They were always playing tricks on each other. They'd rigged up a huge shanghai on the verandah, with a big fork from the branch of a tree, you had to use both hands and your feet on the verandah rail to pull it back. Every time they emptied a bottle, they'd put it into this giant catapult and shoot it down the hill, and time how long it was before they heard the crash...

'Dick Davis was a real night owl then. When he wasn't flying, he'd sleep all day and play all night. He'd generally get in about 2 a.m. Qantas used to switch off the power at midnight then, as an economy measure, and bring it back on at 4 a.m., so that those involved in the 6 a.m. flight departures could rise and get started. So Bruce Fawcett, Ian Taylor and Ross Biddulph decided to work the ultimate practical joke on Dick Davis.

'Ian Taylor had the biggest hi-fi loudspeakers in Lae. One night when Dick Davis was out on the prowl, these blokes shoved the speakers under his bed, slapped a record of Souza marches on the record player, and set it at full volume to start when the power came on at 4 a.m. They ran the leads behind the wall panelling and put the player in the next room, so Dick couldn't get at it, then locked their own bedroom doors so Dick couldn't get at *them*.

'Dick came in at his usual time, turned in, and at 4 a.m. was hit with a blast of sound that blew him out of bed. Woke up the whole mess area, too. Dick went nearly crazy trying to stop the racket, and yelled at the top of his voice, "Anyone who could think of a thing like this must have a brother down at the university!" 'You reckon, Dick?' called Ian, and Dick yelled, "Yes, in a bloody bottle with two heads!" And he was 'Two Heads' Taylor forever...'

Dennis O'Malley, who flew with Qantas in PNG from January, 1950, to May, 1956, remembers House 14 with rueful affection: 'Every Boxing Day, the single boys in House 14 used to invite their married colleagues — of whom I was one — for a drink. Dick Davis always organised a met balloon to be tethered on as long a rope as possible above House 14. This thing stretched up 500 feet or so [about 150 metres], and the met. department always cooperated by issuing a Notam — Notice to Airmen — advising a danger to aircraft should they be flying anywhere in the region of House 14, Qantas mess, Lae...

'On this particular Boxing Day, I went up for my usual drink with the mad people who lived in this house. I was met by Dick Davis as host at the top of the stairs. In those days, the *de rigueur* correct rig to be seen in for any social gathering was like a cricket outfit — white slacks, white shirt, white socks. Dick took one look at me, said, "Happy New Year," then hit me with a mud pie...'

When Bill Forgan-Smith returned to Qantas operations in Australia in 1953, Bruce Fawcett became chief pilot at Lae, reponsible also for pilot basings at Goroka and Madang. He flew mostly on DC3's, and had the distinction of flying the last DC3 into Bulolo before the old strip was closed to these aircraft, following pressure from the Australian Air Pilots' Association New Guinea branch, in May, 1951.

The pilot's union was to become militant in PNG aviation in later years, but in the early 1950s industrial action by pilots was unheard-of. However, as Bulolo Gold Dredging Ltd continued to step up its rehabilitation programme operations into Bulolo aerodrome multiplied, and pilots became increasingly aware of the dangers posed by The Pimple.

In the early days, DCA left the landing and take-off procedures at Bulolo to the discretion of the pilot, but eventually a Notam was issued, requiring 'no wind' landings to be made into the north-west end of the strip, and take-offs into the south-east.

The standard method of landing a DC3 was from The Pimple end, as described by Dennis O'Malley: 'The northern end of the strip pushed up against The Pimple,

*A total of fourteen awards were made to whites and Papua New Guineans following the Mount Lamington eruption. Sadly, Tony Taylor died in 1972 of a heart attack following an inspection on foot of the active crater on Manam Island. He was a true hero, and an unassuming one.

a 500-foot [about 152 metres] hill directly in the flight path. The circuit was done by overflying the strip to see if it was clear, flying a left-hand circuit losing sight of the strip on the down-wind leg, turning on to final approach by lining up on The Pimple, the strip still out of sight, over The Pimple, full flap, no power, and a dirty dive at the deck to flare and touch down. Pretty interesting first time...'

The new Notam was hotly debated by the pilots. Ron Ballard wrote to DCA: 'It is far safer to take off into the north in a light aircraft, as it is sufficiently manoeuverable to follow the river around The Pimple to dropping ground. I have found landing into the south by far the most satisfactory direction...'

John Spiers agreed. 'This Notam requires pilots to land downwind and take-off uphill *and* in this season, downwind...a most dangerous procedure...'

The Pilots' Association called for the immediate closure of Bulolo to DC3 aircraft. Ian Taylor, the secretary, wrote: 'There is a period during the take- off run towards The Pimple of about fifteen seconds where safety is dependent *solely on the operation of both motors.....*'

Bulolo was closed to DC3 operations, but remained open for light aircraft. Argument continued among the pilots as to the safest way to use this strip.

Qantas operations in PNG continued uneventfully following the Mount Lamington disaster. The reputation of New Guinea Internal with the travelling public stood high. Then, in July, 1951, came the first of three fatal accidents for Qantas, the first for the airline since March, 1946, when a Lancastrian vanished on a flight between Colombo and Cocos Island, with five crew and five passengers on board.

On 16 July, DHA3 Drover, VH-EBQ, crashed into the sea off the mouth of the Markham River whilst on a flight from Bulolo to Lae. The pilot, Johnny Spiers, and his six passengers were all killed.

Qantas had employed the DH84 in New Guinea internal operations only with reluctance, and because no better aircraft were readily available. When the de Havilland Aircraft Company's Australian subsidiary began the design of a Dragon replacement, Qantas was immediately interested.

The aeroplane that finally emerged from the drawing boards was the DHA3 Drover, a robust, all-metal aircraft, with fixed undercarriage, simple wing flaps and variable-pitch propellers driven by three Gipsy Major 10 Mark 2 engines, each of 145 horse-power. It was designed for durability, and a safe performance after the loss of an engine. As originally designed, it could accommodate eight passengers, or carry a freight load of some 862 kilograms. On paper, the Drover looked to be exactly the aeroplane for New Guinea operations, and Qantas ordered four, with financial assistance from the Australian Government. TAA also ordered six Drovers, and it was hoped that there would be considerable export sales. The poor initial record of the Drover in service was to dash these hopes, and only twenty were eventually constructed, although Qantas increased its order to seven. The Drover proved a bitter disappointment for de Havilland's. By the time the early problems of the Drover were overcome, the reputation of the type was so bad that it was not deemed economic to continue with its manufacture. There are many parallels in the Drover story with the fate of the magnificent Comet jetliner, also a product of the remarkable de Havilland Aircraft Company.

The Drover project was under the direction of Martin Warner, and all of the prototype and certification flying was done by de Havilland's chief test pilot in Australia, the famous ex-RAAF identity Brian Walker, better known as 'Black-Jack'. Walker had a long association with de Havilland's. He was seconded to the company from the Air Force in July, 1944, to assist in overcoming problems that had developed with the manufacture of the Mosquito, and after he was demobilised, joined de Havilland's as their test pilot. In a flying career that spanned forty-eight years, Brian Walker flew some 130 different types of aircraft.

Development of the Drover was relatively trouble-free, apart from persistent trouble with the propellers. The first production Drover was registered as VH-EBQ, and was officially handed over to Qantas at Mascot Airport in September, 1950.

This first Drover was basically a freighter, but was also fitted with twelve light-weight collapsible canvas seats. It was ferried to Lae by Lionel Purkiss, and was soon in operation between Port Moresby and Kokoda, piloted by Dennis O'Malley. Later, the aeroplane replaced Dragons on the service from Lae to Wau and Bulolo, and then on the Highlands runs.

The second Qantas Drover, VH-EBR, was handed over in April, 1951, by which time operational problems were beginning to emerge. In PNG service, the Drovers generally flew short trips, climbing steeply to cross mountains and descending equally rapidly on the other side. Many of the airstrips it worked out of were short, and at high altitudes. A considerable strain was placed on airframes, engines and propellers. The pilots frequently experienced carburettor icing, and vapour locks in the fuel lines. There were constant maintenance problems with the variable-pitch propellers, caused, it was considered, by the constant changes of pitch that resulted from the type of flying done in New Guinea. Finally, DCA directed that Drover propellers had to be returned to Bankstown after 150 flying hours for overhaul, instead of the 300 hours originally laid down. This, naturally, increased the cost of operating the Drover.

Opinion on the Drover among the pilots varied. Lionel Purkiss is quite blunt about them. 'Greatest monstrosity of a thing that was ever foisted on the Australian flying public...you couldn't coax the Drover into the air. It had to have its flying speed, otherwise it just wouldn't fly. But once it had its speed, it would climb up the side of a hill...not a kind aeroplane for New Guinea work'.

Shortly before the Drover was introduced, Gibbes Sepik Airways had managed to capture most of the lucrative work of flying out rubber from the big Kienzle plantation at Kokoda. The Qantas Dragons were no match for the GSA Norsemans. Len McNeil developed a technique to handle the Kokoda-Moresby leg in a heavily-loaded Dragon: 'Straight climb to 7,205 feet [2196 metres] to the gap. Then down to Kokoda. Bales and bales of rubber, and Christ, it was hard getting your altitude straight from Kokoda to get through the Kokoda Gap. You'd get a bit cunning, have a look at the wind, go across the wall of the valley that the wind was hitting and you'd do your circles there to get the lift. If you went the other side you'd never get up. And it was full bore from the moment you took off until you got through the gap, then downhill into Moresby...'

The Norseman, with its powerful engine, had no trouble. The situation changed when Qantas got the Drover. Dick Davis recalls:* 'With the Drovers, we were able to do more trips each day, and carry more rubber at a better price, and more dependably than anyone else. With the Drover you could get up a lot higher and get over the gap and work right through the bad weather of the afternoon in some instances, whereas the Dragon would have given it away hours before. We had quite a battle, but won all Kienzle's contracts back again.

'The Drover had variable-pitch electrically-controlled propellers, and the pilot had three little toggle switches on the side of the cockpit which you worked with your right hand. You had a separate control for each one and you played these switches like a flute to get the propellers all doing the same RPM, otherwise they were out of synchronisation, and used to howl like mad...

'The Drover was built like a tank, and was covered with rivets. There was no bend or give in the wings at all, they were as stiff as girders. The method of changing a tyre out in the New Guinea jungle was to jump up and down on a wingtip, rocking it until one wheel left the ground, and then somebody would put a brick under the axle, leaving the wheel off the ground. You could then change the wheel, and kick the brick away, ready to taxi off again.'

On 16 July, 1951, Johnny Spiers took off at 8.35 a.m. from Bulolo in Drover VH-EBQ, bound for Lae. On board were six passengers: the Bishop of Alexishafen, Right Reverend Stephen Appelhans; Father William Bachus of the same Order; J.W. Palmer of Lae, and his fiancee, Miss Helen Connor; Miss Frances Tindale, a friend of Miss Connor, and K.C. Macdonald, a BGD employee. Also on board was gold bullion in bars, estimated at different times to be worth between £30,000 and £45,000.

The weather was very bad. It was raining, and visibility was poor. Spiers flew low along the Snake valley and into the Markham, and at 9.06 a.m. gave his position as being at the mouth of the Markham River. No further word was received from him. Shortly afterwards emergency procedures were instituted. Floating wreckage was located that afternoon, about six kilometres south of Lae, and the bodies of Miss Connors and the Bishop were recovered.

The main wreckage of the Drover was located on 25 July, in sixteen fathoms of water. An attempt was made by a DCA salvage trawler to raise the aircraft, but the trawler's gear was inadequate for the task, and only the starboard wing and a portion of the cabin flooring could be recovered.

*Quoted in G. Goodall's unpublished MS on the Drover, which has been extensively drawn upon for this chapter.

DCA officially abandoned attempts to raise the Drover on 30 July. Qantas, however, was determined to recover the wreckage. A world-famous salvage expert, Captain J.W. Herd, was sent to Lae by the insurance underwriters, for as well as the Drover, a small fortune in gold was at stake. The wreckage was again located, and on 14 August was raised, and towed by barge to Lae. The body of John Spiers was still strapped into the cockpit, but there was no sign of the other victims, or of the gold. (Captain Herd finally recovered the gold bullion in November, 1952).

Almost the whole of the Drover had now been recovered, with the exception of the cabin door, and the two outer engines. The bits and pieces were painstakingly pieced together by Aircraft Surveyor J.M. Simpson of DCA, and Bill Burns, representing Qantas. It was found that the centre propeller was missing a blade, but it was considered at the time that this had been detached by the force of the impact. No other faults could be found. In March, 1952, the official DCA findings were published: 'The probable cause of the accident was an error of judgment by the pilot in attempting to continue the flight with visibility so low that flight in sight of ground or water could only be maintained at a dangerously low altitude'.

And so the dead Johnny Spiers carried the sole blame for the loss of the Drover. Pilot error!

Qantas came under immediate public criticism when the finding was released. Why was the Drover permitted to fly in such bad weather? All that the company could do was point out that while their operations officers could certainly prevent an aircraft from taking off, once it was in the air, the pilot was in command, the decision whether to continue with the flight or turn back, his alone.

Scarcely had Qantas recovered from the loss of the Drover than there was another fatal crash, this time of a DH Dragon. The pilot, F.G. Barlogie, a first officer, was the sole occupant.

Fred Barlogie joined Qantas in May, 1950. He was transferred to New Guinea Internal in August after being given a DC3 endorsement in Sydney. In accordance with normal Qantas practice, he spent his first three months as co-pilot of DC3's under experienced captains, being familiarised over the different routes. He made his first flight in command of a Dragon in January, 1951, and over the following six months was carefully checked by seasoned DH84 pilots over Highlands routes. As was always the case with Qantas, his training was painstaking and methodical. During his time in PNG, Barlogie's instrument flying was checked and was considered to be 'above average'. In fact, at the time of his death approval was being sought from Qantas headquarters for Barlogie to carry out DC3 command check piloting. He had accumulated 1,745 flying hours when he crashed. However, in March, 1951, Barlogie had been reprimanded for permitting a DH84 to swing during a landing run, resulting in airscrew damage. The cause was assessed by DCA as 'poor technique'.

Bill Forgan-Smith had a good opinion of Barlogie's ability. In evidence before John Arthur at the subsequent investigation, he said: 'My experience with Barlogie for the whole time he was under my supervision was that he was a steady and reliable person, and a most careful pilot. His approach to flying in New Guinea was always one of caution, and never gave the impression of a pilot who would press on regardless...'

Early on the morning of 21 September, 1951, Fred Barlogie took off from Lae in Dragon VH-AXL on an RPT flight to Wabag, although his aircraft was carrying freight, not passengers. Visibility in the Highlands was much reduced by smoke haze and low cloud, and at around 7.30 a.m. the Dragon crashed on a ridge approximately five kilometres north-east of the Karanka airstrip. Barlogie died instantly. He was believed to have been flying on instruments when he crashed, and the DCA finding was that the probable cause was 'an error of judgment' on his part, in attempting to continue a flight under conditions of visibility so bad that flight in sight of the ground could only be maintained at a dangerously low altitude.

Qantas accident investigator, Bill Burns, beat DCA to the scene of the crash, and was severely criticised by DCA for carrying out an inspection of the wreckage without specific authorisation. There was considerable amusement among the pilots when John Arthur, the DCA boss, crashed in a DCA Auster, VH-CAM, at Fisherman's Island on 5 December. The aircraft was destroyed, and both Arthur and his passenger, S. Elliott-Smith, District Commissioner of the Central District, were injured.

John Arthur had 831 hours up at the time. He held a private licence endorsed for DH82 and Auster aircraft, and was in the process of qualifying for his senior commercial licence.

The ostensible reason for the flight was to examine the progress of work on the new Fisherman's Island airstrip, off Port Moresby. Elliott-Smith had not yet seen the work, and went along for the ride, and not in his official capacity. On taking-off from the rough coral surface of the new strip the engine cut dead at low altitude, and Arthur was forced to land on a coral outcrop. The impact ripped away the belly-tank, which burst into flames. The Auster somersaulted, and the engine was torn from its mountings. Arthur suffered two fractured ribs and lost two teeth; Elliott-Smith's right foot was badly shattered. The Auster never flew again, and DCA replaced it with a Tiger Moth, fitted with radio, long-range tanks, and wheel brakes.

A whisper quickly raced around the Territory: the real reason for the flight to Fisherman's was not to inspect the strip, but to give Elliott-Smith a free flying lesson! This juicy little morsel reached the ear of the Director-General in Melbourne, who thereupon wrote to John Arthur: 'There was a suggestion in the Press that Elliott-Smith was undergoing instruction for a pilot's licence, and that the flight to Fisherman's had something to do with that. I would be glad of your assurance that this is not the case...'

Arthur, of course, gave the required assurance. He had also to submit to the normal DCA investigation of the crash, which was carried out by C.R. Hibbert, and E.W. Eberbach. They concluded that Arthur was a skillful light aircraft pilot, who on this occasion demonstrated airmanship of a high standard. The cause of the accident was fuel starvation, caused by vapour locking of the fuel pump.

Fisherman's Aerodrome was being constructed as an alternative to Jacksons, for use when bad weather closed that strip. It had originally been intended to locate the alternative strip near Papa village, twenty-four kilometres north- west of Port Moresby, but it was concluded that this would be economically unsound. Work began on the Fisherman's strip in July, 1950, by Hornibrooks Constructions Ltd. The project was costed at £12,000, but expenditure soon soared to £80,000 and beyond, which the Director-General 'viewed with concern'. Although Qantas had been pushing for an alternative strip since 1946, because of weather delays to the Bird of Paradise service, Fisherman's was not popular with either pilots or passengers, and was little used after it was opened in June, 1953.

Barely a week after John Arthur's accident at Fisherman's Island, another Qantas Dragon crashed in the Highlands, with the loss of three lives. The pilot was First Officer S.W. Peebles.

On 13 December, 1951, Peebles took off from Mount Hagen in VH-URV, bound for Ogelbeng, Tremearne, Baiyer River, Wabag, Wapenamanda, Yaramunda and Madang. With him as passengers were Airport Inspector Keith Kear, and Assistant District Officer Ken Earle. The Dragon was on charter to DCA, the purpose of the flight an inspection of airstrips, particulary the new one under construction at Yaramunda.

Peebles was twenty-seven, a married ex-RAAF pilot, who joined Qantas after the war as second pilot on DC3's and Lancastrians. He was checked out on Dragons in May, 1950. At the time of his death he had the considerable total of 6,000 flying hours, 900 as pilot-in-command, but he had had very limited Highlands experience. He was a sound pilot, and had never previously been involved in an accident. But he was known to be over-confident.

It was over-confidence that killed Syd Peebles and his passengers. He made a run over Yaramunda strip at a very low altitude, and was forced by the terrain up a gorge from which his Dragon lacked the performance to escape. He crashed, and the aeroplane burned.

It was a particularly sad, and unnecessary, crash. Peebles was popular with his fellow pilots, but his over-confidence was well recognised. Lionel Purkiss will never forget that crash: 'My worst day in New Guinea ever was the day that Syd Peebles killed himself. I was stopping overnight at Goroka with Dennis O'Malley. We were sitting there, wondering how long it would be before Syd killed himself. We didn't know that he already had, that afternoon. I got into the air at daybreak in a Drover, and DCA was calling up for somebody to go to the area. I called them, identified myself, told them I was unrestricted New Guinea, where did they want me to go. I had to go and pick up the coroner and investigators, and that's when I heard what had happened...

'I flew into Nondugl, tried to get to Banz for fuel, but it was fogged over, went up to Wabag and picked up the blokes there, back to Wapenamanda, then back to Nondugl, and Banz for fuel...landed through fog, one prop jammed in fine in this blasted Drover, and had to take off with one pulled back. I identified the bodies wben they brought them in, flew the bodies to Madang, went back in to take Nep Blood from Nondugl to the coast. I just started to cry on the way back, I'd had it...

'Sheer bloody over-confidence. Bloody stupidity. He was doing a beat-up, uphill and downwind, then got into a valley. If he'd said, "OK, I'm in strife," and put it on the ground — it was the only way out — he'd probably have broken a leg at most, but he kept trying to climb and climb, throwing typewriters and things out, until he hit the top of a tree, killed himself and two good blokes...'

Bill Forgan-Smith agrees: 'Peebles should never have been there. He completely disregarded company instructions. He didn't know the Western Highlands, was only relieving at Madang while Jock McKee was on leave. He was only checked into Goroka. He went over to do Madang-Goroka runs, this DCA trip came up, and I told the station manager that when the arrangements were confirmed, I'd send Bill Hosking in to do it. But Peebles got the message instead of the manager, and went in himself. Flew the wrong way up the valley, uphill, trying to outclimb those mountains. In a Dragon! Jesus! If he'd have ditched the thing he'd have been all right. He flew on until he hit the hill and caught fire, and burned everybody up. Should never have been there. I was in a Doug nearly down to the Solomon Islands when I heard about it...'

The verdict of the accident investigators, C.W. Hibbert, Superintendent of Air Navigation, and Rube Murray, Divisional Aircraft Surveyor, was inevitable: pilot error.

Ten days after this sad affair there was yet another crash of a Qantas Dragon, VH-URD. The pilot, First Officer R.J. Davis, was slightly injured, but the aeroplane was completely destroyed.

The Dragon was under charter to the Administration, carrying mails and freight to the outpost at Menyamya, in the wild, mountainous hinterland of the Morobe District. Dick Davis had only recently been checked over the route.

On the approach, Davis encountered strong winds from the south-west, which rocked the Dragon. A mighty downward air-current seized the aeroplane, and forced it into the valley of the Tauri River, where it crashed.

Qantas was criticised by DCA investigators for failing to ensure that Davis was thoroughly checked over the Menyamya route.

This accident happened on Christmas Eve, 1951. It was a sorry Christmas present for Dick Davis, and for Qantas. Three Dragons and a Drover lost in less than six months! Never in its history had Qantas suffered such a run of misfortunes. The pilots of the four aircraft were all first officers. The Drover accident was clearly in a special category, but there were points of similarity in the Dragon crashes.

Two days after the loss of URD, Bill Forgan-Smith called an emergency meeting of all pilots to discuss the run of accidents, and the adequacy of Qantas training and checking procedures.

It was generally agreed that there was nothing wrong with the system of training, which was adequate in all respects. In his report on the meeting, Forgan-Smith said: 'All unanimously agreed that the series of fatal accidents had definitely been due to the attitude of pilots "pressing on", and that this would not be under actual control — it would be up to the pilots.'

The shortcomings of the Drover came under much discussion. Experienced pilots agreed that under bad weather conditions, it was extremely hazardous to fly Drovers into the Highlands. The pilot's seat was located so far back from the windshield that vision from the cockpit forward and to the right was quite inadequate. Certain features of the controls and instruments were also criticised.

A later Qantas inter-departmental inquiry concluded that there was indeed something lacking in pilot training and checking. Pilots were normally sent to New Guinea to obtain their Dragon training, and the inquiry recommended that initial training should in future be given in the kindlier environment of Australia, before pilots were sent up.

C.R. Hibbert and Rube Murray had investigated the crash of Davis's Dragon at

Menyamya, and ironically, Hibbert was himself involved in the dramatic crash of another Drover, a few months later. It was an accident that was to lead to a vindication of the reputation of the dead Johnny Spiers, and it happened less than a month after the official DCA report blaming Spiers for the loss of VH- EBQ was issued.

DCA had been involved in the Drover project from the beginning, and ordered two of the new aeroplanes for departmental use. The first, VH-DHA, was test-flown by Brian Walker as early as January, 1948, but was not transferred to the Papua New Guinea Region until 1951.

On 16 April, 1952, C.R. Hibbert — at the time acting regional director in the absence of John Arthur, on leave — took off from Wewak in VH-DHA on a flight to Momote, Manus Island. With him as passengers were Major A.L. Sheedy, of the Attorney-General's Department, and Tom Drury.

As previously recounted, Tom Drury joined MAL after a disagreement with Sam Jamieson of Guinea Air Traders: 'MAL was a different kettle of fish. They'd operated in New Guinea before the war, knew the country, and all the short cuts. Much more professional that GAT. Had better financial backing — Carpenters owned copra plantations, stores, shipping lines, all sorts of things. Doug Elphinstone and Brian Carpenter ran a very good outfit, gave me a pretty fair spin...

'I was with MAL for six months when Qantas offered me a job. Doug Elphinstone said, "You'd be a fool not to take it." Again, Qantas was a totally different ball-game. Harold Hindwood welcomed me into the fold — great bloke, Harold. Qantas sent him down to Darwin soon afterwards. He was like a fish out of water there; like me, he never wanted to leave New Guinea. But they made him go, and he did an excellent job at Darwin. Then they wanted to move him again! Wouldn't tell him where to...anyway, Harold walked out of a Qantas Board meeting in Sydney, rang Jack Sedgers, who was down at the time, and Jack signed him up immediately. Harold came back to Lae as manager of Qantas's opposition, MAL, and he did a great job...

'John Arthur had been after me for a long time to join DCA as Examiner of Airmen. Said I was just the bloke they were looking for, so eventually, after having a squabble with Qantas, I left them and joined DCA, at Lae...

'I started as an Air Traffic Controller. That left me ample time for flying, so I flew DC3's for MAL at ten bob an hour, in my spare time. John Arthur heard about this, so he sent for me. "I understood you were working for DCA," he said, "What's this about flying a DC3 for MAL? As a public servant, you can't do that without permission from the chief officer." 'Who's that?' I asked. "Me!" he said — "you're nuts to be flying for ten bob an hour — ask for more!"

'I went from temporary Air Traffic Controller "Grade Nothing" to permanent Grade Four in about nine months, and then was given the job of Airport Manager, Lae. Then they made me Acting Safety Officer, and then Superintendent of Air Navigation, and I had to go to Port Moresby. A couple of months later I was Deputy Regional Director, in an elapsed time of a little over a year from the time I joined DCA. I wasn't popular with many senior DCA bods in Australia — I'd broken into their circle as a complete outsider, and went up the ladder far too quickly...the luck of the game.

'After our rocky start, John Arthur and I got on like blood brothers. He was the best man they could have got for District Superintendent. Didn't have a lot of background in the early stages, but he eventually got some pretty crafty characters around him, who kept him on the rails, and administratively he was excellent. Without doubt, the hardest-working public servant I have ever known. We were always short of officers with the right experience to do the total job. A few of us at the head of affairs were overloaded to blazes, we were in a state of nervous exhaustion every day in the week. People thousands of miles away in Canberra and Melbourne never realised the amount of work John actually did for the department...

'I could say a lot about the accident with Hibbert and Sheedy in the Drover. It was the only one I had in New Guinea. The Director-General had given de Havilland's half a million pounds to produce an Australian bush aeroplane, on condition that DCA got the first two off the line. The ones we got were VH-DHA and VH-BMU. Clarrie Hibbert flew the first one, DHA, up from Melbourne, and I met him on his first trip over to Lae. Not overly impressed with the overall appearance of the Drover.

'Hibbert was unfamiliar with the country, and I was given the job of navigating the aircraft around New Guinea. The object of the exercise was to take Major Sheedy, who was regional director of ASIO, around a great number of outlying ports for discussions with the District Commissioners and other responsible people on security matters. Had a great heap of documentation with him...

'I wasn't endorsed on the Drover — had never been in one — so Clarrie flew it; to Madang, the Highlands, and Wewak. Next stop was Momote, over the sea. I'd realised by this time that there was nothing much in the way of emergency equipment on board, so I arranged with Tom Deegan of MAL for the loan of a rubber dinghy, and Leo Simpson, our OIC at Rabaul, sent me a Gibson Girl transmitter, and some Mae Wests.

'We set sail for Momote with Hibbert at the controls, far too low for my liking. He went over the Bismarck Sea at 2,500 feet [762 metres]. Nobody in their right minds should have gone across that distance of ocean at 2,500 feet, but he was flying it...

'I was using the radio for him back in the cabin, giving him the DR positions. I'd just given him one at about the mid-way point, when there was a great explosion at the front of the aeroplane. Papers started to fly about the cabin, and the aircraft started into what was to become a vertical dive from 2,000 feet [610 metres].

'I raced up front. There was a great hole in the side of the aircraft. Hibbert looked up at me with dazed eyes, then put his head down...his foot looked to be almost severed. I said, "Quick, out!" I steadied the thing out of the dive with one hand and grabbed him and helped him out of the seat with the other. I got in the seat, and nearly went out the hole in the side. When I recovered from the dive, I suddenly found I had three throttles in my hand — and they came off. That left us with minus nothing.

'I said to Sheedy, "Look after Hibbert — we're going in!" Sheedy was terrific — no panic at all. Hibbert had managed to get out a Mayday call, and I yelled to Sheedy, "Jettison the top hatches and the door! I'll call out the heights!"

'I had no airspeed indicator, all the lines had been cut by the propeller. I had a quick look at the port engine. The propeller and crankshaft had gone completely. The engine had blown itself to bits — pistons, wires and things dragging over the port wing. Fuel spraying all over me from the cut fuel balance lines under the floor. I started to call out the heights at about fifty feet [15 metres] — fifty, forty, thirty, twenty, ten, going in...the undercarriage went, and I ditched her, tail first.

'The thing started to sink, because of that great, gaping hole in the side, and the hatches and door had gone. She filled up with water in a matter of seconds...'

Working feverishly, they launched the rubber dinghy. Just before the Drover vanished, Drury swam to the rear door and got the emergency Gibson Girl transmitter in its pack, carrying it to the dinghy between his teeth, 'like a retriever dog'.

Clarrie Hibbert was in great pain, and Drury had banged his head during the ditching and was suffering from concussion. It was almost 5 p.m., and they had no way of knowing whether or not the Mayday call had been picked up. While Sheedy applied a torniquet to Hibbert's mangled foot, Drury sent up the Gibson Girl's copper aerial with the hydrogen-filled balloon supplied with this equipment, and started to wind, sending out emergency transmissions: 'I had been trained on the Gibson Girl during the war. We had a lot of trouble then with the aerials — found that if you let all the wire up, the aerial just floated off into the sky with the balloon if the nylon security clip at the bottom had rotted — and a lot of them did. Sure enough, the Gibson Girl we had in the dinghy was an old wartime set, and the nylon clip *had* rotted — so Sheedy had to sit there with two laps of copper wire around his arm, all night. I didn't realise it, but every time I wound the transmitter it generated a tremendous amount of heat in the first ten feet [three metres] of the aerial, and it burned right into his biceps, and he never flinched. I must have wound it 500 times...'

The sun went down, and darkness settled over the empty sea. Sharks appeared, and they threw over repellent. Then, coming up over the horizon, they saw a light approaching, low and fast. It was a searching aircraft. They began to haul down the aerial, and a DC3 roared overhead, homing in on the Gibson Girl. As it flashed past they could see the aerial balloon, wavering in the slipstream in the moonlight.

The pilot of the DC3, VH-EAL, was Captain I.D. Ralfe. The Mayday signal had been received, and emergency procedures had swung smoothly and efficiently into action.

Ian Ralfe was on relief duty for New Guinea Internal, based temporarily at Lae. He was a wartime pilot, and had survived a ditching and an escape from behind enemy lines in New Guinea. He and his first officer, Rex Cuthbertson had completed a routine flight from Lae that afternoon, and after refuelling the DC3 for the return flight next morning and putting the aeroplane to bed for the night, were enjoying a beer in the officer's mess at the RAAF base when the Mayday was received. It took just nine incredible minutes for Ralfe and Cuthbertson to be airborne, after a rush to the airstrip in a madly bouncing jeep at nearly 100 kph. 'It was a ride I won't forget,' Ralfe wrote in a letter to the author in 1984: 'Having found the dinghy, in my excitement I made a low pass with my landing lights on, to see who had survived. I completely forgot the hydrogen balloon and aerial, and had to do a steep turn at 100 feet [thirty metres] when I noticed this large white disc above and ahead of me. Sheedy had the aerial wire wound around his arm...one shudders to think of the consequences of hitting that wire. Of course, it was at night...

'We had to go back to refuel, and our attempts to communicate with the dinghy by Aldis lamp were only partly successful, but they did get our message that, like MacArthur, we would return...

'An RAAF crash boat was on the way, but from memory it had no DF facilities, and would have trouble finding the dinghy. So we loaded a number of flame floats on board, with some RAAF personnel, and flew back. Relocated the dinghy at 1 a.m. and dropped the floats, and lit up the Bismarck Sea like Pitt Street. We contacted the crash boat and gave the skipper a magnetic course to steer. The boat reached them at about 2.30, and took them on board...'

When the Mayday signal was received (by Aeradio Stations Madang, Wewak, Lae, Moresby and Momote), Lae Air Traffic Control cleared virtually all of South- East Asia from Ralfe's radio communications frequency in a matter of minutes after he took off. The Rescue Co-Ordination Centre at Townsville, normally responsible for search and rescue procedures in PNG, was advised of the emergency, and a long-range Lincoln bomber of No. 10 Squadron was hastily prepared for take-off. The Lincolns were aging aircraft, with a poor serviceability record, and as an additional measure, Qantas was asked to put a Catalina on stand-by.

Captain Fred Fox and his wife were entertaining a number of fellow pilots at their house that evening. Fox arranged for flares and emergency equipment to be loaded into VH-EBU, just in case. He remembers: 'We kept the soiree going, but realised we couldn't drink too much with a possible job on. When we heard that a Lincoln was coming, the blokes chortled, "If that Lincoln gets airborne, we'll buy a lottery ticket...and win..."

'Before midnight, we heard that they'd located the dinghy, and that the DC3 was circling it. About midnight, they told us the Lincoln had turned back with engine trouble, and that the 'Cat' would have to come into action.

'Ken Montagu, my first officer, and I planned the flight to be over the Purdy Islands at dawn, so we could at least see to land, because it would be an open- sea landing. Took off at three in the morning. The 'Cat' was fairly light, had fuel for ten hours' flying — I didn't believe in stinting myself with fuel — and there were a lot of high mountains in our path, so I went to 14,000 feet [4267 metres]. No oxygen of course, but we were fit blokes, and watched each other to see if one was going to flake out. Went past Mount Yule, and headed direct for Madang, knowing we were then clear of all mountains on the New Guinea side. The weather was completely clear, a dark night, no problems, all beacons on, all assistance we needed in communications. Interesting flight — there was no night flying then, of course.

'Started to let down over Lae. Got to Purdy Islands, and knew by this time the air-sea rescue boat had reached the dinghy, and taken them on board. The quickest way to get them to medical attention was for us to take them to Port Moresby, which had the best-equipped hospital. So we landed at Purdy, cut the motors — no wind at that early hour — and they brought Drury, Hibbert and Sheedy out. Hibbert was drugged, and strapped into a stretcher...took off, and flew back to Moresby.

'It was a classic case of how a rescue can be made when everything clicks...'

Immediately following this crash, DCA grounded all Drovers pending an examination of their propeller barrels and retaining nuts, for it was obvious that the crash had been caused by a failure of the port propeller. It was now suspected that the loss of VH-EBQ the previous year might have been caused by propeller failure, rather than pilot error, for when the wreckage of that aeroplane had been recovered from the sea, a blade of the centre propeller had been missing. The propeller assembly of EBQ, minus the missing blade, was recovered from a scrap-metal dump at Lae, and sent to Melbourne for expert examination.

De Havilland Australia, seriously concerned about the implications of the failure of the port propeller of VH-DHA, requested the parent company in England to carry out strain tests on the type of de Havilland propeller fitted to the Drover. A strain gauge specialist was sent out to the Bankstown factory, and a test programme was set up.

While testing was still in progress, a third failure of a Drover propeller settled the matter.

The third Qantas Drover, VH-EBS, had been delivered in September, 1951, and after twelve months in service was now due for a C of A renewal overhaul in Brisbane. The aeroplane was ferried from Lae by Rex Cuthbertson. On 11 September, 1952, Cuthbertson commenced his take-off run from Mackay, Queensland, on the final leg to Brisbane. Suddenly, a blade flew off the centre propeller, and the nose engine was wrenched out of its mounting. Fortunately, Cuthbertson was able to stop the aeroplane on the runway, so avoiding what could have been a bad accident.

The blade and retaining nut were recovered. The propeller had logged only sixty-four hours since its last inspection, and there was now no doubt that the villain of the piece was the variable-pitch propeller.

DCA immediately grounded all Drovers again. Results of the de Havilland strain tests now proved conclusively that there was an excessive vibration-induced stress with the engine-propeller combination fitted to the Drover. The propeller blades were held at the root by a retaining nut and ring assembly, and because of metal fatigue caused by the vibration stress the nut was wearing down the thread against the propeller root. The Drovers remained grounded until 26 September, when operations were again permitted subject to propeller inspections every 100 hours, and reduced engine settings.

While the strain tests were still under way, intensive efforts were being made to achieve the seemingly impossible: locate and recover the missing propeller blade of Spiers' Drover, EBQ, from the sea bottom off Lae (shades of the needle and the haystack!). On 11 October, 1952, the blade was recovered. Expert examination revealed that the propeller blade retaining nut had failed.

The DCA and Qantas accident reports on the loss of EBQ were now altered. The official DCA finding was that the crash was probably caused by 'the loss of control resulting from failure of the centre propeller in flight'. The stain on John Spiers' professional reputation was thus, posthumously, removed.

The DCA report on the loss of VH-DHA, issued at the end of October, stated that 'the cause of the accident was that a portion of the port propeller became detached and caused damage to the fuselage to such an extent that the remaining two engines were rendered inoperative'. The possible causes of the failure of the port propeller were a fatigue crack of the propeller barrel or retaining nut, poor machining of the nut itself, and the detachment of the spinner due to failure of the mounting attachment. DCA strongly recommended that all Drovers be fitted with different propellers.

By this time, Qantas had almost had enough of the Drover. Five of the type had been placed in service in PNG — VH-EBT was delivered in November, 1951, and EAZ in July, 1952 — and the company decided that enough was enough. To the distress of de Havilland Australia, Qantas cancelled its orders for the last two of the seven Drovers originally ordered, after EAZ was delivered.

Following the DCA report on the loss of DHA, de Havilland's developed a modification kit to convert all Drovers to fixed-pitch propellers. This had to be done — both Qantas and TAA — the other major operator of the type — were exceedingly dissatisfied with 100-hourly removals of the variable-pitch propellers, which were naturally expensive. TAA had had few problems with Drover propellers, but their aircraft operated under much more favourable conditions that did those of Qantas.

Qantas had continued to be plagued by minor troubles, with engines and brakes as well as propellers. In August, 1952, Jack Stammer — who had just joined Qantas from MAL — flew VH-EBR from Brisbane to Lae after a C of A overhaul: 'I had a devil of a lot of trouble with the aircraft, and I ferried it most of the way on a combination of one or two engines...the trouble was being caused by fuel starvation. The Drover had a small fuel filter made out of bronze. The fuel tank sealant had been overdone, and it choked the fuel filter...we found the answer at Cairns, and from then on there were no more problems...'

Jack Stammer loved the Drover. 'It was a lovely little aeroplane, despite all the troubles it experienced.'

In the following month, Clyde Johnston had a brake failure and an uncontrollable central engine in VH-EBT, and a week later Jock McKee had to make a forced landing in EBR when his centre engine began to smoke, and lose revolutions. The cause was found to be overheating in the propeller box assembly. Qantas engineers were already performing top overhauls on the engines every 300 hours, to maintain economy and efficiency. The company, too, was dissatisfied with the payload of the Drover, and had already had discussions on the matter with de Havilland Australia.

Re-equipping the Drover with fixed pitch propellers could hardly be expected to improve the performance of the aeroplane, so new double-slotted high-lift Fowler-type flaps were designed to replace the original flaps. With these modifications, the aeroplane became the Drover Mark Two. The four surviving Qantas Drovers were in turn converted to the new specifications, and one, VH- EBR, was fitted with new windows and a raised roof in an attempt to answer pilot criticism of bad visibility in poor weather.

The Drover Mark Two was not a success in PNG operations. The climb performance was now clearly unsatisfactory, and the payload was reduced into some strips to only 544 kilograms — little better than that of the ancient Dragon the Drover was designed to replace.

By April, 1953, Qantas was fed up with the Drover. The company had by this time invested more than half a million pounds in New Guinea Internal, and in a letter to the Director-General, C.O. Turner, the managing director, wrote: 'The financial position of our New Guinea Internal operations is causing us a great deal of concern...'

He went on to request a contract payment of £79,000 to cover the current year's developmental flying for the Administration, and continued: 'We are up against a most difficult problem in regard to Drover operations, and unless some alleviation of present restrictions can be achieved we are likely to lose a great deal of money on their operation.'

The Drover was at that time costing £44 per hour to operate, although the charter rate — all that the market would bear — was only £27/10/-. The Drover was almost as expensive to maintain as the DC3, which cost £65/5/- per hour to operate and had almost four times the payload. C of A repairs, maintenance and engine overhauls on the Drover cost approximately £17/10/- per flying hour. Operating Drovers in PNG was clearly a ridiculous economic proposition, made worse by the low utilisation of the Drovers, which averaged only 528 hours per annum.

Drovers were by now being used in a restricted role, mainly into the Highlands from Lae and Madang. Although the charter rate was well below the operating cost, customers were reluctant to employ them since the Dragon could carry much the same load, at a lower cost. Qantas began to lose business to its competitors.

Despite the problems of the Drover Mark Two, some pilots liked it. Mal Shannon, an ex-RAAF pilot who joined Qantas in May, 1951, after some exciting times flying cropdusters, thought the aeroplane was 'great': 'I was checked out on 27 April, 1954. It was the coastal trials of this new monster, EAZ, with the Fairey-Reed fixed props and double-slotted flaps like a modern-day Junkers. Old Brian Walker flew it up, and he said to me, "This'll do the job, son!" I thought, Oh hell, once you get those damned flaps out, you've got to get them back. They had solved the propeller problem, but you'd get no revs out of them...

'But I thought the Drover was great. Wizard. Used to be able to play with those three little buttons like a Wurlitzer, you know? You'd wind those Gipsy Majors to about 2,550 revs on take-off, 2,450 for the climb, 2,250 for the cruise, and she'd go like a beauty. Controls were beautiful. Like a DC3, only more responsive. After the coast, I went with Brian to the hills. At Goroka, Hagen, Wabag, Kainantu you'd lose power, but on the downhill runs you'd accelerate quickly. Take-off weight was

6,500 pounds [2948 kilograms], and we were limited to 5,750 [2608 kilograms] at Kainantu, and about 5,600 [2540 kilograms] at Hagen...'

T.E.W. Howes was to fly more hours on the Drover than any other Qantas pilot. Tom Howes was an ex-RAAF bomber pilot/navigator, who was awarded an immediate DFC in Europe while flying with 70 Squadron, Royal Air Force. He joined Qantas in March, 1945, as a navigator on the Indian Ocean run, and then became a first officer on the Bird of Paradise service to PNG. In April, 1951, he was posted to New Guinea Internal. Howes was engaged to be married, and the ceremony was performed just before his scheduled departure. This made Qantas, rather than Tommy Howes, responsible for his wife's passage to Lae!

Tom Howes flew both models of the Drover, and he, too, champions this unlucky aeroplane, which was almost so good, and could have been a world-beater. Howes recalls: 'Frankly, I really enjoyed the Drover, even with its problems. For example, on take-off from Wau on a sports charter with fourteen passengers on board, an essential component on the port engine decided to shear. Fortunately, it was a Mark One, so I put the pitch into fully coarse on the failed engine, which was as near as I could get to feathering. Due to the steep slope at Wau, it was possible to continue to lose height after take-off, describe a circle and re-land on the strip again — at a much lower altitude than I began at! Half-way round, I shouted to the somewhat confused passengers, "We're going back in." I wasn't quite sure if it shouldn't have been, "We're going in!"

'The Mark Two's propeller blades didn't fly off like the Mark One's were prone to do. However, being only fixed pitch and with the extra drag of the flaps, it was rather like driving a car with second gear only. The result was that, particularly on a soft strip, it would not get up and go, and once in the air had hardly a dazzling climb. This sometimes impressed observers as one disappeared below the level of the strip after take-off in an endeavour to achieve enough flying speed to make the necessary turn away from the hills safely...'

In September, 1954, Howes flew a full load of labour recruits out of Menyamya. John Wakeford, the ADO, watched the take-off with two of his patrol officers. The Drover roared down the strip, and vanished over the end into the gorge of the Tauri River. It was long seconds before it appeared, climbing away from the river towards the steep, grass-covered ranges of this forbidding Kukukuku country. Wakeford wrote to Howes: 'There are going to be no more recruits of that number taken out in that cursed plane, not whilst I am here at any rate. My God, I thought you had gone this morning. All three of us were waiting for the crash...'

The end had almost come for the Drovers. In 1954, Qantas sold three to Fiji Airways. The first two, EBR and EBT, were ferried to Sydney by Dennis O'Malley, and Tom Howes delivered the third, VH-EBS, in May. The sole surviving Drover, EAZ, had been the last to be converted to Mark Two standard, and it continued in limited service out of Lae, flown solely by Tommy Howes, until mid-1955. During this year, the Drovers were replaced by DHC2 Beavers, strong bush aeroplanes designed and built by de Havilland Canada.

What irony! The twin-engined Dragon had been replaced because of its poor single-engined performance by the tri-engined Drover, which was in turn replaced by the single-engined Beaver! Some humorist produced an 'Ode To A Terrified Airman', which summed up the feelings of the pilots quite nicely. It was sung to the tune of the old wartime favourite, 'Bless 'Em All', and it went, in part:

> 'Though I'm only the first of the few,
> I'm quite sure that one will not do,
> I'd rather look over, the nose of a Drover,
> If one goes, I'm still left with two!
> Just think of a Wau, the Watut, the Snake,
> Shove in a Menyamya as well,
> I don't care one iota, I like a tri-motor,
> You can damn all your Beavers to hell!'

The Beaver, however, was to be highly successful, and well-liked by its pilots. And Qantas never had another fatality, or serious accident, in PNG.

88

90

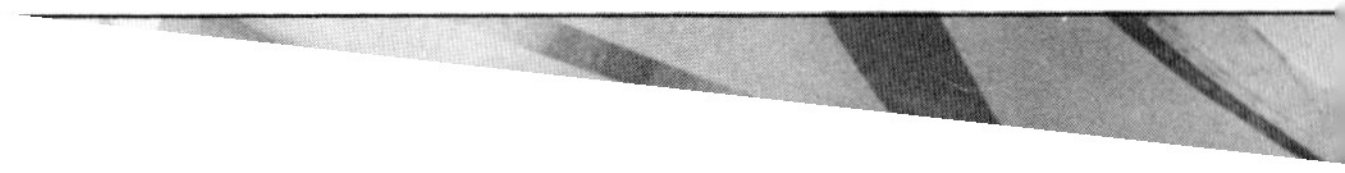

9

92

9

88. Lae aerodrome, 1954. *Photo: Qantas*
89. John Arthur in the DCA Auster, VH-CAM. *Photo: Juin Arthur*
90. Tom Howes in a Qantas DC3. Mt Lamington may be seen in the background. *Photo: Tom Howes*
91. Qantas DC3 at Rabaul. *Photo:* Pacific Islands Monthly.
92. John Arthur and Bob Green in VH-CAH. *Photo: Alf Lane*
93. Qantas apprentice working on tailplane of DC3, VH-EAQ at Lae. *Photo: Qantas*

94

95

96

97

98

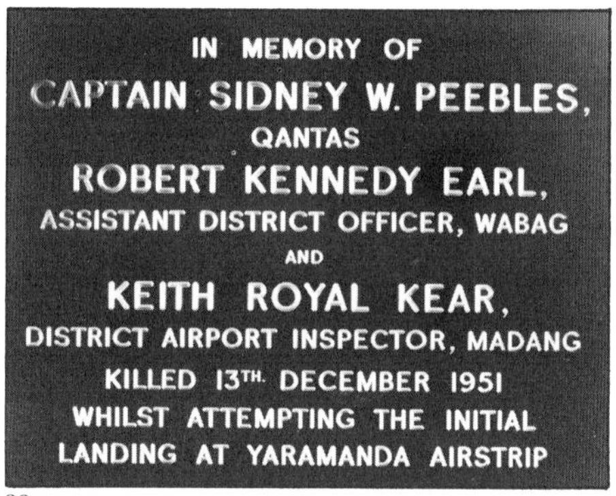

99

94. Qantas Drover flying over Sydney harbour. *Photo: Lionel Purkiss*
95. Qantas Drover, early 1953, shows modified screen with flat planes for better pilot visibility. *Photo: Qantas*
96. Three Qantas Drovers and a GSA Norseman. *Photo: Mal Shannon*
97. Mal Shannon and a Qantas Drover at Minj. *Photo: Mal Shannon*
98. Pilot Roger Wilson in a Qantas Dragon. *Photo: Roger Wilson*
99. Memorial erected by Qantas following the crash of the Dragon. *Photo: Qantas*
100, 101 & 102. Parts of Qantas Drover, VH-EBQ at Lae, 1951. *Photos: Bill Burns*

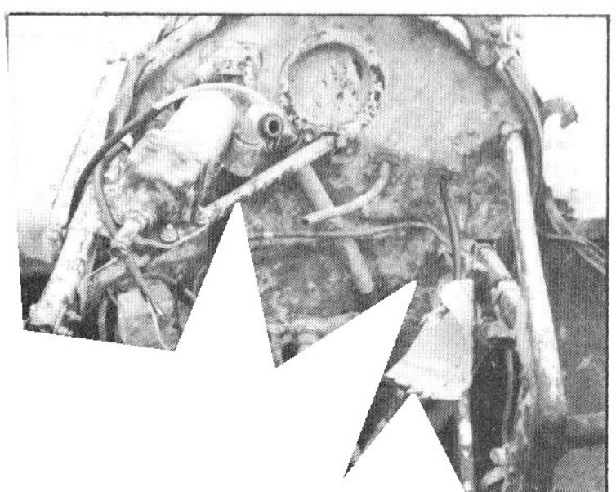

100

101

102

103

4

Flying High

Success And Failure

Although Qantas and MAL dominated the PNG aviation scene at the beginning of the 1950s and Bobby Gibbes was starting to make his presence felt, there were other, smaller, operators at work. Of the individuals and companies who bravely bid for their share of the rapidly expanding air transportation market, few survived. One who did was Laurie Crowley.

As detailed in Chapter 2, New Guinea Internal, Crowley had originally come to PNG as a licenced engineer for Guinea Air Traders. After the Second World War, he obtained his pilot's licence, but he did no flying for GAT.

Laurie Crowley had no intention of starting what would become a notably successful airline operation when he arrived in Lae. But there was little to do in one's spare time in the war-battered little town, and a group of eighteen GAT men — of whom Crowley was one — decided to chip in £20 each and purchase a light aircraft, for purely recreational flying. Crowley was given the job of finding a suitable aeroplane. He remembers: 'First weekend I flew to Wewak, because I had heard that Dr John McInerney was thinking of selling his old Tiger Moth, VH-BLT. I approached John in his "surgery" — a room littered with Gipsy Major engine parts. Not being much of an engineer, he had removed each part from a spare motor and laid them out one by one around the room — he said he was going to start at the other end and reassemble the bits, and if when he got to the starting point no pieces were left, the job had to be satisfactory.

'Anyhow, John suggested we could not carry out an important deal like selling an aeroplane in his surgery, as the BMA would hardly approve, so we retired to his "main office", the Sepik Club. The deal was completed, and I hurried back to break the good news to my seventeen partners — but, to my dismay, sixteen of them had changed their minds, with varying excuses like, "My wife thinks I shouldn't fly," and, 'There's some talk of the company closing down, and I may need the dough...'

'My empire had crashed around my ears — how to face John McInerney? The situation needed some serious thinking over, so I rustled up the lone survivor, Ray Stockden, and we proceeded to the old Hotel Cecil for a summit meeting. My first question to Stocky was, "How much have you got?" He said, 'About 150 quid.' I checked with my wife, Betty, and she said she could let me have a hundred pounds, so all in all we raked up the £350 we needed, and I set off post-haste to Wewak, paid John, and tore off to the airstrip to gloat over our first aeroplane. There are no words in the dictionary to tell you how it feels...Sleep was out of the question, so I got out of bed at 3 a.m. and was airborne at first light. After refuelling at Madang, was on my way to Lae. Arrived just before last light, feeling like Charles Lindberg when he arrived in Paris. I was jerked back to reality when hardly a soul bothered to give our plane a second glance, but the look on Stocky's face restored my faith in human beings...'

Crowley and Stockden purchased BLT in June, 1949. Raymond John Stockden was also an engineer/pilot, with a licence in category D only, which allowed him to work on engines, not airframes. After purchasing the Tiger, the two men were almost broke. All they had were their jobs, and by this time it was obvious that GAT wasn't going to last much longer. They were refused credit for fuel, but an old friend, Pappy Duff, came to the rescue. Duff drove the Mobil fuel tanker at Lae, and he found that by filling his tanker in the cool of the morning and dispensing the contents during the heat of the day, a balance of fuel remained in the evening after he had accounted for his daily uptake. It wasn't much, but the Tiger had a modest thirst. There was, too, still a lot of old 100-octane fuel from the war at various disused wartime strips in the Morobe District which was just useable, and these two sources of supply allowed Crowley and Stockden to fly their Tiger at

103. Captain Fred Fox in the cockpit of a Qantas Sea Otter. *Photo: Qantas*

weekends.

Then GAT folded, and the partners were out of work. Again they retreated to the sanctuary of the Hotel Cecil to discuss the situation. Neither wanted to leave Lae: 'Neither of us could stand post-war Australia, with its black markets, key money for flats, and tipping waiters to serve you, so we decided to form Crowley and Stockden Airways, and have a go. We moved into Joe Taylor's old hangar, half way down the Lae strip.

'We managed to keep our heads above water by buying vegetables at Boana, and selling them in Lae, and doing what charters a single passenger plane could do...the local indigenous population didn't mind sharing the front seat of a Tiger with a couple of others to keep the cost per head down!...

'The vegetable business began to boom, and in no time we were flying from first light to last light to cater for the public. BLT was flying every daylight hour, the maintenance being carried out at night by self, Stocky and Kandilop. On rainy days we'd catch up with major items.'

Kandilop had arrived at the GAT workshop one day in November, 1948, as a young boy, just in from the bush and covered with the severe skin infection, *tinea imbricata,* called *girili* in Pidgin English. The engineers cured him, and a grateful Kandilop repaid the favour by donating his time in cleaning up the workshop. When GAT folded, Kandilop went with Crowley and Stockden. Neither Laurie Crowley nor Ray Stockden took wages from their little business at first, and Kandilop was paid a small wage. Says Crowley: 'Kandilop worked for me until the end of his life. We hardly had need for words. I would be working on an engine, and be short of a spanner and look around, and there would be Kandilop, holding it for me. In the early days, Ray and I and Kandilop were the whole staff, two-thirds of us unpaid, and sometimes it would be necessary for us to work twenty or thirty hours without sleep to get the planes back into the air after major breakdowns. Quite often I would notice Kandilop's eyes begin to droop, and I'd say, "Go and have a sleep," and his reply was always the same. *'Em i orait, Masta, maski...'*

'The big, black cloud looming over camp in the beginning was what to do when the annual overhaul for C of A came due. We asked for, and received, one extension of thirty days or 100 hours, whichever came first. We nearly had enough money to buy another aircraft, so Stocky was despatched south to get a job, and look around for another plane.

'In the meantime, Rube Murray arrived in Lae, took one look at BLT, and said, "My God! This machine looks like it has flown a couple of thousand hours since last year's overhaul." With that, he began to push his big thumb through the fabric in all sorts of places. I had witnessed Rube despatching the Ansons to the Busu graveyard a year earlier, and I was terrified BLT was going to share the same fate. But Rube said, "New fabric all over, and a complete overhaul, and I'll be over to inspect the final job."

'So there we were, vegetables to be collected and passengers depending on our arrival, and no plane. Fortunately for us, Jack Gray had a Tiger Moth, VH-AQS, which we were able to hire for a few weeks. Next day, good news from Stocky. He bought a Tiger for £250, and had enough money for fuel for the flight back, with eighteen shillings left over. By sleeping under the wing and eating cheaply, he made it to Lae, with two shillings to spare.

'The new Tiger was VH-ARQ. Beautiful, red and blue, with leather upholstery. By midday next day we were flying both Tigers, but in due course had to give AQS back to Jack Gray. By working nights, we got BLT back into the air, and we then got the bright idea that if we sold BLT, we could think about buying a plane with a cabin, and offer the public a ride inside, out of the rain...'

Fortunately, there was a buyer at hand for BLT; W.L. Nesbit in partnership with E.J. Desmond. The deal was concluded and the purchase price of £300 paid over on 10 October, 1949, just before Laurie Crowley was about to take off for Boana in the fully-loaded aeroplane. He recalls: 'Before we had time to get the cargo out of the plane, Lloyd was in the cockpit, pulling on his helmet. As soon as the last box was out he called for prop swing, and in no time was airborne, much faster than a rusty pilot should have been. Stocky said to me, "Don't worry," and patted the wad of notes in his shirt pocket...

'Ten minutes later, we heard BLT's motor change note, and were surprised to see Lloyd Nesbit doing aerobatics over the Butibam River, just south of Hospital Hill. Soon a small crowd had assembled on the tarmac to watch. After a few very rusty loops, Lloyd climbed back to about 4,000 feet [1219 metres] and put the aircraft into a left-hand spin — a manoeuvre that is impossible to execute badly — and after three or four turns the non-flying spectators were very impressed, while the flying bods were no longer smiling, and were murmuring things like, "About time he pulled out of that spin," and, 'I hope he knows what he's doing...'

'There was a deathly silence as Lloyd continued through a few more turns, and eventually disappeared behind Hospital Hill. We waited for the dull thud and the black smoke, we were too stunned to talk...Then somebody pointed towards Buster Purnell's chicken farm, and said, "I don't believe it." Sure enough, there was Lloyd making a left turn towards the runway. By the time he landed, a few more people had arrived. He taxied up, and we all waited patiently for his explanation. His face was very pale, and he was silent for quite a minute or so. The silence was getting embarrassing, so to break the ice I said, 'What happened, Lloyd?'

'He said that the plane wouldn't come out of the spin by using all the proper control movements. "How did you recover?" I asked. He said, 'I knew it was all over, so I panicked and let all the controls go, and next thing I was out of the spin..."

In March, 1950, Laurie Crowley applied for a charter licence on behalf of the partnership, to cover 'flights within the Territory of New Guinea' with a single DH82, VH-ARQ. Crowley advised that six months' supply of general maintenance spares was held. The licence, No. 66, was issued on 16 May, 1950.

The Tiger Moth was by no stretch of the imagination a commercial aircraft in the normally accepted meaning of the term, although Moths were used in a wide variety of commercial roles, in particular crop-dusting, in the early post-war years. The Crowley and Stockden charter licence application resulted in a great deal of argument and discussion among senior DCA officers in Melbourne, over the question of what operating weight limits should be applied to the Tiger Moth in New Guinea service. No altitude/weight charts for the DH82 existed. Not even the manufacturer, de Havilland's, had one. The largest user of the type, the RAAF, had no performance charts on the Tiger; the use the Air Force had made of the aircraft had 'never justified knowing its performance', as one DCA officer noted. Consideration was given to carrying out a series of exhaustive tests on the performance of the Tiger, using a departmental DH82.

The famous aviator, and now DCA officer, Stuart Campbell, who had flown in Papua for the Oroville Dredging Company before the war, was loath to allow the tests. He noted: 'I don't want to start doing performance trials on DH82's at this stage, as once we do we will have to continue with all other obsolete and light types. Surely we can work out a theoretical relationship between weight and altitude which will be accurate within the limits of the performance and handling technique of a DH82 in New Guinea.'

Stuart Campbell settled the matter, after further internal discussions: 'We're not going to waste time doing performance trials on DH82's. Their altitude accountability must be left to the pilot.'

And that is how the matter was resolved. The Director-General advised John Arthur that no attempt would be made to specify operating weights for Tiger Moths at high density aerodromes, since owing to the low performance of the aeroplane, variations in rigging and fabric, and the condition of individual engines, it would be impossible to apply them realistically.

'It is decided, therefore, that the appropriate weight shall be determined at the site in the light of the operator's experience and knowledge of local conditions...'

This decision allowed Crowley and Stockden an unusual degree of flexibility in the use of their Tiger Moth. The little aeroplane was flown hard and to the limit of its performance into high altitude strips, but it was used carefully, and maintained meticulously.

The sale of BLT to Nesbit and Desmond provided Crowley and Stockden with funds for the purchase of a light cabin aircraft. Unfortunately for the partners, new aeroplanes were expensive, and well beyond their financial reach. The aircraft that was finally purchased was of a lineage more ancient than that of the Tiger Moth. A Curtiss Robin.

The Robin was designed in 1927, and first flew in 1928. A three-seat cabin monoplane, the Robin was produced by the Curtiss Aeroplane Company, one of the oldest names in aviation, and was originally equipped with a 90 hp Curtiss OX-

5 radial engine, a relic of the Great War! The strut-braced wings were of wood, and the fuselage of fabric-covered steel tube construction. It was cheap to buy and fly, and for a short time during 1929 Curtiss produced Robins at the rate of seventeen per week. Many different engines were fitted to the later models of this enduring classic, and production totalled some 325. It was in a Robin B that Douglas Corrigan flew from New York to Ireland in 1938, instead of New York-Los Angeles as flight-planned, thereby earning for himself the nickname, 'Wrong-Way'.

The Robin purchased by Crowley and Stockden, VH-UJE (the registration was later changed to CSB) was one of the earliest constructed, in 1928. It was imported into Australia by S.L. Tyler of Sydney and fitted with the OX-5 engine. In 1939, the Robin was acquired by C.H. Fischer. During its long years based at Mascot, the little aeroplane was constantly flown, and it was a long way from its original state when permission was granted for its export to New Guinea. The engine was now a Gipsy Major, and the propeller was of wood, and also of de Havilland manufacture.

UJE was flown from Sydney to Lae in four days in March, 1950, with Crowley and Stockden alternating at the controls. The single remaining passenger seat was occupied by Ray Stockden's new bride. The Robin was soon flying hard. With a 145 kph cruise and a 227 kilogram payload, it was able to operate off the shortest airstrips in the Morobe District, which was still the principal area of Crowley and Stockden operations. The Robin was considered by DCA to be in the same category as the DH82, and John Arthur was advised: 'DH82 and Curtiss Robin aircraft will not operate to PK charts in New Guinea. There are no rules beyond the operator's assessment as to whether he can get into and out from aerodromes.'

So carefully did the partners operate that their aircraft never had accidents or went missing.* But Crowley and Stockden were about to split up. In 1951, the firm was dissolved, and was de-registered as a limited liability company in New Guinea. Laurie Crowley formed a new company, Crowley Airways Limited. Ray Stockden had no interest of any kind in the new company. In March, 1952, Crowley applied for an extension of his charter licence to include an Avro Anson. He purchased his first Anson, VH-BAC, in November. Laurie Crowley was on his way.

Ray Stockden eventually joined Qantas. He retained his love of the Tiger Moth, and in May, 1952, purchased VH-ARQ from Laurie Crowley. He had earlier been one of a syndicate of three which had owned another Tiger, VH-BLX, for a brief period between October, 1948 and January, 1949, when the aeroplane overturned on landing at Slate Creek. The other syndicate members were John J. McCann and Lloyd Nesbit. BLX had originally been owned by pilot R. Devenish-Meares, who had come to New Guinea to fly for MAL, but who later joined DCA. After the Slate Creek accident, Meares bought back BLX, which was later acquired by George Larum, an administration officer of Lae.

In April, 1954, Ray Stockden crashed in VH-ARQ in the mountainous country between Lae and Finschhafen. Four ground parties and a number of aircraft were involved in the search for the missing aviator, which ended when Stockden arrived at Lae on board a Lutheran mission vessel. The engine of the Tiger had failed, and the aircraft had crashed into high trees. Miraculously, Ray Stockden was unhurt, and walked for fifty-six kilometres through the mountains to a mission station. ARQ, however, was a write-off.

A company that commenced operations soon after Crowley and Stockden and which was destined for a short life was Charter Aviation Service, also known as Charlier's Air Service. This company operated out of Lae, and for a time had an aircraft stationed in Popondetta. Another was Amphibious Airways, based at Rabaul. A rather fascinating and certainly enterprising figure was concerned with both companies: Eric E. McIlree. McIlree was the proprietor of Self Drive Cars Pty Ltd, Sydney; U-Drive Pty Ltd, Mudgee; and Aircraft Disposals Company, Mudgee, whose cable address was 'Loneragan, Mudgee'.

Eric McIlree was to achieve fame in the business world as the man who brought Avis rental cars to Australia. Long before this, he was heavily involved in the purchase of war-surplus aircraft from the Disposals Commission, and in aircraft operations. He specialised in Avro Ansons, and the Vickers Supermarine Walrus amphibian. By 1950, McIlree had purchased all Walrus aircraft then available, and had accumulated between sixteen and twenty Ansons.

In September, 1950, it was announced that a new air charter company would

*In 1951, the author flew with Laurie Crowley in the Robin on a long survey flight over the then little-known mountain country at the head of the Tauri River and towards the Kratke Range. This flight lasted for over two hours and the old Robin never missed a beat, although it was noisy and draughty, and vibrated to an extent that was worrying to the passenger, if not to the pilot.

commence operations out of Lae. The company was Charter Aviation Service, wbich took over the remaining Guinea Air Traders facilities there. The directors of the company were R.W.J. Charlier and his brothers, Roger and Ronald, and Robert Shute. Two Ansons, VH-BMN and BMO, were initially employed and a third, BMP, joined the fleet in March, 1951. The aircraft were supplied by Eric McIlree, who informed the Director-General of Civil Aviation that he was 'sponsoring' the formation of Charter Aviation Service. Pilots for the new company were Robert Shute, Frank Minjoy and Ernie Reid.

Reginald William John Charlier was aged thirty-one, and had flown in PNG with the RAAF during the war. Before coming to New Guinea, he had been a principal in Charterflite Aviation Service, based at Bankstown.

Charlier arrived in Port Moresby in November with his brothers and the rest of the staff of the new company in the two Ansons, bound for their headquarters at Lae. The departure date for Lae had to be postponed when weather conditions deteriorated, and Charlier asked for and obtained accommodation for his party at the DCA Aviat Mess.

There was an impromptu dance at the mess on the evening of 10 November, and during the evening an argument developed between Charlier and others, which ended in blows being struck. Charlier was hit on the head, and died before reaching hospital. A post-mortem examination revealed an abnormality in the blood vessels of his head; blows that would hardly have worried the average man proved fatal to Charlier.

Two men who were seen to strike Charlier, Robert Clancy and Victor Carter, were put on trial for unlawful killing. Because of the evidence of Charlier's abnormal condition, both men were convicted of assault causing bodily harm, and were each fined £200, in default a year in gaol.

It was an ominous beginning for Charter Aviation Service. As mentioned in the preceeding chapter, the company was active in the Mount Lamington relief operations. Bob Shute was actually the first pilot to land on Popondetta strip after the eruption, and he made scores of trips in his Anson into that strip, Embi and Kokoda during the emergency. On 8 February, 1951, he was taxying for take-off from Jacksons in VH-BMO, heavily loaded with emergency supplies for Kokoda, when his brakes failed. The Anson ran into an embankment and was damaged. Shute and his passengers escaped injury.

BMO was quickly repaired and back in the air again, but Charter Aviation Service was soon in financial difficulties. BMN was returned to Eric McIlree later in 1951. BMO was returned the following year, and the last of the trio, BMP, was sold to Papuan Air Transport* in February, 1953. Charter Aviation Service had ceased operating long before, and one of its pilots, E.J. Reid, announced in January, 1953, the formation of another company, Goilala Air Services, which would operate mainly between Tapini, in the Central District, and Port Moresby. Reid announced that his company had acquired three Ansons. The first was already in PNG, and others would follow. He said: 'Our two- engined Anson is fully radio equipped, and fitted with radio compass. We believe healthy competition will be achieved by small companies giving better service to the public.'

The first Anson was VH-BDD. It had been purchased by Eric McIlree from the Commonwealth Disposals Commission in December, 1946, and sold by him to Theiss Brothers Pty Ltd, Brisbane. Like many of the Ansons sold by McIlree, BDD had been overhauled and prepared as a feeder airliner, with modified engine cowls, Avro electrical installations, and comfortable cabin trimmings. It was registered in the name of E.F. Reid on 12 December, 1952, and departed for Port Moresby the following day, with Ernie Reid at the controls. A second Anson, AVY, also supplied by McIlree, arrived in January, 1953.

Although BDD was registered in Ernie Reid's name, the actual owner was Dr Alexander L. Casselberg, a retired doctor who had settled with his wife at Tapini, where the cool climate favoured his health. Casselberg became managing director of the new company; Reid acted as manager and chief pilot. The second pilot was R.K. Pilling, who for the previous fourteen months had flown for Gibbes Sepik Airways. Ernie Reid was a licenced engineer, and was responsible for overall maintenance of the Ansons.

Reid applied for aerial work and charter licences, and submitted operations and maintenance manuals, plus details of a loading system, with his application, which was received by John Arthur on 28 January, 1953. Everything looked to be in order,

*See Chapter 4, Patair

and Arthur recommended the granting of the licences, and gave Reid a temporary authority to operate pending a decision by the Director- General.

While the application was still being considered in Melbourne, disaster struck. At 6.55 a.m. on 4 February, Ron Pilling left Popondetta in VH-BDD bound for Port Moresby. On board were two girls, Leani and Erica Healy, daughters of the ADO Popondetta, Clarrie Healy, and nine Papuans. Pilling was not familiar with the route, and when dense cloud forced him to turn away from the Kokoda Gap, he decided to return to Popondetta, but he was soon lost. He called the Port Moresby Aeradio, and described the country over which he was flying. At 10.52 a.m. Pilling called again, and advised that he was flying at 762 metres, was short of fuel, and was about to land on a patch of *kunai* grass.

At Port Moresby, Tom Drury — at the time, regional safety officer — and other experienced DCA men pin-pointed on the map the probable location of the missing Anson, and diverted a Qantas DC3 en route from Lae to Moresby to the area to search. The Anson was found ten minutes after Pilling had put it down. That afternoon, food, clothing, camping equipment and firearms were airdropped to the stranded party, all of whom were unhurt.

Pilling had forced-landed at Embessa, near an old wartime airstrip. On the following morning, Clarrie Healy led a rescue party to Embessa, and in due course they all paddled down the Musa River in canoes, to an Administration trawler that took them back to Popondetta.

The affair was widely reported, and DCA immediately suspended consideration of the licence applications pending a full investigation.

This was carried out by John Arthur and Tom Drury, and proved highly damaging to Goilala Air Services and Ron Pilling. The finding was that the forced-landing was caused by 'a shortage of fuel resulting from the pilot getting lost due to poor navigational technique and his failure to comply with VFR requirements'. The pilot had allowed the aircraft to be overloaded, had failed to carry the necessary maps, had filed a flight plan which did not show the correct amount of fuel carried, had flown in IFR flight conditions without an instrument rating, had been in command when not familiarised over the route, and had carried out pre-flight inspections of BDD without being properly authorised.

Ron Pilling was thirty at the time of the incident. An ex-RAAF pilot of limited experience, he had obtained a commercial licence in 1947. This lapsed, and was revalidated in 1951. He joined GSA in November of that year, and spent most of his time as co-pilot to Captain Francis Wiza, flying a Lockheed Lodestar, VH-FAC, that Bobby Gibbes had unwisely leased from Fawcett Aviation Pty Ltd of Bankstown, in anticipation of the granting of a licence to operate a Sydney-Port Moresby service — a licence which was, predictably, refused by DCA. The Lodestar was an embarrassment to Bobby Gibbes from the beginning of its New Guinea service, in December 1950, to its ignominious end in November, 1952, when it was grounded by DCA because of the poor condition of the airframe. During his time in PNG, Pilling accumulated 140 flying hours, mostly as second pilot. His total hours were only 729. The Embessa flight was his first in command of a charter aircraft. His licence was suspended for six months following the inquiry.

Ernie Reid, too, came under fire. He was another ex-RAAF pilot, with 420 flying hours. In 1945 he obtained his commercial licence, which was suspended in 1948 following two accidents which showed 'a lack of responsible airmanship'. He received his Aircraft Maintenance Engineer's Licence in Divisions C and D in 1947, and this, too, was suspended in 1950 after an incident which revealed a lack of maintenance supervision. The inquiry found that Reid had modified the seating of the Anson without DCA approval, and had improperly endorsed Pilling on the Anson.

Eric McIlree figured in the inquiry. He had earlier formed another company, Amphibious Airways — whose operations will soon be described — and John Arthur was fed up with him, and all his ways. Reid had been associated with McIlree in the formation of Amphibious Airways, and Charter Aviation Service. Arthur wrote: 'Mr Reid has exhibited all the hallmarks of the highly unsatisfactory type of operations conducted by Mr McIlree and his aviation organisations in this Region...'

Reid's own commercial licence was suspended for six months, and his engineering licence was cancelled. The interim authority for Goilala Air Services to operate was suspended, pending the re-organisation of the company. Dr Casselberg began to look around for another pilot and engineer, and asked DCA for approval for Ernie Reid to continue as manager.

The suspension was lifted in April, and Goilala Air Services continued to operate, at least for a time. Amphibious Airways, however, failed to survive.

Amphibious Airways was Eric McIlree's own particular baby. On 6 June, 1950, he lodged an application for a charter licence with DCA Melbourne, to operate 'approximately twenty' modified Avro Ansons on interstate flights within Australia, and to PNG. He estimated that his aircraft would operate for 5,000 flying hours per annum. The aircraft would be equipped with Marconi TR1196 radio and Gibson Girl emergency transmitters. They would be maintained by Marshall Airways, Bankstown; Airflite; and J.S. MacConnell of Camden, New South Wales. Two supervising engineers would be employed.

To handle the PNG end of his proposed operations, McIlree stated that he would purchase the hangar, workshop and equipment from the now-defunct Guinea Air Traders at Lae. The original application said that the PNG organisation would be called 'Papuan Airways Ltd', but in September the name was changed to Amphibious Airways Limited.

A considerable amount of correspondence between McIlree and DCA Melbourne and Port Moresby followed. McIlree advised that he planned to use Walrus amphibians in conjunction with Ansons — which had modified wingspars and were authorised to operate in Australia at 3990 kilograms — to provide a service to outlying plantations in the New Britain, New Ireland and Bougainville Districts.

A charter licence was issued to Eric McIlree in December, 1950, which allowed him to operate his Ansons within Australia subject to conditions, but which restricted their use in PNG. Passenger operations over the water, out of gliding distance from the land, was not permitted. This would render their use on such routes as Lae or Madang to Rabaul, and to Bougainville, impossible. McIlree protested, pointing out that his Ansons all had modified wingspars, and had been streamlined by the removal of astro-domes and gun blisters, the faring of bomb-bay doors, the improved faring of windows and the fitting of Cheetah 25 improved-type engine cowls. He wrote: 'In view of the foregoing, it is requested that at the weight of 8,200 pounds [3720 kilograms], permission to carry passengers over water be given. This company has invested a great deal of money in an effort to bring about a speeding up of native labour from the Highlands of New Guinea to the plantations in New Britain by the combined use of the improved Anson together with the civilian Walrus. This is an enterprise of a highly hazardous nature from a financial point of view, and yet one which has a very tangible benefit for the areas it proposes to serve...'

An argument also developed over the use of the Walrus in New Guinea. McIlree pointed out that he had five of these ex-naval amphibians, 'together with a great portion of the residual stocks for this type of aircraft, formerly the property of the Royal Australian Fleet Air Arm. These stocks of spare engines, airframes, engine and airframe component parts, overhaul kits, tools, jigs, and manuals have been acquired over the past four years. It is the possession of this considerable supply of equipment which gives to the founders of Amphibious Airways the courage to tackle the remaining difficulties.'

McIlree requested permission to operate the Walrus at an all-up weight not exceeding 3833 kilograms for both land and sea conditions, because of 'the exceptional strength of the aircraft as a whole and its phenomenal take-off performance due to the fitment of the Pegasus Mark VI engine which gives 775 hp available for take-off.' He asked for approval to make water crossings for destinations up to 161 kilometres, to fly the aircraft with a single pilot, and to extend the period between complete engine overhauls by an additional 250 hours.

Approval was given to use the Walrus over sea up to a distance of about ninety kilometres from the nearest land, subject to the carriage of specified safety equipment, including a self-powered radio transmitter. No flight where the aircraft was likely to be more than thirty-two kilometres from the land could be commenced until flight details had been provided to the nearest Air Traffic Control centre, which had authority to veto the flight should weather conditions be such as to make it unduly hazardous. Stuart Campbell advised McIlree that this was because 'Search and Rescue facilities in New Guinea waters are far below the standard which we would desire'. Maximum all-up weight of the Walrus was restricted to 3560 kilograms.

In view of the conditions imposed on the use of the Avro Anson, McIlree decided against their use in PNG by Amphibious Airways until the question of their operation was resolved to his satisfaction. He would establish his company using only Walrus amphibians. He discussed his plans with John Arthur, who wrote to the Director-General: 'Mr McIlree has informed the writer that in an endeavour to obtain some return from the capital he has invested, he decided to favour the New Guinea Region with the unwelcome presence of Avro Anson and Walrus type aircraft. He states that he sponsored the formation of Charter Aviation Services Ltd, who will operate the Ansons in New Guinea. I understand from McIlree that he formed Amphibious Airways Ltd...for the purpose of operating Walrus amphibians as a financial experiment...'

Amphibious Airways Ltd was registered at Port Moresby, with a nominal capital of £20,000. McIlree left Sydney for PNG at the end of March, 1951, with two Walruses, VH-BGP and BLD. He flew one, and pilot C.N. Bell the other. The aircraft arrived at Daru on the night of 1 April, after radio communications had been shut down. When they failed to arrive at Port Moresby at the time advised, DCA began emergency procedures. The amphibians arrived the following day.

John Arthur was not amused. McIlree had neglected to include any licenced engineers in his party, which meant that the aircraft could only be used for private flights until engineers were engaged. The two amphibians flew on to Rabaul, which would be the company's base. McIlree announced to the Press that the aeroplanes would be used 'to survey territory for a new airline'.

On 14 April, McIlree took off from Rabaul in VH-BGP on a private survey flight to Kavieng and return, via Namatanai. With him were Bell, as co-pilot, Rabaul businessman J.L. Chipper, who was considering an investment in Amphibious Airways, and three others. It was McIlree's first flight since his arrival in Rabaul.

After landing on Kalili harbour, McIlree took off for Patlangat Plantation, where he landed, at Logon Bay. After forty-five minutes on the water, McIlree prepared to depart. He made an attempt to take off in the open sea, but encountered rough water before he reached flying speed, and broke it off. He taxied further back, and commenced his second take-off run. He had reached a speed of 64 kph when the amphibian hit a slight swell, and bounced into the air. It fell back, and a bigger swell threw the Walrus six metres up, but flying speed had not been attained and back on the water it crashed, several times. The starboard float was torn off, and the fuselage was heavily damaged. The aeroplane was in danger of sinking, and McIlree asked two of the passengers to stand on the port wing to keep the amphibian in balance while he slowly taxied to a sheltered bay one-and-a-half kilometres away, where he beached it.

The Air Accident Investigation that followed disclosed a number of breaches of the regulations by McIlree. He had departed Rabaul without submitting a flight plan or obtaining ATC clearance, and had in fact informed the duty aeradio operator that he was 'not in the habit of making out such documents'. The investigating officers, Tom Drury and Rube Murray, 'experienced difficulty in obtaining the cooperation' of McIlree. The Certificate of Safety covering the Walrus was invalid. McIlree had been involved in nine reported air safety incidents since 1946. During the investigation, McIlree estimated his total flying hours as being 'in the vicinity of 3,000 as pilot-in-command', including 180 hours on the Walrus. His claims were politely doubted by the investigating officers, and he was asked to submit his logbooks to support his claim.

The inevitable verdict was pilot error, and John Arthur recommended that McIlree be prosecuted. The Director-General, however, let McIlree off with a stern letter of warning, and he was permitted to continue the operations of Amphibious Airways.

They were not to continue for long. BGP was severely damaged during the Patlangat incident, and was shipped to Lae for possible repair. There it lay for many months, before going back into service. In January, 1952, a severe storm at Rabaul blew the Walrus into a Qantas DC3, which was slightly damaged. The repairs to BGP were estimated to amount to £3,000. She was again shipped to Lae, where on 21 March, 1952, her Certificate of Airworthiness expired. She was advertised for sale 'to the highest bidder', but there were no takers. The second Walrus, BLD, was returned to Sydney when her C of A was cancelled in December, 1951, 'due to owners inability to maintain the aircraft in accordance with the department's requirements'.

The Walrus did some good work in the New Britain-New Ireland region while it lasted. The ability of the type to land on water, then taxi up on any firm sloping beach won a lot of support from plantations and outstations. According to a report in *Pacific Islands Monthly* the sophisticated locals of Buka Island were 'rocked' at their first sight of the ugly, lumbering Walrus. 'When they saw the amphibian settle on the water, shake itself, put down its wheels, and solemnly clamber ashore, they ran away, howling, and yelled, "Go to 'ell 'ere — this plurry airplane *savi tumas*!" ' But the Walrus was a military design, produced for work in cold climates, prodigal of fuel and difficult to maintain. As APC and Qantas had already discovered, it was not the aeroplane for New Guinea.

Amphibious Airways owned several Avro Ansons, but only one appears to have been actually employed. This was described as an 'Avro Feeder, licenced to carry sixteen passengers, with a range of one thousand miles [1609 kilometres]'.

The troubles of the company increased. A well-qualified aircraft engineer resigned in December, 1951, after only fourteen days with Amphibious Airways. He made an adverse report to DCA on the management, conduct of operations and general safety standards of the company. McIlree advised DCA that the operations of the company had been 'temporarily suspended' pending the arrival of replacement engineers. John Arthur wrote to the Director-General: 'Mr C.R. Hibbert is at present in Rabaul investigating unsatisfactory matters which have come to notice. If reports amounting to hearsay are confirmed, a recommendation will be made to cancel the charter licence in the interests of public safety...'

The licence was not cancelled, but Amphibious Airways was nevertheless finished. A DCA report noted that the company was 'in a chaotic condition, with rapid changes of all personnel including management and ownership'. The charter licence expired, and although McIlree asked for a renewal in August, 1952, this was opposed by John Arthur on the grounds that there had been no operations 'for the past twelve months'. The Director-General agreed, and the renewal was refused.

Eric McIlree went on to his outstandingly successful business career in Australia, where his imagination and drive won him wealth and standing. There were no regrets in the New Guinea region when he departed those shores!

The demise of the Amphibious Airways' Walrus amphibians almost corresponded with the retirement of the two Vickers Supermarine Sea Otters, VH-AJN and AJO, operated by Qantas for the owners, Australasian Petroleum Company. These were virtually modernised versions of the Walrus, but fitted with tractor engines. It will be recalled from Chapter 2, The Flying-Boats, that they came into operation early in 1950, as replacements for the little Grumman Widgeon amphibian originally flown by Phil Oakley. Both were purchased, second-hand, in England in September (AJN) and October (AJO) 1949. They were crated to Australia, assembled at the Qantas marine base, Rose Bay, and flown to Port Moresby. AJN was in service in March, 1950, and AJO in May.

The senior flying-boat captain in charge of operations out of Port Moresby when the Sea Otters came into service was Fred Fox. Although he mainly flew the Catalinas, Fox also piloted the Sea Otter on APC work from time to time. Few of the handful of pilots who flew the Sea Otter — Fox, Oakley, Ron Roberts, Ken Montagu and Struan Robertson mainly — liked the aeroplane, but Fred Fox did. He remembers: 'The Sea Otter — that was quite a machine. Typical of British naval design, very rugged, extremely solid in every way, access to engine components and refuelling pretty primitive. Engine itself was pretty good — a Bristol Mercury — but when you think of an aircraft weighing about 12,500 pounds [5670 kilograms] with an 850 hp engine and with a cruise of only about 70 knots, you get some idea of the drag. Like flying a barn door through the sky.

'With flaps down, it had a fantastic angle of glide. Coming over the top of Hospital Hill in Port Moresby, you'd cut back to idle, and the aircraft would touch down at the edge of the moorings; really down, in a steep flight path. If you were heading downwind and came back to idle at 1,000 feet [305 metres], you could just complete a 180-degree turn to the left in time to land on the water. It had no great climb performance when heavily laden, of course. Strong, though — I had interesting events when I was training people to fly on it. It required careful handling. It could land on three points quite happily, on the two main gear and the

104. From left: Bruce Braithwaite, Ron Roberts and Charles Wade. *Photo: Ron Roberts*

104

tailwheel, could do a nice tail-up wheels landing, but my God, you had to be on the ball, otherwise it'd bounce to high heaven. Ultimately, the amount of time we were spending on Otter training required Hugh Birch to come up, so we could get on with our other flying...'

The first pilot that Fred Fox endorsed on the Sea Otter was Phil Oakley. The aeroplane was AJN, the place was Jacksons, and the date was 23 April, 1950. Says Fox: 'It was hilarious, with a touch of drama. Phil put her down just a little too hard, she commenced to bounce, came down and we caught it before it touched, but it went off the runway into a ditch. The damage to the aircraft was just a slight cut along the keel — but it smashed a rock! Shows how strong it was...'

Phil Oakley remembers this incident: 'We were doing circuits and bumps as part of my endorsement training. The water operations had been relatively simple. The Sea Otter was a good rough-weather boat, but on the aerodrome it wasn't that hot. The braking system was the old air-brakes, with a single control on the column. You got differential braking by pushing on the right or left rudder. The narrow-track undercarriage, a tail wheel with the nose up in the air and a big keel surface meant that she was hard to keep straight. On this occasion, she dived off to the right, and it was just not possible to get enough braking and rudder on the left wheel to stop it running off the runway.

'But the Otter did a pretty good job. Served the same sort of areas as the Widgeon had done, but it carried a much bigger load. Our problem was serviceability. Finally, we just couldn't operate them. There were hideous ignition problems — the whole motor would cut out. Nobody locally could do much about it. The engines were always overheating. They sent an expert up who brought a whole heap of modifications for the carburettor, but we continued to have trouble...'

Faulty air brakes almost caused the loss of VH-AJO in September, 1951. The aeroplane was on the slipway in Port Moresby harbour with the engine running, in preparation for taxying down the ramp into the water. One of the cables controlling the brake valves let go, and the Sea Otter quietly rolled forward, went over the side of the ramp, and nosed down eight or ten feet [about 2.5 to 3 metres] onto rocks. AJO was hauled back onto the ramp, with the head of one rivet on the keel missing — but the rocks were badly damaged.

Ken Montagu had first-hand experience of the unpredictable nature of the Bristol Mercury engine fitted to the Sea Otter when he was converted onto the type by Fred Fox, in December, 1951.

Unlike the other pilots who flew the Sea Otter, Ken Montagu was not RAAF-trained. He spent five years in the Australian Army during the Second World War, and failed in several attempts to transfer into the Air Force. He obtained his private licence with the Royal Aero Club of New South Wales after the war, and later his commercial licence. His first job as a commercial pilot was with Trans Oceanic Airways (see later), flying Sunderland Mark II flying-boats from Rose Bay to such destinations as Hobart, Vila, Santos and Noumea. After twelve months with TOA, Montagu joined Qantas, in June, 1951. He was converted onto Catalinas by Hugh Birch, who was then Flight Captain Flying-Boats. At the same time, Phil Oakley — then a senior first officer — did his Catalina command check flying with Birch.

Ken Montagu began his service in PNG in August, 1951, as second officer to Captain Ross Treadgold on the Catalina service to Rabaul. He stayed on the Catalina, under the supervision of Fred Fox, until December of that year. Then Fox decided that Montagu was ready to take command of a Sea Otter, from Struan Robertson, 'a great lift for a relatively new and junior first officer,' as Montagu recalls today: 'The Sea Otter was not at all a suitable aeroplane for what we were using it for — much too heavy and much too costly and much too complicated. As for incidents — I had one, the only major one in the whole twelve months I flew the aeroplane.

'It happened when I was training with Fred Fox, out of Fairfax Harbour, Port Moresby. As you are aware, you practice forced landings in single-engined aircraft — obviously, the reason for this is that in the event of a complete engine failure there is nowhere to go, but down. Fred would close the throttle and announce that I had an engine failure at various positions in and around the Moresby area, and I then would have to get it down safely...anyway, I had completed two forced landings this day satisfactorily, and was proceeding around the circuit for an ordinary landing when I had a complete engine failure...I then proceeded to force-land the aircraft on the harbour. Having had a couple of practice shots, it

105. Ken Smith carrying out repairs at Samarai while Ron Roberts waits to lend a hand. *Photo: Ron Roberts*

105

wasn't any great problem...'

Says Fred Fox: 'The Bristol Mercury had hollow exhaust valves filled with sodium crystals. When these became hot they became liquid, and helped to cool the exhaust valves. Had a habit of breaking. That's what happened to Ken and I. Valve stem had broken, valve head dropped inside the cylinder and hammered the piston to pieces. I have the piston now, as a doorstop to my front door...

'APC called all sorts of odd jobs for the Sea Otter. Once I went up the Purari, the geologists wanted to have a close look at a cliff-face along the edge of the river. Long flight, three hours...refuelled on the Kikori River. I didn't care much for refuelling. You had to pull down a ladder from the end of the nacelle, open a hatch and climb up behind the engine, open another hatch on top and climb out on top of the wing, which was generally pretty slippery with oil. Not a popular thing to do en route, and we did it on most flights...'

The APC empire in Papua expanded with incredible speed in the 1950s, and reliable air transport was of paramount importance to the company. The serviceability record of the Sea Otter was so appalling that it was privately considered a blessing for all concerned when VH-AJO sank at its moorings in Port Moresby harbour on 21 January, 1952.

The Port Moresby district had just emerged from one of the worst dry seasons in living memory, and then was hit by gale-force winds and driving rain, generated by a cyclone off the coast of the Northern Territory. The storm- battered town came through the ordeal with only minor damage, but AJO was unable to ride out the tempest. The force of the wind capsized her, and she filled and sank. Her mooring-lights stayed on for some time, shining through the clear harbour waters. Qantas station engineer, Les Weekes, was told of the sinking by telephone late that night, and he ran out onto his verandah, which overlooked the harbour. He saw the lights of the sunken Sea Otter, and told his caller, 'No, she's OK — I can see the lights!'

AJO was beyond salvage — or perhaps APC was happy to leave her there, at the bottom of the harbour. There were dark rumours that perhaps some human agency had assisted the welcome work of the storm! Certainly, the pilots were less than upset at the loss of the Sea Otter. Said a report in *Pacific Islands Monthly*: 'They, and their engineers, have long regarded it as a bomb. They recorded their feelings on the mess notice board: "The Sea Otter passed away peacefully at its moorings this morning. R.I.P." '

Unreliable or not, the sinking of the Sea Otter came at the worst possible time for APC, for AJN had been out of action for the past four months, awaiting spares from England. In the meantime, the company was now forced to rely on sea transport, and the occasional use of a Qantas Catalina. The fate of the surviving Sea Otter was sealed. Ken Montagu tells the story of its last flight: 'The Otter was taken out of service on 21 June, 1952. Apart from the frequency with which it would go unserviceable for all kinds of reasons from radio to little bits and pieces, nuts and bolts, it was always a problem if the spares were not on hand — they had to be shipped or flown out from England — very time consuming, and very expensive. Then I noticed a crack developing in the root of one airscrew blade. These were wartime airscrews, and only available in England from the Rotol company. There was much technical discussion, and many engineering brains viewed this thing and finally it was decided a new airscrew would be required.

'APC had just about had the aircraft by this time, and they said it wasn't worth the effort, the cost would be prohibitive, so my last flight in AJN was to Fairfax Harbour from Jacksons. The aircraft was towed in to Badili, and subsequently broken up on the APC lot...'

The death of the Sea Otter coincided with the final failure of another air company serving PNG, and also using waterplanes: Trans Oceanic Airways Pty Ltd.

TOA was formed in 1947 by a group led by ex-Squadron Leader Bryan W. Monkton, a highly-experienced pre-war light aircraft pilot, who did a great deal of combat flying in Catalinas during the Pacific conflict. The original business manager of the company was Douglas Lindsay. In April, 1947, Lindsay announced that TOA would employ five converted Sunderland Mark III flying-boats, crewed by widely experienced ex-RAAF pilots, within and without Australia on a charter basis. The Sunderlands would be fitted with two-way radio and advanced navigation aids, and with seating accommodation for twenty passengers. Steamships Trading Company was appointed PNG agent for the new company.

106. Refuelling a Sandringham flying-boat. *Photo: Ron Roberts*

106

Over the next two years, TOA developed services to Lord Howe Island, New Caledonia, New Hebrides and the Solomons, and pioneered a coastal service from Sydney to Grafton, and a direct service to Hobart. These services were operated under charter licences, as TOA was refused airline licences 'on grounds of government policy'.

Early in 1950, Bryan Monkton applied for an airline licence to operate a once-weekly service between Sydney and Port Moresby, in direct competition with Qantas. Captain P.G. Taylor was by this time associated with Monkton. TOA's plans proposed the use of three Sunderland Mark II boats with seating for thirty passengers, and a cargo and mail capacity of 1814 kilograms. Monkton stated that six spare engines would be held in serviceable condition. TOA had a total of forty-two engines in store, and employed thirty-four full-time aircraft maintenance personnel. The company's flying staff consisted of three captains, three first officers, four flight engineers, two radio operators, and four flight stewards and hostesses.

The TOA application was refused. It was held by DCA that the service that Qantas was now providing on the Bird of Paradise route with Douglas DC4 and DC3 aircraft was adequate for the business offering.

In his comments to the Minister on the application, the Director-General, Air Marshall Williams, noted that it raised the question of policy as to whether the future use of flying-boats should be officially encouraged. Flying-boats — any seaplanes — were in general much more expensive to operate than landplanes of similar capacity. Salt water corrosion meant that maintenance costs were high. Both Pan American Airways and the British Overseas Airways Corporation had ceased the use of the flying-boats they had operated so extensively on their international routes before the war, and had gone over completely to landplanes. Within Australia, the flying-boat base at Darwin had been reduced to facilities sufficient to handle the RAAF Catalina Air-Sea Rescue Flight, the base at Bowen had been shut down, and Rose Bay was being maintained only because of the flying-boat service to New Zealand. A base was maintained on the Brisbane River to meet the needs of Barrier Reef Airways, but the Queensland Government maintained that this was interfering with the movement of shipping, and was pressing for its removal. There was no flying-boat base north of Brisbane, and one would have to be established probably at Townsville if TOA were allowed to operate their service. Since an airways system for the operation of landplanes already existed along the route, it would be an unnecessary, and heavy, expense to maintain additional flying-boat bases.

This argument was accepted by the Minister, T.W. White, and it is really surprising that TOA *was* later granted a licence, in the face of opposition from Qantas, and the DCA attitude. But there were now some substantial backers for TOA. Several shipping companies, and in particular Clan Steamship Line, had invested considerable capital in the company, and the Townsville Chamber of Commerce strongly supported the new service.

Following an interview in September, 1950, between the Minister, the Director-General, P.G. Taylor and Bryan Monkton, approval was given for the Sydney-Moresby service. TOA announced that a once-weekly service with Hythe flying-boats would commence in mid-October, leaving Rose Bay every Sunday morning for Port Moresby with calls at Brisbane and Townsville.

A survey flight over the new route was made on 29 October, and Townsville Harbour was used as an alighting-place, with no problems. The new service commenced immediately. The Hythe boat, *Pacific Star* captained by P.H. Mathieson, made one return flight, but then came another check: DCA ruled that TOA flying-boats could not land at Townsville harbour when there was more than a 155-millimetre swell there — a common occurrence. TOA was stopped dead, and announced that the service was cancelled until further notice.

Qantas had operated the C Class Empire boats through Townsville for some years before the war, but had never considered it a satisfactory base, and just after the war had transferred to Bowen on grounds of safety. The RAAF also considered the harbour unsafe for large flying-boats. It was possible to land outside the harbour in favourable conditions, but in the event of a swell this could be dangerous. During the war, the RAAF lost an Empire boat and a Catalina outside Townsville, with heavy loss of life, and had refused to operate Catalinas there for the rest of the war. The manufacturers of the Hythe and Solent flying- boats, Short Bros, themselves considered a 155-millimetre swell the outside limit for safe operation. Although the Townsville Chamber of Commerce strongly protested, the DCA decision stood.

Bryan Monkton and his associates had now to reconsider their whole approach. The Hythe could not carry sufficient fuel to fly Brisbane-Port Moresby non-stop with a payload, but the Solent could. Monkton decided to replace the Hythe with the Solent III. The Solent was the most modern flying-boat available in the world. It was a civil version of the Short Seaford, equipped with four great Bristol Hercules engines each of 1,680 hp and with accommodation for thirty- nine passengers on two decks. Monkton went to England to take delivery of the first Solent Mark III and planned to commence the return ferry-flight on 26 January, 1951. It was hoped to commence the long-planned New Guinea service on 18 February. To provide for travel beyond Port Moresby, TOA entered into an agreement with Mandated Airlines. It looked as though Qantas would soon be facing some real competition.

Alas for Bryan Monkton and his hopes! On 29 January, the new Solent crashed just south of Malta on the flight back to Australia, shortly after take-off. There were thirty-nine passengers on board, French building workers going to a special construction job. All were rescued by an RAF crash launch, except for one man, who was presumed drowned.

There seemed to be a jinx on the operations of TOA. This latest disaster would have discouraged most men, but not Bryan Monkton. He returned immediately with his crew to the United Kingdom, where Short Bros were converting another Solent to TOA specifications. He found that work on the new flying-boat had stopped, because of an industrial dispute! However, the dispute was settled, and Short Bros advised that the Solent would be ready for delivery on 28 March. It was now planned to commence the Sydney-Port Moresby service on 28 April.

The new Solent, bearing the British registration, G-AKNP, and christened *Star of Papua*, reached Sydney in late April, 1951. She was later re- registered as VH-TOB. The inaugural flight to Port Moresby was made on 27 May, with Bryan Monkton at the controls. Advertisements in both the Australian and local press claimed that this new 'Chieftain' service offered 'the last word in luxury travel'. The *Star of Papua* boasted a cocktail bar, a scenic lounge and promenade deck, an electric grill-room, writing tables, a library, electric razors for male passengers, and dressing-tables for the ladies, and a flight steward and hostess. These were luxuries that could not be matched by the Qantas DC4's — or, indeed, any of the landplanes then flying the international routes.

The inaugural flight was attended by the kind of public relations ballyhoo common today, but rare then. Four beautiful models from Sydney were among the invited guests. It was reported that 'they were the target for both movie and still cameras when, clad in floral summer frocks, topped by beach straws, they were surrounded by a curious and friendly crowd of grass-skirted native girls at Koki'. Free goodwill flights were given to 'prominent residents of Moresby and their womenfolk'. It was in all a great show, and the new service was warmly welcomed by the *South Pacific Post*, the newspaper which had begun publication on 20 September, 1950. Said the *Post*: 'Links with Australia were strengthened this week by the entry of Trans Oceanic Airways into the Australia-Papua air service. It also marked the beginning of a fight for air passengers and freight which, in addition to providing additional services, already has resulted in lower fares...'

As if the attractions of its spacious new luxury flying-boat were not enough, TOA had lopped ten per cent off the Qantas fare on the Bird of Paradise service. The return fare Sydney-Moresby was set at £70/13/-. Qantas immediately reduced its fares to match those offered by TOA, but Hudson Fysh complained to the Director-General. Qantas had been put into 'a most embarrassing and damaging position by the fact that TOA commenced operations by undercutting the DCA-approved Q.E.A. fares and thus forcing us to lower our fares, which must result in great support for the new company in New Guinea'. He pointed out that it was the practice of 'all countries' to accredit an operator to a route, and only licence a competitor if a careful investigation revealed that the original operator was not providing an adequate service. Fysh continued: 'I believe that no such detailed consideration was given before this additional operator was granted permission to operate to the Territory of Papua, and now there is talk of this new operator increasing frequencies and extending into new Territories where, undoubtedly, the greatest embarrassment will be caused to your accredited operator, Qantas

107. Engine change for a Crowley Airways Curtiss Robin; note the elastic band shock absorbers. *Photo: Laurie Crowley*

107

Empire Airways...'

TOA had announced that it would increase its service from one return flight a week to two, from 20 June. Bryan Monkton was soon to depart for England to take delivery of the second of the three Solents TOA were purchasing.

Air Marshall Williams did not accept Hudson Fysh's arguments. He pointed out that world practices in regard to the authorisation of airline operators varied greatly. Moreover, it was Commonwealth Government policy not to favour a private, or government, monopoly of air transport, which implied 'some degree of competition'. The TOA fare structure had been set at that of Qantas when it was first planning its New Guinea service. During the delay in commencing the service, Qantas had increased its fares, without advising TOA.

The hoodoo that had dogged the efforts of TOA, however, was still active. On 18 June, *Star of Papua*, south-bound, struck a tugboat on Hamilton Reach in the Brisbane River. The captain, P.G. Taylor — vice-chairman of TOA — was taxying for take-off. The flying-boat collided with the tugboat, and *Star of Papua* then ran aground. The hull was holed forward, and the flying-boat was out of action.

The passengers were sent to Sydney on an ANA aircraft, and TOA had once more to assess its position. It was decided to carry on, using Sandringham boats from its other Pacific operations, while *Star of Papua* was being repaired, and pending the arrival of the second Solent, G-AHIV (VH-TOC).

In retrospect, it seems that the Brisbane River accident was the beginning of the end for TOA. The Sandringham, while outwardly similar, was slower, noisier and less capacious than the Solent, and less popular with passengers. It was a makeshift service, and passengers realised this. Passenger numbers dropped alarmingly. The second Solent, *Star of Hobart* arrived in August, and TOA announced that a twice-weekly service would begin from 5 September. On 28 October, *Star of Hobart* hit a dredge at Colmslie on the Brisbane River just after taking off for Port Moresby! The flying-boat was about six metres airborne when the collision occurred. The starboard wing hit the superstructure of the dredge, and over four metres of the wing, and the float, were torn off. The pilot, P.H. Mathieson, immediately cut the motors, landed, and ran the flying-boat into a mud bank. Nobody was hurt, but the damage to *Star of Hobart* was severe.

The third Solent, G-AHIO (VH-TOD) was delivered in December, 1951, but the company was virtually finished. On 15 April, 1952, TOA announced the cessation of the Australia-Papua service. Increasing costs had made its continuance impossible. The company went into voluntary liquidation. Ansett Airways purchased the last of the assets of TOA, and continued with the Lord Howe Island service.

A combination of extraordinarily bad luck, rising costs and determined competition from Qantas beat TOA. Rather amazingly, at the very moment TOA was expiring, Qantas was itself planning the introduction of Sandringham flying- boats on the Bird of Paradise service, and within PNG.

The Catalinas had done a great job for Qantas within PNG, but their payload was relatively limited, and they were becoming very costly to maintain. The company decided to explore the possibility of replacing the 'Cats' with the much larger Sandringhams. On 30 March, 1952, a Sandringham landed at Port Moresby, under the command of Captain J.W. Solly, to begin a survey to test the suitability of the aircraft over the routes flown by the Catalina. The Qantas area manager for PNG, who was now Captain Orm Denny, and the assistant area manager, Captain E.C. Sims, were on hand to welcome Solly and his crew. Over the following fortnight, the Sandringham made a series of flights over the Catalina routes, which were considered to be successful. Outstation people in general welcomed the possible introduction of Sandringham services, for, reliable as the Catalina undoubtedly was, it could not match the bigger aeroplane in speed, comfort and capacity.

On 9 May, the general manager of Qantas, C.O. Turner, wrote to the Director-General: 'The tests proved the Sandringham is quite satisfactory from the operational standpoint, plus the ability to carry greater loads...we propose to substitute the Sandringham for the Catalina.

'The Sandringham is to be based in Sydney, and despatched fortnightly to New Guinea from Sydney via Brisbane, and thence operate from Port Moresby. As each group of services are completed in twelve days, the Sandringham is to return to Sydney, completing a total of sixty-two flying hours in that period...the services will be the same as the Catalina, except for an additional extension from Samarai to Rabaul...we propose to despatch the first Sandringham from Sydney on Saturday, 31 May...'

Qantas had four Sandringhams based at Rose Bay — VH-EBV, EBY, EBX and EBZ. A fifth had been lost in the New Hebrides in 1951. The first two were Mark V Sandringhams and the others, Mark IV. The substitution of the Sandringham for the Catalina made financial sense to Qantas, and while in PNG, Captain Solly instructed the Catalina pilots in the handling of the big flying-boats. However, it was planned to transfer these pilots to Sydney, where the Sandringhams would be based. The Qantas station manager, Port Moresby, J. Samuel, announced that the Catalinas had had their day, and would be sold by public tender.

But PNG had not yet seen the last of the Catalina. Both the Administration and the Australasian Petroleum Company were considerably disturbed at the prospect of losing the 'Cats'. Both depended on the Catalina for servicing isolated outposts and field parties. Qantas had announced that the Sandringham would be unable to service Abau, Inus and the Trobriand Islands. The Sandringham could not be used for airdropping, and with the retirement of the Sea Otters, APC was in a particularly vulnerable position. Finally, Qantas agreed to leave one of the two Catalinas at Port Moresby, primarily to work for the Administration and APC.

On 18 August, 1952, Captain Fred Fox took off from Port Moresby harbour in VH- EBU on his way south. With him were First Officer Marsh Burgess and Flight Engineer Tom Mitchell. With the departure of Fox, Phil Oakley took over as senior captain, Port Moresby. One Catalina remained, VH-EBC. In January, 1953, EBC was flown to Sydney for overhaul, and was temporarily replaced by VH-EBD.

Fred Fox had proved the unique worth of the Catalina in New Guinea service on many occasions. In April, 1951, he had successfully airdropped supplies to an Administration patrol led by ADO Ted Hicks and PO Ken Chester at Lake Tebera, 113 kilometres north-west of Kikori. Lake Tebera is cupped in a hollow between 305-metre-high ridges, enclosing an area about eight kilometres long by two- and-a-half wide. The only possible drop-site was a tiny island in the middle of the lake, 183 metres long by fifty-five wide. It would have been impossible to even attempt an airdrop into such a confined area in the ponderous Sandringham, but Fox managed it — only just — in the Catalina, and with the Administrator, Colonel Murray, as an interested and apprehensive passenger. Fred Fox remembers: 'We couldn't land on the lake — which was actually more like a swamp — so we had to drop. It was like an elongated crater, just enough room between the mountains to turn the aircraft. Had to fly a kidney-shaped pattern. It was a tight kind of dropping area. Colonel Murray came along for the ride, and I put him in the right-hand seat, while the first officer went down to watch the cargo over the side.

'We made the first drop, and I banked to starboard and flew over towards the mountainside, then rolled the "Cat" into the start of the 180-degree turn. Poor old Colonel Murray shrank back in his seat. I think he thought we were going into the trees. The old "Cat" was pretty responsive, though, and it was quite safe, provided you knew what you were going to do before you started...it was a good drop, 95 per cent recovery. One of the more precise drops I ever had to do in PNG...'

The Catalina made sixteen runs over the drop-site in forty-five minutes of flying, and not one of the bags and bales missed the little island. Fred Fox did many similar jobs for the Administration, and on one occasion, accompanied by ADO David Fenbury, located the village of the mysterious Mokolkol people in the mountainous hinterland of New Britain. Fenbury subsequently led a patrol to contact this tiny band of fabled warriors. Just weeks before he left PNG, Fox made a successful drop of live poultry to another patrol in the Southern Highlands, led by ADO Syd Smith and PO Des Clancy.

Smith and Clancy and their patrol were constructing an airstrip at Tari, in the heart of one of the last remaining uncontrolled and partly-explored regions of Papua. Such work would have been impossible without the support of aeroplanes. The two officers wanted to establish a poultry run as a supply of fresh meat in this wild, remote country. It was just about impossible to carry in live poultry; an aerial delivery was the only way — if it was feasible.

There was a poultry run at Lake Kutubu, lorded over by a particularly cantankerous old rooster. Said Smith to Fox, per radio, 'Take him up and chuck him out — if he can't fly, he's dinner, if he does we'll know it's possible to drop poultry.'

Fred Fox obliged: 'Problem was to get it out of the aircraft into the slipstream without stripping its wings. Chooks are stupid birds, not like pigeons. If you drop a pigeon, he knows he's going too fast, so he keeps his wings tucked in until he slows down. So I decided to hold his wings in, wrap him loosely in two sheets of the *Sydney Morning Herald,* and stick his head in a sickbag. By the time the paper all blew off in the slipstream he should have slowed down to the point where he could fly.

'So we wrapped him up, took off, circled over the station at about 100 feet [about thirty metres], and dropped him. Sure enough, he unwrapped and fluttered down, probably the world's first flying rooster. A few weeks later we dropped a dozen or so, four at a time — no problems.

'Also dropped a big dry-cell battery pack for their radio at the Tari base. We wrapped it up well, and dropped it at the base of long *kunai* grass, which slowed it down until it hit earth. Heard after there was only one crack in the outer casing of the pack.'

Of no lesser importance was the work the Catalina did for APC, particularly to parties in the field: 'We were often engaged in airdrops of supplies to oil search people, making magnetometer surveys way out in the bush. This required the dropping of dynamite in storepedoes. We'd launch these through the tunnel hatch and the blisters. It wasn't the weight that was the limitation, but the bulk. Capacity was the problem. This was low altitude flying. They'd clear an area about fifty yards square [nearly forty-two metres square] in the forest. At first we dropped from the tunnel hatch, then we tried to reduce the flying time — and APC were paying on time — by dropping storepedoes two at a time, from the blisters. We got pretty adept at this. The storepedo had a parachute, and you could drop it much closer to the site than free packs, which had a trajectory...'

After Fox left, Phil Oakley and Ken Montagu did a lot of work for APC with the surviving Catalina. Oakley recalls: 'We did a lot of APC charters, particularly to places in the delta, wherever they opened a new drilling site. We'd indicate an alighting area we could use, and we'd have them make up moorings for us — we'd become fairly sophisticated. Made them out of forty-four gallon [200- litre] drums, suitably treated, with a tyre around each end, chain, concrete blocks or even cylinder blocks, anchors — whatever would hold on the bottom. Rough rule of thumb was, three times more chain than the depth of the water. That would hold up pretty well. We'd have a small drum trailed by a long rope from the main mooring for ease of mooring-up.

'We also did a lot of supply drops, to people like old G.A.V. Stanley, out for four months at a time. We became very expert. Free drops were done from as low as possible — fifty feet [fifteen metres], if you could get down to it — and the cargo was bagged in forty-pound [eighteen-kilogram] loads, inside a large copra sack. We dropped through the blisters. We finally had wooden slides made — load up with a dozen bags, then all you had to do was push. This was Ken Montagu's idea. Rigged up a system of red and green lights activated by a switch in the cockpit in case the intercom broke down. It worked extremely well. Used the tunnel hatch for storepedoes. From the pilot's point of view, provided you got down low and assessed the drop-site, the wind direction and so forth, it wasn't difficult.'

The success of free-dropping using double bags eventually enabled APC to abandon the use of the expensive storepedo, which cost £10 and which could seldom be recovered and used again.

In October, 1953, Captain Ross Treadgold free-dropped 2268 kilograms of rations to an APC engineering party surveying the Hathor Gorge, high up the Purari River. So narrow was the gorge that the Catalina could not drop there. A small island in the Purari, two days' walk from the gorge, was selected, and Treadgold and his crew, Ken Montagu and flight engineer Phil Heenan, dropped the urgently-required food stocks in double bags in nine runs, from a very low altitude. This was an important survey, preliminary to an ambitious scheme to harness the waters of the great Purari to generate hydro-electric power for a possible aluminium industry, using bauxite mined in Australia's Northern Territory — a scheme still being mooted thirty years later.

The Catalina continued to serve the needs of the Administration and APC, while the Sandringham took over its role in scheduled services within PNG, although the Catalina did continue to fly a scheduled route west from Port Moresby as far as the Aramia River and Lake Kutubu. During the three weeks that followed the departure of Fred Fox in VH-EBU, the remaining Catalina, EBC, made three hazardous mercy flights to remote regions under the command of Phil Oakley to bring out seriously ill patients to Port Moresby hospital. During one day, the

Catalina was in the air for fourteen hours.

One consequence of the replacement of the Catalina was a sharp increase in the amount of the Administration subsidy paid to Qantas for the scheduled services along the Papuan coast, and to the islands. In the 1950/51 financial year, a subsidy of £18,535 was paid; in the following year this soared to £75,000, a reflection of the much greater cost of operating the Sandringham.

The Sandringham was well liked by outstation people, and in operation was generally reliable. The great strength of the aeroplane was well demonstrated in October, 1953. Captain Robert Fairservice landed at Lindenhafen, and hit a submerged reef. A gash 152 millimetres long was ripped in the hull, and the massive keel was heavily dented. Water poured in. So slight was the impact that Fairservice did not notice it. He was taxying to the mooring buoy to pick up two passengers when a crewman reported that water was pouring in.

Fairservice immediately took off, and advised Port Moresby by radio that he would be making an emergency landing. It was witnessed by a crowd of more than 200 people, and two DCA crash launches were standing by, together with a crew from the fire station. But Fairservice put the big flying-boat down cleanly and efficiently, and she was then beached, and repaired. Fairservice estimated that two-and-a-half tonnes of water had poured into the Sandringham by the time he took off from Lindenhafen.

In 1953, Qantas decided to withdraw the faithful old DC3's which were still sharing the Bird of Paradise work with the big DC4 Skymasters, and substitute Sandringhams. The Sandringham had 'much greater passenger appeal', said the company. The last DC3 flight through to Rabaul was EM300/792 ex-Sydney on 19 June. The last south-bound DC3 Bird of Paradise flight left Rabaul on 15 June. With the introduction of the Sandringhams, the Bird of Paradise service now terminated at Port Moresby, with operations to Rabaul a part of New Guinea Internal.

The Sandringhams did not last for long. Passengers certainly preferred their solid, spacious comfort to the cramped conditions of the old DC3, but the cost of operating heavy flying-boats in the decade of the 1950s had increased alarmingly. There really were few routes where flying-boats could still be gainfully employed. From being a major user of flying-boats, Qantas was now relying almost exclusively on landplanes for overseas services. A new Lae- Hollandia once-monthly service was begun in September, 1953, and it was flown by the DC3. A subsidised service, the frequency was increased to once fortnightly in July, 1954.

Moreover, the Sandringhams were beginning to develop faults. In February, 1954, an oil leak in one engine forced the return to Port Moresby of a Sandringham when half-way across Torres Strait. A fortnight later, a broken stay-wire on a wing-float caused a delay of twenty-four hours to another flight. In May, a Sandringham was delayed at Cairns for three days while an engine was changed. There was another twenty-four hour delay early in June. Later that month, John Arthur held up the departure of another Sandringham, following complaints by the crew that they had not been allowed an adequate rest period. An inquiry was conducted by Qantas and DCA officials, but by the time the dispute was resolved, deteriorating weather conditions forced another twenty-four hour delay. Passengers, who had already boarded, were upset. Rising public dissatisfaction with the service to Australia prompted Qantas to provide an extra weekly Skymaster service.

However, it was the cost of operating the Sandringhams, not their reliability, which finally forced Qantas to withdraw them. The company was losing £55 per hour on Sandringham operations, and this was a financial haemorrhage that clearly had to be staunched. Qantas applied to DCA for permission to replace the Sandringham on the Bird of Paradise route with the Skymaster. The venerable Catalina would be brought back to service the internal routes.

Permission was granted. The last Sandringham services were flown in December, 1954. One Sandringham, VH-EBX, had already been sold to Ansett Airways, and the remaining three were broken up in 1955. A sad end for these majestic flying- boats.

108

109

110

111

108. Sea Otter VH-AJO which foundered in Port Moresby harbour in January, 1952. *Photo: APC*
109. Sea Otter VH-AJO at Port Moresby, September, 1951. *Photo: Bill Burns*
110. Qantas Catalina VH-EBD at the marine base, Port Moresby. *Photo: Qantas*
111. Pilot Ray Stockden in Crowley and Stockden's Tiger Moth at Kaiapit, 1949. *Photo: James Sinclair*
112. Crowley Airways Stinson L5, VH-CSA. *Photo: Laurie Crowley*
113. Crowley Airways Curtiss Robin, VH-UJE at Lae. *Photo: Laurie Crowley*

112

113

Territory Airlines

In 1952, two of the most successful air transport companies ever to operate in PNG were born. One, Papuan Air Transport Ltd, formed by the ex-Guinea Air Traders/Gibbes Sepik Airways engineer, Cliff Jackson, and ex-GSA pilot Frank Goossens, became dominant in third level operations within the Territory of Papua. The other grew from puny beginnings to pre-eminence in the Territory of New Guinea and, ultimately, PNG. This was Territory Airlines Ltd.

The two men who were principally responsible for the formation of Territory Airlines — generally known as TAL — were employees of Mandated Airlines: engineer Jack Gray, and pilot Ray Harris.

John Arthur Wilson Gray was born in Mullumbimby, New South Wales, in January, 1920. His parents were banana planters on the north coast before moving to the Mandated Territory of New Guinea in the mid-1920s. There Gray's father became manager of a copra plantation, on the Gazelle Peninsula. Jack Gray and his elder brother, Les, received primary education at Rabaul. When their parents were divorced, Les Gray was apprenticed as an electrician with the Administration. Jack accompanied his mother to Lae, where both were employed by Guinea Airways, the mother in the mess, and the son as an apprentice aircraft engineer.

Jack Gray worked on the maintenance and repair of the Junkers W34 and G31 and the Ford Tri-Motor aeroplanes that were the mainstay of the Guinea Airways fleet. Guinea Airways employed excellent engineers, Australian and German, and young Jack received a thorough training. In 1939, he learned to fly, on DH60G Moths hired from Guinea Airways and Jerry Pentland, and Tiger Moths in Australia. He was granted his private licence, No. 3655, in October, 1939.

When the Japanese struck, Jack Gray assisted in the evacuation of the white women and children of the Mandated Territory to Australia. Like most of the other fit young men of the Territory, he joined the New Guinea Volunteer Rifles, but was soon released, for aircraft engineers were sorely needed by the Department of Aircraft Production in Australia. Gray's Notice of Release read: 'Reason for Release — Reserved Occupation, Munitions, Clyde Engineering Co, Sydney'.

114. Joan and Jack Gray at Goroka, December, 1953. *Photo: Joan Gray*

114

Jack Gray was working at Clyde Engineering in 1942 when he met his future wife, Joan. In October, 1944, Gray and others were sent to the New Guinea theatre, and to the Philippines, as engineers on American small ships taking up supplies to forward areas as the tide of war swept closer to the home islands of Japan. He was sent home on leave in September, 1945, and married Joan. Almost immediately, he was away again, and did not return for another year.

The war was over, but Jack Gray considered New Guinea his home. He had no intention of remaining in Australia, and joined MAL as an aircraft engineer. He arrived in Lae in October, 1946. Joan followed in January, 1947. She still remembers the spartan living conditions in that shabby little town: 'Our first home was an American army shed, alongside the Lae airstrip, 100 feet by 25 feet [30 metres by 8 metres]. Rough concreted floor, walls of corrugated iron, wire netting windows, sisalkraft internal walls, and parachute-silk ceilings. At the side of our dwelling, two big Caterpillar diesel engines housed in an arc-mesh shed boomed out day and night, no silencers at all. These machines supplied DCA, just near us, and the town of Lae with electric power. I brought a grocery order with me from McIlrath's, and a 'Silent Knight' kerosene refrigerator and a beautiful American 'Kitchencook' petrol stove...

'For quite some time, we had the wings of Jack's Tiger Moth, VH-AQS, in our lounge room, supported by forty-four gallon [200-litre] drums. Jack worked on them in his spare time. Jack wanted me to sew the fabric for the wings, but my machine would not take the strong aircraft thread, and I made him take it down to Chinatown, where they had much stronger machines.

'Those were the days when everything was mostly out of a tin, and one longed for an onion or tomato to season a dish. The refrigerator mainly contained water, which had to be boiled before being bottled, and tinned butter. The local bread made from old army flour just rolled into a dough-ball in one's mouth. I occasionally received meat — green and slimy — which I would wash well in salt and vinegar before cooking...

'But we were very happy. We made lots of friends, and all in all made the best of everything...'

After two years in the army hut, the Gray's purchased a block of land on the Nadzab road where they built a temporary home using war-surplus materials.

In 1948, Jack Gray underwent two unsuccessful operations at Lae Hospital for a fistula, and had to go to Australia for further treatment. Upon their return to Lae, Joan fell pregnant. She had already suffered a miscarriage, and she was now advised to go to Australia for her confinement. Jack stayed in Lae, and it was while he was living the life of a bachelor that he conceived the idea of purchasing two old DH84 Dragons which MAL had for sale: VH-BDC and APL. Their hours had expired and they were without Certificates of Airworthiness.

This was late in 1951. BDC and APL were among the batch of Dragons purchased by W.R. Carpenter and Co. from the Commonwealth Disposals Commission in 1945. BDC was first registered to MAL in June, 1947, and APL in April, 1946. Gray approached the MAL accountant, Vic Nagle, about a possible purchase of the two aeroplanes. The price quoted, £750 each, was certainly cheap, for with C of A, Dragons were fetching up to £4,000 each in Australia. It was, however, beyond Gray's financial resources. What followed was told to the author in an interview with Ray Harris at his home in Mapleton, Queensland, in August, 1982: 'Jack got together with me, and we decided we'd buy these two Dragons, work on them at weekends, then I'd fly them south to sell them. We had no idea of starting an airline. The Dragons had run out of C of A, and MAL put them up for sale, rather than overhaul them.

'So we bought them, and took them over to the DCA hangar on the opposite side of Lae strip, and worked on them there. We got BDC ready for flying. I was going through Goroka in a DC3 soon after that, and George Greathead, the District Officer, approached me. He'd heard that we had a couple of aeroplanes, and he said, "How about bringing them up here?" I said, 'I can't do that, Bobby Gibbes is here.' "That's the trouble," said George, "Bobby hasn't got an aeroplane in the air, and we don't know when he will." George promised to give us all the government work from Goroka if we'd start up. Said that even when Bobby got going again, there'd be plenty of work for both of us.

'So I flew off and thought about this, and talked it over with Jack. We both had good jobs with MAL, and we weren't that interested. It took us a fair while to think about it. Of course, the story got around, and people like Lloyd Hurrell used to talk to me about it when I flew in to Menyamya. "You'd make a fortune," he'd say. Les Gray, Jack's brother, had the same idea. Vic Smith, who was then flying with Qantas, too — George Greathead got onto him, I think, urging me on. So we decided to form a company, and give it a go...'

It is ironic that it was the misfortunes of Bobby Gibbes that provided the opening for a new air transport company at Goroka, for Dennis Buchanan, who was to eventually control it, was still working for GSA, although he was stationed at Port Moresby at this time.

The future for GSA had looked bright for Gibbes Sepik Airways when 1950 dawned. Despite the losses of some Austers, previously described, Gibbes had steadily added to his fleet, and during this year purchased three more Norsemans, VH-ASS, BLM and BNT, and aother Auster, VH-KST. His principal base was still at Wewak, but he had a branch at Port Morseby and, from mid-1950, at Goroka. In addition, he acquired a fine block of land, 1,400 hectares in extent, near Mount Hagen, which he began to develop into a mixed farm and coffee plantation. Called Tremearne, after a property near Orange, New South Wales, owned by Gibbes' grandfather, the farm was to absorb more and more of his time, energy and money over the coming years, to the undoubted detriment of his air company. Gibbes encouraged his staff to take up land in the Highlands, and later formed a company called Highland Development Co. Ltd. in which the shareholders were all GSA employees. A property called Kimel was acquired, near Tremearne, which was planted in coffee. It was to prove a highly successful venture.

This was an exciting time to be living and working in what was still called the Central Highlands District, a vast region that included the whole of the present Eastern Highlands, Western Highlands, Southern Highlands, Simbu and Enga provinces, together with the Telefomin and Aiome sub-districts. This enormous expanse of territory, much of it still uncontrolled and some unexplored in 1950, was administered by a single district officer from headquarters at Goroka. It was an impossible task for one man, aided only by a few assistant district officers and patrol officers. Huge areas were perforce left untouched. Almost the entire area of the Southern Highlands was uncontrolled, although as previously recorded the old pre-war station at Lake Kutubu had been re-opened, and during 1950 the station at Mendi was established as the headquarters of what became the Southern Highlands District in September, 1951, following the partition of the gigantic, unwieldy Central Highlands District in that month. At the same time, the title of the officer in charge of each of the new Highlands districts, and the other districts of PNG, was changed from 'district officer' to 'district commissioner'.

The decade of the 1950s saw the virtual completion of the exploration, pacification and initial development of the Highlands. Townships grew, and government stations were opened. Patrols went out, probing into the last unknown corners and bringing the government peace to warring tribes. Hardy souls began to develop coffee plantations. Roads were constructed, and airstrips built. The work was made possible only by the aeroplane, for the road link to the coast, the Highlands Highway, was not fully opened to heavy traffic until the mid-1960's (although three Seventh-Day Adventist missionaries travelled by jeep from Lae to Mount Hagen in three-and-a-half days in November, 1953, the first time the trip had ever been done). The opportunities for companies like Gibbes Sepik Airways in the Highlands appeared to be almost endless.

The troubles that came close to finishing GSA in the 1951-53 period began in 1950. Bobby Gibbes, and Jean, had worked themselves into the ground establishing their air company, and the farm, Tremearne. Gibbes had been flying maximum hours — way over the legal limit — ever since GSA started. It was common knowledge that he often ignored DCA regulations governing flight times, and frequently landed in an Auster at Wewak after dark, guided only by hurricane lamps set out by Jean, and dodging the clouds of big fruit-bats that spiralled up from the tree-tops as dusk fell. His was a hardy constitution, but the strain of almost continuous flying, as well as coping with the management of the company, inevitably told on him. Upon the advice of his doctor, Gibbes decided to cut down on his personal flying, hire a manager to look after GSA, and retire to Tremearne and devote more of his time to building up the property.

Bobby Gibbes was always loyal to his comrades of wartime days, and he invited Group Captain Ken Ranger to take over the management of GSA. It was not a successful appointment, and Ranger left six months later. In October, 1950, W.G. Matthew, who had been with Gibbes in 450 Squadron, took on the job. V.W. Cox joined GSA as accountant. The company was prospering, and in June, 1951, another branch was opened, in Lae. The Madang branch was established in August, managed by another old friend of RAAF days — ex-Wing Commander Bill Duncan. 'Prim' Primrose, an ex-3 Squadron man, became manager of the Goroka branch. Each branch was self-accounting, each manager reporting independently to Gibbes.

It was not an efficient company organisation. Overall direction was lacking as Bobby Gibbes became more and more immersed in his farm. Large book debts began to accumulate. The effectiveness of each branch depended largely on the efficiency and energy of the branch managers. In some regions, GSA began to earn a reputation for unreliability — a dangerous brand for an air charter operator to wear. A reorganisation of the company was becoming imperative.

Again Gibbes turned to an old wartime friend for assistance. A.W. Barr, MC, DFC and Bar, had been one of the RAAF's leading aces in the Middle East, shooting down eight enemy aircraft in his first thirty-five operational flying hours. His career as a fighter pilot was brilliant, but brief. In June, 1942, Nicky Barr was himself shot down, and ended up a prisoner-of-war.

Barr was an engineer and an accountant as well as an aviator, an unusual combination which well fitted him for the task of devising a tight management structure for an air company. He was, too, a successful investor. He made an investment in

GSA, and became financial director in mid-1951.

GSA was actually a partnership, owned jointly by Jean and Bobby Gibbes. This was no longer an appropriate structure for a fast-expanding organisation. After a close examination of the affairs of GSA, Barr recommended the floating of a limited company to take over the assets of the partnership. Since GSA was now operating all over PNG, and not just in the Sepik, Barr felt that the name should be changed to 'South Pacific Airways', to more accurately reflect this fact. Barr also considered that a central financial reporting and consolidation system should be established at Wewak, to provide firm overall control.

In August, 1951, a statement of affairs was drawn up, showing the nett worth of GSA as £49,000, and on 1 January, 1952, Gibbes Sepik Airways Ltd was born. At the end of that month, Nicky Barr resigned and returned to Australia. He had had differences of opinion with Bobby Gibbes over the future direction of the company, which he felt was limited unless there was an immediate large injection of capital. For his part, Gibbes was apprehensive of the effects of Barr's management proposals. It was an unfortunate falling-out between two good friends, and one which did not last.

Following Nicky Barr's departure, Vic Cox became assistant general manager. The financial position of GSA was deteriorating alarmingly, and Don Levy was appointed secretary/accountant in March. He commenced proper accounts on the double-entry system, but GSA was by this time almost paralyzed by a lack of cash-flow. Disaster was averted by appealing to the company's debtors, which included the Administration — traditionally a poor payer — and by a rapid debt-collecting tour by Peter Manser in an Anson. This resulted in a breathing- space for the company, but the relief was short lived. GSA Norseman began to suffer a series of engine failures. By the end of 1952, the entire fleet of seven Norsemans were out of action, crippling GSA and providing Ray Harris and Jack Gray with their chance to get started.

At this time, GSA did not have an engine overhaul workshop. The company was under-capitalised throughout the whole of its existence, and overhaul workshops, properly equipped, are very expensive. GSA engines were shipped south, to Sydney, for overhaul by a certain well-known specialist company, at considerable expense. The overhaul work was badly performed. One engine failed after only five hours operation; none lasted longer than 130 hours.

It will be recalled that VH-BNL came to grief in long *kunai* grass near Goroka in July, 1950, with Frank Goossens at the controls. The second Norseman to be lost was VH-ASR, which Gibbes put into service in April, 1951. On 6 October, Gibbes was en route from Madang for Mendi in ASR. With him was a newly-joined pilot, Tom Lumme, who was being familiarised over the route. They were flying at an altitude of 914 metres, some sixteen kilometres west of Amele mission station, when the engine began to vibrate severely. Gibbes immediately stopped it by drawing the mixture control into idle cut-out, preparatory to making a forced landing. The terrain below was not suitable, so he re-started the engine, but a minute later it stopped abruptly and the propeller shaft broke, shearing the propeller. Gibbes proceeded to make a forced landing, using a technique that he urged upon his pilots. In a statement to DCA, he said: 'I alighted in some tall *pitpit* in close proximity to the Gogol River. To avoid the danger of turning the aircraft on its back, I deliberately applied the port rudder and dropped my port mainplane in the *pitpit*, immediately making contact with same, causing the aircraft to drop its left wing and stall to port, striking the ground while travelling sideways, and thus avoiding the danger of the engine coming back, or of the cargo coming forward...'

Gibbes and Lumme were slightly bruised, but the cargo remained intact, apart from the breaking of a few eggs. ASR was so badly damaged that it was a write- off. Gibbes offered a reward of £5 for the recovery of the lost propeller, and, amazingly, it was later found.

GSA lost another Auster, in 1951, VH-KSX. The pilot was Peter Manser. It was his first accident in 1,715 hours of flying, 1,166 of them on Austers.

As noted previously, Peter Manser had joined GSA as an accountant, but his true love was flying. He had by this time forsaken his desk for a cockpit, and over the weeks preceeding this accident had been flying constantly. On 29 March, he departed Hayfield for Dreikikir, with 179 kilograms of freight and one native passenger, who had never before been in an aeroplane.

115. Peter Manser at Goroka, 1956. *Photo: Sten-Erik Molker*

115

Arriving over Dreikikir, Manser made a circuit, and commenced an 180-degree turn in to the strip. Suddenly the engine spluttered and stopped. The Auster was at a height of about 213 metres, and after a few seconds Manser realised that he was not going to clear a heavily-timbered ridge about 400 metres from the end of the strip. He yelled, in Pidgin, to his passenger who was sitting quite unconcerned in his ignorance, 'If we hit those trees, jump out!'.

KSX clipped the top of a tall tree, and plunged down into the forest. Manser was knocked unconscious, and the passenger flew in a beautiful parabola through more tree-tops to the ground, ninety metres ahead. He was shaken, but unharmed, and promptly decamped, leaving the bleeding pilot to extricate himself from the wreckage as best he could.

The accident investigators concluded that the engine failure was caused by fuel starvation. Dreikikir was a nasty little strip and was classed as an emergency landing ground, where operations were permitted at company risk. Tom Drury suggested that Peter Manser, while not having exceeded his flight time limitations, may have been suffering from flying fatigue, following 'hazardous and arduous flying duties', and it was noted that he had not been on leave for twenty-one months. A medical examination, however, found him to be perfectly fit.

The series of engine failures that were to prove so damaging for GSA continued in 1952. In June, PNG's first woman airline pilot, Miss Patricia Graham, arrived. She was an ex-hairdresser, and only the fifteenth woman to qualify as a commercial pilot in Australia. Pat Graham was engaged primarily as second pilot for the leased Lockheed Lodestar, VH-FAC, which was still encumbering the GSA fleet, but soon began instruction on the Norseman.

Barely a week after her arrival, Pat Graham was accompanying Rinus Zuydam, then the GSA chief pilot, in VH-BLM on a flight from Wewak to Angoram. The Norseman was fitted with a newly-overhauled Pratt and Whitney R1340 Wasp engine, which had done 130 hours since overhaul in Sydney. The engine stopped dead, giving no prior warning, and Zuydam forced-landed in heavy *kunai* grass. The pilots and their passengers — none were hurt — walked back to Wewak. A new engine was carried in to BLM and fitted, a rough strip was cleared and the Norseman was flown out six weeks later.

On 27 June, Zuydam landed in VH-BNE at Popondetta, and the engine seized up on touchdown. It had completed nine hours since overhaul. One by one, the remaining Norsemans were grounded. Only one, VH-ASN, was still in operation by Christmas, and a week later ASN, too, was grounded. In Port Moresby, John Arthur announced that the whole Territory was affected by a shortage of light aircraft, in particular the Sepik and Madang districts. The grounding of the Norsemans was having a serious effect.

Gibbes was overseas with Jean when the engines began to fail, looking for suitable cheap heavy aircraft to supplement the Norsemans. Bill Matthew, the GSA manager, was in an impossible position. With aircraft failing one after the other, the cash-flow began to dry up. Customers turned to other operators, and the reputation of GSA was badly tarnished. Reconditioned Wasp engines were held by the Sydney overhaul company, but they wanted cash before they would release them. At his wit's end, Matthew wrote to the Minister for Civil Aviation, H.L. Anthony, on 13 August, outlining the plight of the company and asking for Commonwealth financial assistance to save GSA, offering as security the Norseman fleet. He concluded: 'If we do not get this assistance, we are faced with the necessity of dismissing the majority of our staff — at present twenty-eight Europeans — abandoning our branches at Moresby, Goroka, Lae and Madang, and retiring to Wewak with minimum staff, where we originally commenced four years ago...'

The wheels of government began to turn with customary leisureliness, and in the meantime Bobby Gibbes — advised by Vic Cox that company debts exceeded £40,000, and collapse was imminent — started for home, alarmed at the development since his departure.

By the time Gibbes returned to PNG — towards the end of 1952 — the company that Ray Harris and Jack Gray had decided to form was alive and well, flying hard and taking full advantage of the lack of competition over many Highlands routes. As previously mentioned, Qantas was phasing out its Dragons but were having troubles with the replacement, the Drover. MAL was still recovering from its problems, and had only a few Dragons still flying. The company had turned to the de Havilland Dove.

Territory Airlines Ltd was born during a meeting at Les Gray's house in Lae, on 17 January, 1952. Those present were Jack Gray, Les Gray and his wife, Vic Smith and his brother, Jim, and Ray Harris and his wife. Absent were Lloyd Hurrell and Bill (Darky) Johnson, who had both agreed to invest in the new company.

Several names for the company were considered and discarded, including 'Highland Air Transport Company', 'Fleet Air Service' and 'New Guinea Air Service', before there was unanimous agreement on 'Territory Airlines Limited', or TAL for short.

Ray Harris recalls the steps that led up to this meeting: 'There were seven of us who agreed to form a company — Jack Gray, Les Gray, Lloyd Hurrell, Vic Smith and his brother, Jim, Darky Johnson, and myself. Seven of us, and I think it takes seven to register a company. We put in £450 each, except Darky — he put in £280. And that's what we started the company on. Only £2,980 capital. Jack and I had outlaid £750 each to buy the two Dragons, BDC and APL, and we got back £300 each...'

The overhaul of APL was still incomplete, and it was agreed that Jack Gray — who was appointed chief engineer of TAL (he was the only engineer!) — would be paid ten shillings per hour while he was working on her. Gray's private pilot's licence had expired — he had accumulated a total of seventy solo hours only — and he agreed to obtain its renewal, and an endorsement on Dragons. The meeting decided to look into the possibility of obtaining fire insurance cover of £3,000 each for BDC and APL, and £1,500 for a third Dragon, VH-AAC, which was also being purchased from MAL.

This Dragon, AAC, was a real veteran, with an interesting history. It had been built by de Havilland Aircraft Co. at Stag Lane Aerodrome, Middlesex, England, in 1933, and became the personal aircraft of the Prince of Wales. It was sold to W.R. Carpenter & Co to be operated by MAL in the Mandated Territory of New Guinea in 1938, and in 1941 was impressed into the RAAF. In 1948, it was acquired by Joe Taylor, and in 1949 passed to Guinea Air Traders when that company took over Taylor's Air Transport. When GAT died, it returned to the ownership of MAL.

AAC passed briefly into the hands of J.A. Corrigan, who operated the short- lived Wewak Air Transport, in May, 1950, but was returned to MAL in November. In January, 1951, the aeroplane crashed when approaching the dangerous Slate Creek strip in the Morobe District. The pilot, Tom Lumme, and his European passenger were unhurt, but AAC was badly damaged. MAL engineers dismantled the Dragon, and stored it at Lae, pending a possible rebuild. It was sold to TAL in its dismantled condition, for spares.

The first general meeting of TAL was held in June, and the first directors were appointed, Jack Gray, Les Gray and Jim Smith, with Ray Harris as managing director. Mrs Joan Harris was appointed as temporary secretary. Shares were later allocated to each of the principals, 400 to all but Johnson, who requested 200. E.B. Davis, public accountant of Lae, was appointed company auditor.

116. Ray Harris. *Photo: Ray Harris*

116

As mentioned earlier, Les Gray was an electrician, and Jack Gray and Ray Harris had worked for MAL as engineer and pilot prior to resigning to start TAL. Lloyd Hurrell was an assistant district officer with the Administration, and was the man who established the outpost of Menyamya, in 1950. Darky Johnson was a veteran Qantas aircraft engineer, and Vic Smith was currently a Qantas pilot.

Vic was the first of three brothers who came to New Guinea after the war. They were from Mudgee, New South Wales, home of the Loneragan and McIlree aviation enterprises. As recounted in Chapter 3, The Trials of Mandated Airlines, Vic Smith joined MAL as a pilot in 1947, and in mid-1948 commenced flying for Qantas (where he was to remain until his retirement in 1976 as a Boeing 707 captain, with 19,450 flying hours). When he left MAL, his place was taken by his brother Elwin, an RAAF-trained pilot. The third of the brothers, Jim, was a fitter and turner by trade, although he had spent some months as a flight engineer with Trans Oceanic Airways, on the Pacific runs. Vic and Elwin convinced Jim that there was a future for him in PNG, and Doug Elphinstone agreed to employ him in MAL.

So, for a few months in 1950, the three Smith brothers were together in Lae, Vic with Qantas, and Elwin and Jim with MAL. Elwin Smith remembers: 'In those days, a lot of people couldn't tell us one from the other, and this used to cause a certain amount of consternation. When they decided to form TAL, Jim went in with them, and Vic, but not me. They both put money into it, but I didn't — I was on the verge of returning to Australia. I think they wanted me to fly. That was the thing that talked me out of it. I'd just about had enough up there, and they only had this couple of clapped-out Dragons...

'If pilots today realised the conditions we flew under they'd reckon we were bloody mad! No radio, no aids, nothing. But I reckoned the Dragon was a wonderful little aeroplane. No faults, apart from being underpowered. You could get away with anything. Easy on the controls. I came home on one-and-a-half engines a few times, but never with a load — other than that, you could do anything with it.'

Elwin Smith resigned from MAL and joined Trans-Australia Airlines. He flew Fokker Friendships in outback Queensland until 1967, when he abandoned commercial aviation, and finally became a chicken farmer. Vic Smith was unable to fly for TAL — he was too involved with Qantas. However, since VH-BDC was the only operational Dragon when TAL began, only one pilot was initially required. Ray Harris was at the controls of BDC when it made the first revenue- earning flight for TAL, on 11 April, 1952, from Lae to Rising Sun airstrip. He recalls the early days of the company: 'We had a lot of problems and headaches. I often used to think we'd never make it. Our biggest problem was getting money in, and the Administration was the worst payer of the lot. Lloyd Hurrell was a great help here. Being in the Admin. himself, he could get money out of them. Once they owed us £3,000, and I went to Moresby to see Harold Reeve, the Treasurer. I said, "Can you see your way clear to giving us some money?" and he said, 'Oh, what do we owe you?' I told him, and he reached into his desk and pulled out a cheque, and said, 'It's for exactly £3,000.' There it was, already made out — I don't know if he had heard I was coming, or whether it was ready to be posted. It saved the day...

'MAL always gave me a free ride over to Moresby on a DC3 — they were very good. No hard feelings about us leaving them. Doug Elphinstone was great...so the relationship was good. Relations with Bobby Gibbes, too, were friendly all the way. He didn't mind us starting up, even with all his problems. We were all good friends, and operated well together. We'd lend, and borrow, things from each other.

'Our office management when we started was done by my wife. Jack did all the engineering, and I did all the flying. We didn't get APL into the air for quite some time, so only needed one pilot at first. Then we got another pilot, Noel Peacock, and another, Les Taylor, and later another engineer, Frank King. It was a relief when the other pilots came. I was operating way over maximum hours allowed by the Air Navigation Regulations, and I was also managing the show, I was working too hard...'

In the beginning, TAL headquarters were in Lae, and for a number of years annual meetings and many regular director's meetings were held in that town. However, Goroka was always seen to be the home of TAL, and the headquarters were transferred there during 1952 and 1953.

The second pilot to fly for TAL, Noel Peacock, joined the company in June, 1952. Peacock learned to fly at Bankstown, and obtained his private licence after forty hours' dual instruction and solo work on Tiger Moth and Ryan STM trainers. He financed his flying from his salary, and he was twenty years old before he completed 165 hours, and gained his commercial licence.

Flying jobs were beginning to be scarce in 1951 for junior pilots, but Peacock was determined and after failing to find employment with the major airlines, he heard that MAL was recruiting, and that their chief pilot, Tom Deegan, was in Sydney. Peacock tracked Deegan down, and was rewarded with a job. Peacock remembers: 'For an enthusiastic young man, my flight to New Guinea on 20 October, 1951, was a memorable experience. I flew by Solent flying-boat to Port Moresby. The captain was the famous P.G. Taylor, and I was fortunate to spend time in the cockpit talking with him.

'I flew as co-pilot on DC3 aircraft for MAL for eight or nine months, with pilots like Tom Deegan, Wally Pawlawicz, Jim Perry, Jim McMillan, and Keith Cooper...

'I flew with Ray Harris, too, and had great respect for him, and when he left to start TAL I joined him as a pilot on DH84's. TAL headquarters were at Goroka, and

117. Noel Peacock. *Photo: Noel Peacock*

117

their main business was a daily service to Lae, and regular flights to Madang and Central Highlands ports such as Chimbu, Banz, Minj, Hagen.

'The DH84's were stretched to their limit. We had to fly on the windward side of hills in fully loaded Dragons, utilising updraughts to get altitude. Airstrips were short, and cargo weights were critical to performance. It was real 'seat of the pants' flying. Operating heights were lowered by the effect of high temperatures, and aircraft on take-off were frequently bounced into the air, utilising elevators to rock the aircraft free of ground contact. More than once in MAL DC3's we used an application of flap on take-off to lift the aircraft into the air, then hold the aircraft at low altitude to build up air speed before retracting flap...'

Noel Peacock had a narrow escape from disaster soon after he commenced flying with TAL. On 29 July, he was en route from Minj to Goroka in VH-BDC. The Dragon was carrying a light load of freight, plus one passenger, a Seventh-Day Adventist missionary, Harold Rudd. Earlier that morning, BDC's air brake had seized while taxying at Mount Hagen, and the aircraft had swerved off the strip into a shallow side-drain. No damage appeared to result, and Peacock proceeded to take off. However, it is almost certain that the port propeller struck a stone during the swerve, although a close examination had not revealed any physical damage.

Over Chuave, the port engine developed a severe vibration. To Harold Rudd in the cabin, the sound was 'like a bath-full of bricks dropping 200 feet [sixty- one metres] onto concrete'. Peacock throttled back, and over Daulo Pass the port propeller shattered, pieces of it tearing through the side of the aircraft immediately behind the pilot's seat in the cockpit, and exiting through the roof. On its way, the propeller smashed through baggage packed behind the light bulkhead at Peacock's back, and split a main structural member on the port side of the cockpit, crimping the cabin section slightly out of line. Says Harold Rudd: 'A big bang, a big hole about three feet by eighteen inches [about one metre by forty-six centimetres] in the side of the fuselage and another in the ceiling...I was in a front seat, and when the shower of chips and bits had cleared away, I very smartly left that seat, and diverted myself to the rear of all the seats. Bits of propeller and plywood you know, all the bits of paint, they took some time to clear. Now all this, the actual action, took place in about one second. The second second found me behind the seats, with my eyes out on stalks. I noted Captain Peacock trying to locate me, and my whole and only contribution was to wave and yell, "Keep her going, Brother..."

Fortunately for pilot and passenger, it was downhill all the way to Goroka, and Peacock landed safely on his one engine.

Harold Rudd was involved in another, minor incident in a TAL aircraft, the Tiger Moth, VH-AQS, which the company purchased from Jack Gray for £500. On a flight to Minj, Rudd sat in the single passenger seat, with a fifty-kilogram box of nails and twenty-three kilograms of carpenters' tools jammed between his knees, resting on the aluminium floor plate that covered the dual controls. On landing at Minj, the pilot said in a worried tone, 'I don't know, there's something wrong, I've had to row this thing all the way'. Only then did Rudd realise that his heavy boxes had compressed the floor plate, which had fouled the dual controls, making every control movement an intense muscular effort for the pilot.

Rudd, and most of the other passengers who flew in light aircraft in PNG in those days, well appreciated the worth of the men who flew and maintained the aeroplanes that did so much to ease the lot of those on the outstations. Men like the GSA engineer, Cliff Haydon, who was killed when he attempted to extinguish a fire that threatened an Auster. Haydon was doing a tappet check on the engine — an inverted Gipsy Major — assisted by workshop hand Paulus, who was with GSA for the whole of its existence. Underneath the engine was a drip- tray, containing waste petrol used in washing down. Haydon took off the magneto covers, which rendered them live. He pulled the propeller blade to check on the tappets, and the motor fired. The petrol ignited, and Haydon lunged for the tray, obviously completely forgetting the turning propeller, which struck him on the head. He died soon afterwards. Rudd recalls: 'Remember, it was pioneer flying by the finest of pilots, and pioneer maintenance. To remove an engine, a tripod was erected holding it up, and then the earth was shovelled from under the wheels of the aircraft, lowering it until the engine hung free. This was of course before the big DCA hangar was built at Goroka. The engines were overhauled under temporary roofs of galvanised iron or *kunai*. The only reason the aircraft went so well was that the engineers and pilots were dedicated experts. We often laughed to see the glossy photographs of aircraft administration and maintenance buildings in the Australian capital cities, and then look at the little tin shack at Goroka, at the other end of the line. The ten foot by ten foot [three-metre by three-metre] iron shed with its shutters banging in the breeze, and the top hinge of the door broken...'

Noel Peacock had a serious accident in VH-APL on 23 September, 1952. He was on a charter flight with government supplies for Chimbu from Madang, and the Dragon was loaded to the limit. The load had been computed by the then TAL agent at Madang, W.E. Passlow. Bill Passlow was undoubtedly one of the most experienced Dragon pilots in New Guinea, but the responsibility for loading aircraft in accordance with the DCA load chart rested with the pilot, whose duty it was to complete the DCA Form CA26 when loading was completed. Peacock followed company instructions carefully, and secured the load with a heavy cargo net. He was thoroughly familiar with the route, and had made perhaps fifty landings at Chimbu. This was a short airstrip at high altitude, flanked by steep ridges, and had to be approached with great care. The dangerous nature of the Chimbu strip had been officially recognised, and a new one was in the course of construction.

In the DCA inquiry that followed the accident, Ray Harris described TAL standing instructions for the approach to Chimbu in a loaded Dragon: 'After normal fly round to inspect the strip, a normal powered approach, air speed sixty knots, engine RPM 1,000 to 1,500, towards the north following the ridge to the southern end with a slight left hand turn just prior to final approach. The ridge is followed to minimise the effect of updraughts.'

Noel Peacock described what happened that September day in a letter, years later: 'The accident at Chimbu was the result of over-shooting the strip, and trying to go around. The strip was very short, and approach speed was critical. I came over the threshold too fast, contributing factors were updraft and a light tail wind. Unfortunately, with a loaded aircraft and restricted area for turning, the aircraft was on point of stall during the turn, losing height, and finally hitting the side of the hill. My leg was badly broken...'

Peacock was in fact lucky to have escaped with his life after what must have been a terrifying few minutes, trying to keep APL from stalling straight in. He suffered a compound fracture of the left thigh, plus minor lacerations and bruises. APL was smashed to pieces, and never flew again.

Noel Peacock was flown to Lae to hospital. His injuries were so serious, however, that he had to be sent to Australia for further treatment and convalescence. It was eight months before he was able to return to duty.

Fortunately, TAL had acquired another Dragon only days before the loss of APL. This was VH-AON. Qantas had offered two Dragons to TAL for £1,200 — AON, and VH-AIA — both without Certificates of Airworthiness. When Ray Harris approached Qantas, only AON was immediately available. It was sold to TAL on 18 September, and was registered to the company on 17 November. AON was in relatively good condition, and little work was required to restore it to C of A condition. VH-AIA was finally acquired from Qantas in March, 1953.

With Noel Peacock out of action for an uncertain period of time, another pilot was required. This was Les Taylor, who joined the company late in 1952. At a meeting of the directors on 28 September, Lloyd Hurrell — who had been appointed as the fifth TAL director — moved that no further flying be carried on until all personnel were covered by worker's compensation insurance. This was carried, and cover was arranged for sums ranging from £1,500 to £1,000 for Harris, Peacock, Jack Gray and Bill Passlow.

Salaries paid to Ray Harris and Jack Gray at this were quite reasonable: £1,975 and £1,816 per annum respectively. Peacock was paid £1,300. It was at this September meeting that J.V. Knight was appointed secretary/accountant of TAL.

Spare parts were always a problem for Dragon operators in the early 1950s. Although they were available, they were very expensive if purchased from de Havilland's and Ray Harris was delighted when an opportunity came to buy a large inventory from Qantas, cheap!

'I was down in Lae, and Eric Sims, who was acting manager at the time, said to me, "We're getting rid of a lot of our Dragon spares. Are you interested?" We were, of course, so he said, "Make us an offer. There's a hell of a lot of stuff there. Make us an offer of £800. You might get it." Nobody else wanted the stuff, so we got the lot for £800, and Qantas flew it to Goroka for us. We got a huge amount of stuff — you wouldn't believe it. Qantas was getting right out of Dragons. We had bought one of their aeroplanes, and this helped them to make up their minds to sell us their spares. So we had no problems — still plenty of spares when I left the company...'

As additional Dragons were acquired, so Jack Gray's workload steadily increased. His only assistants were three untrained natives. One, however, rapidly became an expert on various aspects of Dragon and Gipsy Major maintenance and repair — Sam Wanoi Buri, a young lad from the Kaiapit district in the Markham valley. Sam Buri was hired by Ray Harris in 1952 as an 'engine boy'. Perhaps fifteen years old, he was to remain with TAL all his working life, and was still a trusted, and much respected, employee of the company when this book was being written. He was — and is — known to all as Sam 'One-Eye', although he possessed two very keen eyes indeed.

TAL facilities at Goroka in the beginning were primitive. Bobby Gibbes had erected a big hangar/workshop of bush timber and *kunai* grass at the side of the airstrip, and was pressing DCA to erect a permanent building. The department had no funds available, and it was not until March, 1953, that the Department of Works announced that it would soon commence to erect a big Pentad hangar for DCA, forty metres by thirty metres. The component parts of this hangar were flown in to Goroka in chartered DC3's in June, and the hangar was soon erected. Goroka was becoming an important centre for light aircraft operations, which handled freight for the Eastern, Western and Southern highlands, flown in from Madang by Qantas and MAL.

As an interim measure, TAL also built a huge, sprawling hangar/workshop near the top of Goroka strip of bush materials and *kunai* grass. Here Jack Gray and his helpers carried on the maintenance of the company Dragons. Early in 1953, another licenced aircraft engineer joined TAL: Frank King.

King was an engineer of extraordinary versatility, qualified on an amazing variety of aircraft. At one time or another he was to work for Ansett, TAA, Bush Pilots, Air Niugini, DCA, Douglas Airways, PNG Aviation Services, Helitrans and other companies, and on his own behalf. He had engine and airframe licences on most of the aeroplanes that flew in PNG and Australia during the 1940s to the 1980s from the DH84 and 86 to the DC3, DC4 and DC6, the Cessna family, the Convair, the Viscount, the F27 and the Lockheed Electra. He had electrical instruments qualifications also, and could overhaul, re-wire or conduct major inspections on many different aircraft. But Frank King was a restless man, constantly moving around, and seeking new challenges. He was not a businessman: he simply loved aeroplanes, and aeroplane engines.

Frank King went to PNG to work for Bobby Gibbes, who was then trying to get official approval to operate DC3's. Gibbes never did get DC3's, and King joined TAL. It was the loosest possible kind of employment arrangement, though, as Frank King recalls: 'They were the good days. After I started work for Jack Gray I still did work for Gibbesy, on and off. We used to help each other so much, you didn't know who you were working for. Also, mission planes would come in, and we'd service them. Just Jack and me, and three nationals, Sam "One- Eye", Gengis and Kawon. Good, all three of them, even in those days. They learned on the job. Jack Gray was a very fine teacher, and he knew every stick of that Dragon. He and Ray both knew how to look after that aeroplane, fly it, and look after it like crazy. I didn't know so much about it — but I soon learned.

'That enormous great *kunai* hangar only cost about £100 to build, and we housed all our staff and the Dragons in it. Was quite good, weatherproof, but *binatangs* were always falling from the roof. You'd pull a cylinder-head off, and you had to be quick to get the new one on. Bobby Gibbes got an enormous parachute and put it under the ceiling of his grass hangar, and that fixed them.

'That was the beginning of it all, Gibbes and TAL, we were more or less together. Sometimes you'd get no pay, then double pay. All good fun — tremendous get-togetherness about the whole thing. Particularly pilots and engineers. I'll always remember the tremendous rapport between them. It wasn't a job, you didn't work for TAL or Gibbes, you were there, and part of it all. We were all encouraged by both employers to have our own little coffee plantations. Just about every pilot and engineer took up some form of holding. Government policy was to encourage it, and the employers encouraged it. Everything but give us a bit of dough to do it with, that is...

'Jack and I used to wander around the bush in his jeep, and we found this beautiful piece of land across the Bena, called Kapakamarege. We got 150 acres [sixty-one hectares], put a line down the middle, and we each had half. Jack had a partner, Bruce Blackley, and I was by myself...'

TAL acquired yet another Dragon, the last, in September, 1953. This was VH-ASX, built during 1943 for the RAAF. In 1946, it was purchased by the Flying Doctor Service, Western Australia, and was held in storage under the registration, VH-ASN. The Service had another Dragon, AGJ, which had been involved in a number of accidents, and after one of these, in 1947, parts from ASN including the propellers, tailplane, fin, rudder and portion of a wing, were used to repair AGJ. The resulting hybrid was unfavourably regarded by the Flying Doctor Service pilots, who considered that its flying characteristics were downright dangerous. In 1950, the C of A lapsed. Both AGJ and ASN were purchased by a grazier, Douglas Craig, who intended to build one good Dragon from the two.

VH-ASN was overhauled, and a new C of A was issued. But this registration had been allocated to a GSA Norseman, and the Dragon ASN became ASX. For personal reasons, Craig did not proceed with his plans, and ASX was offered for sale for £1,500, with the dilapidated remains of AGJ thrown in.

TAL purchased the two Dragons. Ray Harris and Jack Gray travelled to Perth to take delivery. AGJ was in a mess, but Gray considered that it was worth having for spares. The fuselage was cut behind the pilot's cabin and behind the rear bulkhead, and the undercarriage, cowlings and elevators were packed inside. The reinforced fuselage was then shipped back to Lae. Harris and Gray flew ASX back to Lae. Ray Harris hated this particular Dragon: 'It was the lightest Dragon we had, and it was the worst. It was a terrible aeroplane. The first day we flew it commercially with a full load out of Goroka, I got lower as I flew. The damned thing — I ended up going around and around over the Bena River. Finally, I used up enough fuel to get me high enough to go back to the strip. The boys knew I was down there, and they could hear me going. They were waiting for me to crash, or whatever. We always called that aeroplane "The Reluctant Dragon"...'

Although all Dragons were built to a standard design, each individual aeroplane varied slightly in weight and flying characteristics, to a much greater degree than is the case with modern all-metal light aircraft. Such things as the state of the fabric and the tension of the flying-wires could have a noticeable effect on handling and performance. The heaviest Dragon operated by TAL was BH-BDC, at 1246 kilograms empty weight. AON weighed 1242 kilograms, and APL, 1221, yet BDC was by far the best performer — Ray Harris considered it 'outstanding' for a Dragon.

'The Reluctant Dragon' lasted less than a year with TAL. Its last flight for the company was made in August, 1954, with Ray Harris at the controls. The C of A expired the following month, and the aeroplane was sold to MAL for use as spares for the two Dragons that company was still operating.

The end of TAL's first year found the little company in a healthy financial position. Despite the loss of VH-APL and Noel Peacock's expensive accident, TAL made a profit of £9,801. The following year a dividend of fifty per cent was paid. Some internal changes in the administration of the company were made, and J.V. Knight resigned as secretary, and was replaced by J. Joynton-Smith, who recently recalled: 'My position was actually company secretary/accountant; however, as I represented 100 per cent of the office staff it also included traffic officer, weighing freight, issuing of tickets, and anything else you can think of...'

At this stage, Darky Johnson decided to sell his shares in TAL. At a director's meeting in April, 1954, Joynton-Smith advised that there had been a 'marked falling-off in the gross takings of the company', due mainly to a decrease in Administration work, and the loss of freights out of Madang into the Highlands to the growing Qantas and MAL DC3 fleets.

The air companies of PNG were starting to complain that more Administration freight was being given to Qantas, rather than being shared as had hitherto been the case. The whole question of Administration support for Qantas was a touchy one to the private companies. Qantas took the view that it had entered PNG internal operations at the request of the Administrator and by direction of the Commonwealth Government, and should therefore be entitled to official backing. Qantas was certainly being paid a subsidy, which covered some of its PNG losses,

118. VH-BDC. *Photo: Noel Peacock*
119. VH-BDC showing gash in the fuselage. *Photo: Noel Peacock*

118

119

but under the current licencing policy had no route protection, to enable forward planning to be confidently carried on. DCA in 1954 had still not finalised a scheme of rationalisation of PNG air services that had been under development since 1952.

An attempt by the Administration in 1951 to reserve all its business for the government-owned Qantas had resulted in such a storm of protest that the Commonwealth Government asked the Administrator to withdraw the direction a week later. Not only the private air companies protested. The then manager of MAL, Doug Elphinstone, had been so enraged by the direction that he told the Administration, 'If you want Qantas to carry all your freight, they can carry all your mails, too'. MAL was then carrying mails twice weekly to Wewak, Madang and Rabaul, and the citizens of those centres complained vigorously at the curtailment of mail services that would have followed an MAL withdrawal. The Administration directed that Qantas should be given preference only when their freight charges were equal to or less than those of their competitors, and providing Qantas was able to offer prompt service. The practical effect of this new policy was that business had been shared fairly equally between Qantas, MAL and GSA.

In July, 1953, three of the best-known residents of Goroka, Jim Taylor, Jim Leahy and George Greathead, wrote to the Administrator to draw his attention to fresh rumours which were 'causing alarm amongst residents of the Eastern and Western Highlands districts'. The rumours claimed that in the near future competitive operations by airline companies between Lae and Goroka, and Madang and Goroka, would cease, and Qantas would take over. Limited MAL services to other PNG destinations through Goroka would be permitted, and a feeder service to all Highland ports would be operated exclusively by GSA.

Taylor, Greathead and Leahy pointed out that were such arrangements to be introduced, air freight charges would inevitably rise. At present, a customer could charter an aircraft as required from any company. By judicious chartering, goods could be imported from the coast for fifty per cent less in freight costs than the normal poundage rates. Qantas applied a flat rate for freight, regardless of its nature, whereas MAL, GSA and TAL charged special, lower rates for perishables. The rumoured new policy, if implemented, would have an inflationary effect throughout the Highlands, and would 'disrupt the lives and businesses of the farmers and settlers of this region, and this in turn would have an ill-effect upon the native population. We believe, Sir, that monopoly by any one company would be an outrage, and we would oppose the creation of such just as strongly, whichever company were selected to hold that monopoly.'

The rumours were not correct, but without doubt DCA planning did involve the rationalisation of existing air services, and it was common knowledge within the industry that changes were on the way. Then came another rumour, that proved to be true. Qantas was to introduce the DHC-2 Beaver, as a replacement for the Drover.

The first public report that the Beaver was coming appeared in early August, 1954, confirming stories that had been floating around for months. A spokesman announced that the Beaver had a 'remarkable record in Canada and America' — true enough — and could be fitted with floats or wheels, which would make it ideal for the Territory. Anticipating criticism of the replacement of the multi-engined Drover with a single-engined aircraft, the spokesman made a rather curious statement: 'While the Beaver has only one engine compared to the Drover's three, there is another way of looking at the disparity. With three engines, you have three times the chance of one stopping. With the Beaver's one engine, you have only the one chance.'

There was no doubt that the Beaver was a most formidable opponent for the Dragon and the Anson, and also the Norseman. It was a robust, all-metal, high-wing monoplane, powered by a 450 hp Pratt and Whitney Wasp Junior engine of undoubted reliability. It could operate into the shortest airstrips in PNG and was stout enough in construction to take heavy floor loads. It could be used as a freighter, or as a passenger aircraft with a capacity of up to seven.

Ernie Reid of Goilala Air Service announced that he was 'scared stiff' of the Beaver. 'I operate an Anson into Tapini and rely on government contracts to keep me going,' he said. 'If I lose these contracts, I fold up. I am considering the threat so seriously that plans I had to expand my service will now have to be abandoned.' Although Qantas had stated that the four Beavers it was going to purchase would be used only on the New Guinea side, Reid was not convinced. He said: 'It is all very well for Qantas to say that because they are going to operate in New Guinea there is no threat to operators on the Papuan side. That could very well be the plan and intention at the moment, but Qantas, being a semi-government show, could go where it likes. The Department of Civil Aviation has only got to say, "Take the Beavers to Papua", and Qantas will do just that. If I spend £10,000 on a new aircraft and the Beavers come here there is no use for the extra aircraft, and there would be no room for any of my planes. I would be left holding the bunny.'

Bobby Gibbes was particularly worried about the introduction of the Beaver, and announced that the very existence of his airline was threatened. He was now back in operation, with six Norsemans flying, and business improving. He had survived the financial crisis of 1952*, had reduced his debts by half, and was soon to introduce the Junkers JU52/3m.

Gibbes had been able to overcome the engine crisis by the purchase, very cheaply, of a number of Pratt and Whitney S1H1-G geared motors with three-bladed Hamilton propellers from the RAAF. These were ex-Wirraway engines, and during 1953 Gibbes engineers, with the assistance of Rube Murray, succeeded in fitting them to the Norsemans, despite formidable problems of stress and balance, for the Norseman had been designed to take a direct-drive R1340 engine and a two-bladed propeller. The re-engined Norseman had a better performance than the original, despite the additional ninety-kilogram weight of the geared S1H1-G To improve the cargo carrying capacity of the Norseman, its empty weight was substantially lowered. The engine cowling was modified, the starboard cabin door removed and the space covered with stringers and fabric, the rear locker removed, the front doors replaced by lightweight duralium structures. Special lightweight Lear radio equipment was purchased from America, and fitted. The cost of the modifications was relatively small, and the gains considerable. Payload was increased to 1089 kilograms. The Norseman now had a greater payload than its competitors, had a greater reserve of power, could operate into high-density strips, and was economical to maintain. Gibbes feared that the Beaver would threaten his investment in Norsemans, and he went to Australia to seek political support.

Bobby Gibbes had come through a traumatic couple of years, and his concern was understandable. The request by Bill Matthew for financial assistance from the Commonwealth had been followed by requests from Gibbes himself. In April, 1953, he asked for a direct loan, or bank guarantee, of £40,000 to keep his company afloat. His request was strongly supported by John Arthur, who wrote to the Director-General: 'The loss of Gibbes' air service operations at Port Moresby and Goroka has already had a serious effect on dependent communities outlying from Moresby and Goroka and, in particular, the new centre at Mendi. It is proposed to utilise Norsemans in the development of uncontrolled areas now being brought under government administration in the Southern and Western Highlands. The disappearance of Norseman aircraft from operations has been acutely felt...'

In a search for a way out of his mounting problems, Gibbes even considered selling his company to Qantas. Negotiations quickly broke down over the valuation of £24,700 that Qantas put on the entire Norseman fleet, claiming them to be 'practically all in poor condition'. Orm Denny recommended to the Qantas management that GSA be purchased at a total cost of £40,000. Gibbes himself placed the replacement cost of each Norseman at £15,000. Needless to say, no sale resulted.

The Minister for Civil Aviation, H.L. Anthony, supported Gibbes' loan application in Cabinet, and a long saga of negotiations, investigations, recommendations and recriminations followed. The loan was granted, subject to the reorganisation of GSA, the employment of a business manager, and a close surveillance by DCA of company operations, but in the event, Gibbes only took up £7,500. In May, 1953, Gibbes paid a deposit of £500 on the last Norseman he was to purchase. This was VH-BHG, then owned by the Zinc Corporation, Broken Hill. The price of the aeroplane was £8,000, and after John Arthur had certified that the Norseman was 'necessary to the efficient operation of GSA', the Commonwealth Government advanced Gibbes £7,500 to finalise the transaction. Gibbes later agreed to repay the advance over five years, plus interest.

This was the sum total of financial help received by Bobby Gibbes, but it was not

*For details, see *Sepik Pilot*, op. cit.

the end of the official assistance given to him. The reorganisation of GSA was done with DCA advice. Headquarters of the company were shifted to Goroka, and Gibbes left his farm, Tremearne, to the management of Peter Maxtone- Graham, and again took over direct control of GSA's affairs. In talks between GSA and Qantas, informal agreement was reached that when the expected airline licences for private air companies were introduced, Qantas would withdraw its light aircraft from the Highlands, and feed in with DC3's from Lae and Madang to Goroka and Minj, with GSA taking the cargo from these ports in to the outstations. MAL proposed a similar arrangement with GSA. John Arthur supported the agreements. This was, of course, the source of the rumours that had caused Jim Taylor, Jim Leahy and George Greathead so much concern.

John Arthur wanted to licence GSA for operations in the Sepik, and the Eastern and Western Highlands districts only. When a new airstrip currently under construction on the Papuan coast near Kikori was completed, Arthur intended the GSA Norsemans to supply the Southern Highlands and western Papuan ports from that strip.

In a memorandum written in August, 1954, to the Director-General covering these matters, Arthur backed Gibbes on the Beaver question: 'I must emphasise our considered opinion that the introduction of Beavers now would be poorly timed...'

He went on to mention the effect of his proposals regarding GSA on the now-struggling Territory Airlines: 'Licencing of Gibbes Sepik Airways Norseman operations would eliminate this organisation. The resignation of the managing director and the expected intention to liquidate, plus the undoubted collapse following the resignation of the managing director and the loss of his experience, makes it certain that this company will cut its losses and cease operations in the very near future. The DH84 is inefficient, and costly to maintain...'

Ray Harris had indeed decided to resign his position as managing director of TAL, leave the company, and join DCA. Following the director's meeting of April, 1954, strict economy measures were introduced to trim operating costs in the face of falling revenue, but the company continued to lose ground.

Harris was tired of the battle to survive. He was still flying long hours — one of the economy measures adopted was that the pilot staff be reduced to Harris and one other — and management worries were increasing. And Ray Harris had another problem: 'I was managing director, and Jack Gray was chief engineer. Poor old Jack was another of the reasons I got out. He wanted to be on the management side of it, instead of work on the aeroplanes. Quite embarrassing at times when a client would come in to talk business and next thing Jack would be sitting in, wanting to be in it, too. It was a bit of a strain. He was such a kind person that it was hard to tell him he was doing the wrong thing. Of course, his money was in it as much as mine, but he couldn't feel that he could leave things to me. He wasn't manager material, but he took over after I left. Wonderful engineer, amazing on fabric aircraft...

'John Arthur had offered me a job as an Examiner of Airmen and Airways Surveyor, so I got out of TAL and joined the department.'

Harris's resignation was effective as from 31 August, 1954. At a director's meeting on 18 August, chaired by Ray Harris, the resignation was received with sincere regret. Much soul-searching followed. The directors realised that the loss of Harris's experience and knowledge would seriously affect the company, and the prospect of Qantas introducing Beavers and probably locating one at Goroka was daunting. A motion was put: 'That the company should go into Voluntary Liquidation as from 31 October, 1954'.

Ray Harris and Lloyd Hurrell voted in favour, Les and Jack Gray against. The motion was carried on Harris's casting vote as chairman.

It was agreed that TAL would carry on with reduced staff until 31 October. Jack Gray took over as manager and Les Taylor continued as pilot. Noel Peacock had earlier left to join Trans Australia Airlines, and Frank King had left. V. Da Vanzo became temporary secretary, and Jeff Joynton-Smith carried on as accountant.

But TAL was not yet finished. A lot of talking went on behind the scenes after Ray Harris left, and it was decided to try and carry on. At a director's meeting on 10 October, the voluntary liquidation motion was rescinded. The appointment of another pilot, 'Mat' Matus — a Czechoslovakian who had previously flown for Gibbes Sepik Airways — was confirmed.

Territory Airlines was still afloat.

120. A DH82 Tiger Moth, similar to the aircraft VH-AQS, operated by Territory Airlines from 1951-6. *Photo: Talair*
121. Norseman VH-ASN flying over the Tari Gap in the Southern Highlands. *Photo: James Sinclair*
122. Aerial view of Mendi, 1958. *Photo: James Sinclair*
123. Goroka as seen in the mid-1950s. *Photo: Joan Gray*
124. Hangar at Goroka. *Photo: Joan Gray*
125. Alberto Buzzi refuelling Norseman VH-GSA. *Photo: Sten-Erik Molker*
126. GSA Junkers JU52 taking off from a Highlands airstrip. Patrol officer Ron Hoad looks on. *Photo: James Sinclair*
127. Qantas DC3 and Jack Gray's Tiger Moth, VH-AQS. *Photo: Qantas*

120

121

122

123

124

125

126

127

128

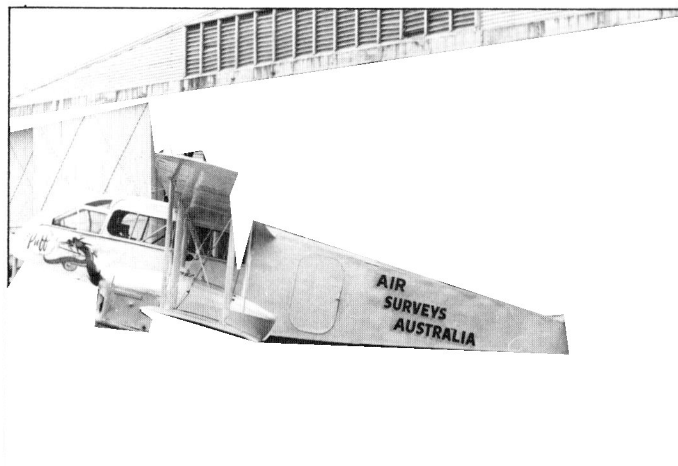

12

130

13

13

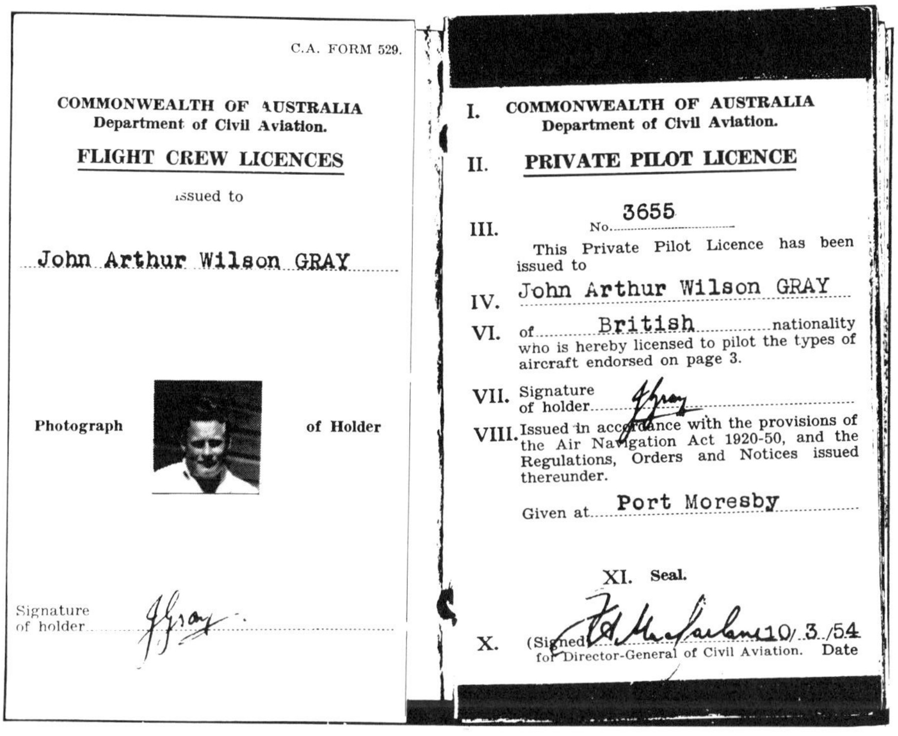

C.A. FORM 529.

COMMONWEALTH OF AUSTRALIA
Department of Civil Aviation.

FLIGHT CREW LICENCES

issued to

John Arthur Wilson GRAY

Photograph of Holder

Signature of holder

I. COMMONWEALTH OF AUSTRALIA
Department of Civil Aviation.

II. PRIVATE PILOT LICENCE

III. No. 3655

This Private Pilot Licence has been issued to

IV. John Arthur Wilson GRAY

VI. of British nationality who is hereby licensed to pilot the types of aircraft endorsed on page 3.

VII. Signature of holder

VIII. Issued in accordance with the provisions of the Air Navigation Act 1920-50, and the Regulations, Orders and Notices issued thereunder.

Given at Port Moresby

XI. Seal.

X. (Signed) for Director-General of Civil Aviation. 10/3/54 Date

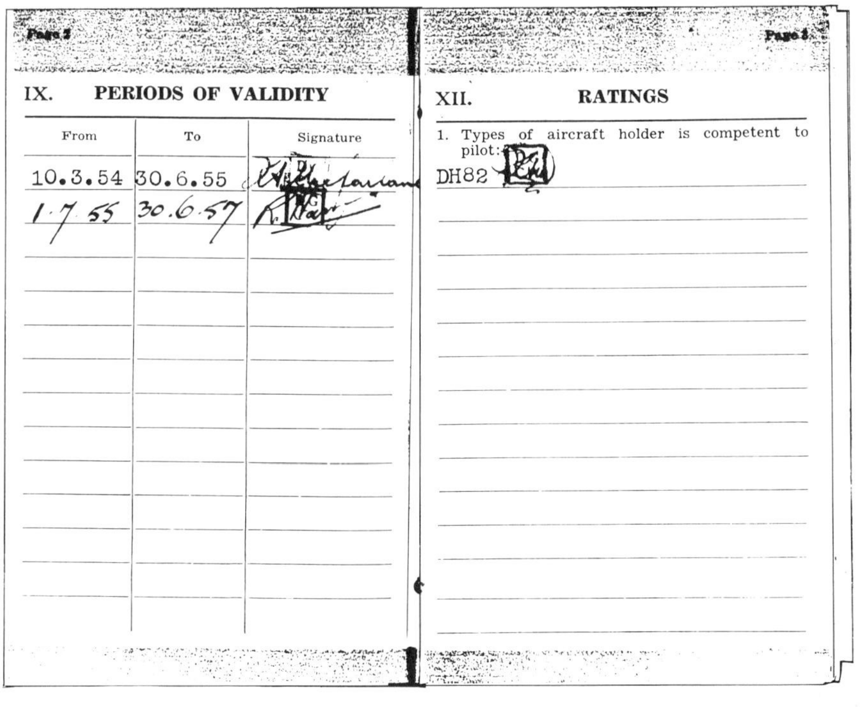

Page 2 — IX. PERIODS OF VALIDITY

From	To	Signature
10.3.54	30.6.55	
1.7.55	30.6.57	

Page 3 — XII. RATINGS

1. Types of aircraft holder is competent to pilot:

DH82

128. Ex-Territory Airlines Dragon, VH-AON. *Photo: Talair (Neville Parnell)*
129. VH-AON which served with Territory Airlines while remaining with Air Surveys Australia. *Photo: Ray Harris*
130. Qantas Dragon VH-AIA at Kainantu, March, 1953. It was later sold to Territory Airlines. *Photo: Qantas*
131. Territory Airlines Dragon, VH-BDC at Minj, 1952. *Photo: Talair*
132. DH84 Dragon, G-ACGG was delivered to H.R.H. The Prince of Wales (later King Edward VIII), in November, 1933. This aircraft became VH-AAC and was one of the first two Dragons acquired by Territory Airlines in November, 1951. *Photo: Talair*

Left:
Pilot's Licence issued by the Department of Civil Aviation, c1954

Following page:
Pilot John Gray's Log Book showing Licence No., dates of validity and fitness report.
Courtesy of Joan Gray
Pilot's Log Book showing flight records

C.A. Form 7

Full Name John Arthur Wilson Gray. Date of Birth 24-2-1920.

Address C/o Guinea Airways Ltd. Lae, New Guinea.

No. of Licence	Date of Expiration	Date of last Medical Board	Result
3655	16-10-1940	16-10-1939.	Medically Fit.
	16-10-1941.	14-10-1940.	[illegible]
	16-10-1943.	10-10-1942	Fit.

RECORD OF FLIGHTS

D.3389.

Date	Aircraft Type	Aircraft Reg. No.	Time in Air Hrs.	Time in Air Min.	Remarks
			65	15	solo
	Brought forward		15	35	Dual.
22-4-47	Tiger Moth	AQS	1	-	Trans (Dual) Lae.
29-4-47	Tiger Moth	AQS	1	-	Cir & Landings (Dual) Lae
1-5-47	" "	AQS	1	-	Cir & Landings, (Dual) Lae
4-5-47	" "	AQS.		20	Cir & Landings Solo.
4-5-47	" "	AQS.	1	-	Lae - Finch. (Dual)
6-5-47.	" "	AQS.		10	Cir & Landings (Check)
6-5-47.	" "	AQS.		40	Cir & Landings. (Solo)
~~7-5-48~~					Cir & Landings
	Carried forward		66	30	hrs solo
			19	35	hrs dual

Patair

Territory Airlines was started by a pilot and an engineer. So was the other major charter company that began in 1952, Papuan Air Transport. TAL was to confine its operations principally to New Guinea and the Highlands in its earlier years, while Papuan Air Transport was to work mainly within the Territory of Papua. The two concerns were thus able to grow without seriously encroaching on each other's preserves.

As mentioned in Chapter 2, New Guinea Internal, Clifford Robert Jackson came to PNG as an aircraft engineer for Guinea Air Traders. At about the time that GAT folded, Cliff Haydon was killed at Wewak, and Jackson moved to Gibbes Sepik Airways as his replacement. As noted previously, one of the pilots then flying for Gibbes was Fernand Jules Goossens, known to all as Frank. Another was John Wells, who built an impressive record of flying into isolated Highlands outposts — he was the first pilot to land at Tari.

Jackson's first posting with GSA was to company headquarters at Wewak, late in 1949. When the new branch at Goroka was established, Jackson was transferred there. The Port Moresby branch had been opened the previous year, with a single Auster piloted by Ted Marriott. The only other light aircraft based at Port Moresby then was a Qantas DH84, with Bill Hosking as pilot.

The Moresby-based Auster was serviced at Goroka as opportunity offered, but as business developed, particularly into Kokoda and Popondetta, the Auster was kept so busy that it was no longer practicable to fly to Goroka, so Jackson was transferred to Port Moresby to service the little aeroplane, and assist with ground management and the commercial development of the branch. As time went by and GSA drifted into financial difficulties, the Moresby business failed to grow as expected. Cliff Jackson remembers: 'During my period with Sepik Airways at Moresby, which ended in June, 1952, I was engaged in servicing whatever aircraft came into the area, but development for charter aircraft was very slow, and a lot of my time was spent selling passionfruit on the corner of a Port Moresby street, which came from Bobby Gibbes' farm in the Highlands. This was not a very lucrative way of earning a living.

'In mid-1952, I decided that it was time for me to have my own aircraft and consequently my own business, mainly for the reason that it sounded like a lot of fun, and I went through the process of seeking a suitable aircraft. Through Kingsford Smith Aviation Services at Bankstown Airport I purchased for £2,000 (all borrowed funds) a 1936 model Waco YKS-6 biplane, fitted with a Jacobs L4MB engine with a huge fixed-pitch wooden propeller. The price included a spare engine and a package of spare parts.

'This aircraft had been imported into Australia in 1936 by Adastra Airways, and was used by them for years for aerial survey camera work. The owner immediately prior to my purchase was a doctor based at Broken Hill, who had fitted along the fuselage on the port side, a long narrow door which was capable of taking a stretcher, which rested on the rear fuselage structure of the metal frame.

'While negotiating the purchase of the Waco, I arranged with John Wells, who was now with GSA at Moresby, to join me as a 50-50 partner in the venture. John was to accompany me to Sydney to fly the Waco back, but on the eve of departure, regrettably John informed me that he had changed his mind. I had an urgent need for another pilot, and contacted Frank Goossens, who was on leave in Sydney. He readily joined me in the partnership, and immediately set about getting an endorsement on the Waco.

'Unfortunately, we had our first setback right at the beginning. We were taxying at Bankstown when the aircraft stood on its nose, and smashed the propeller, the only one of its kind in Australia. We took the remains to George Adams senior who was an approved propeller manufacturer, and he agreed to draw up plans and produce an identical one. This took some considerable weeks to perform.

'We eventually departed for Port Moresby via the route north to Horn Island, across to Daru, and down the Papuan coast. We set up operations having our first charter on 8 August, 1952, which was a half-hour early morning search around Port Moresby harbour for a supposedly sunken submarine. Our gross return for that was £8/10/-, as the charter rate was £17 per hour.

'Our operation started off very successfully, with lots of charter work, and requests from the very few established landing fields to carry in regular food supplies and freight, and we decided from early on to set up regular services to all of those aerodromes in Papua where we could operate if not a daily service, at least two or three times a week. The people in those remote stations and outlying districts came to rely very heavily on our services.

'After a short time, I was informed by the Regional Director of Civil Aviation, John Arthur, that we were to cease operations because we didn't have a charter licence, and weren't registered as a trading company. We had been operating the Waco purely as a partnership...'

Cliff Jackson applied for a charter licence in August, 1952, to operate a 'Waco Ambulance Freighter and an Auster Autocrat'. He estimated a thousand hours' operation for the two aircraft per year. Neither aircraft would carry radio, or radio navigation equipment. The application was in the name of Papuan Airline Transport Ltd. However, the company that was registered on 20 August was called Papuan Air Transport.

John Arthur supported the charter licence application, and wrote to the Director-General: 'The company consists of an experienced New Guinea pilot and an experienced New Guinea engineer. They have purchased a Waco YKS-6 and are negotiating to purchase a J5 Auster. An operations manual has been completed — a copy of the approved GSA manual, suitably amended, and is satisfactory. A satisfactory maintenance manual has been prepared. Maintenance schedules for both aircraft have been prepared and forwarded. I am quite satisfied that it is a fit and proper organisation for a charter licence.

'The company proposes to operate under VFR rules largely between Port Moresby and Milne Bay, Popondetta and Kokoda, to serve numerous plantation and mission interests. The success of this small company is more than reasonably assured...'

Qantas had largely withdrawn from Papuan internal operations at this time, partly as a consequence of the troubles being experienced with the Drovers. Ernie Reid was still operating his Goilala Air Services, but on a very small scale. A gap existed, which Jackson and Goossens were determined to fill.

A charter licence was granted in the name of Papuan Air Transport. Towards the end of 1952, a telegraphic code address, Patair, was registered, and it was by this name that the company was generally known. An ex-Dutch Air Force pilot, Henk Van Santen, joined the partnership. The little organisation prospered, and it became apparent that additional aircraft were needed. The J5 Auster was not acquired, but a DH83 Fox Moth, VH-UTY, was purchased, and in February, 1953, an Anson, VH-BMP. This was one of the Ansons that had been operated by the ill-fated Charter Aviation Service, and was an ex-Eric McIlree aircraft, the first of a number of Ansons to be put into service by Patair.

Cliff Jackson recalls the early days of the company: 'Frank and I had decided that the only way to survive was to offer a regular and reliable service to all outstation 'dromes, and to this end we established a scale of charges for airfreight and passengers and publicised a schedule which we rigidly adhered to. As more aerodromes became available and required regular air services, we added them to our list. We also did extensive charter work, mainly for plantations, hauling native labour and supplies. Of particular interest at that time was Kienzle's rubber plantation at Kokoda, and the government tea plantation at Garaina. Late in 1953, we had our first crash, the Fox Moth...'

133. Frank Goossens pictured after a lucky escape when his Fox Moth crashed near Kokoda on November 17, 1953. The aircraft was a total loss. *Photo:* Pacific Islands Monthly

133

On 17 November, Frank Goossens left Port Moresby in VH-UTY, bound for Kokoda. The biplane was loaded to capacity with mail, freezer meat, bread, Christmas parcels and general cargo. Frank Goossens remembers: 'I had to circle and circle to get over the Gap. I "jumped" over the ridge to get over Lake Myola, and had to tackle one more ridge. Instead of getting an updraught, I was in for a downdraught, tried to turn back and in doing so, dodging around the hills, I fell into a spin, could not get out, and wrapped the Fox Moth around a tree. The plane was a wreck, but apart from a bleeding forehead I was unscratched.

'In those days, we didn't have radios on board to send out a Mayday, so when I didn't arrive at Kokoda the search was on. I decided to find the Kokoda Trail and make my way to Kokoda station. However, once in the thick jungle, I soon lost my way. Once again "Lady Luck" was with me. In the thick scrub I stumbled across what appeared to be a telephone cable. Found out later it was a Japanese telephone line, from the war...

'I was getting pretty panicky, not knowing which way to go, and then I heard an aeroplane overhead. DCA had started the air-sea rescue, and sent out a Qantas DC3 to search for me. I just hung onto the cable, hoping it would lead me to Kokoda, or the lake. I only had shorts and a short-sleeved shirt on, so I got plenty of scratches and cuts, but I wasn't going to let that cable go...

'I finally got to the lake, and with a sigh of relief just rested for a minute. Then I saw the DC3 circling the far end of the lake. How to attract them? I saw that the grass was dry, felt for a match-box. Thank heavens, I had one. Bundled up some grass and lit a fire. Soon I saw the aircraft turn towards me, and I started to wave, running backwards, stumbling on the way. I pointed towards the direction of my wrecked Fox Moth. Because the pilot of the DC3 saw me stumbling and falling, he must have radioed Port Moresby that I had a broken leg. Half an hour later, another aircraft came overhead and dropped crutches, first aid, food, a rifle (that broke in two on impact), cigarettes etc, and a note to say that a patrol was on the way from Kokoda, and to stay put.

'I went back to the Fox Moth, and looking at it realised how lucky I had been. Everything in front of the pilot's cabin was crushed. The engine was pushed right back, and must have just missed my head. I slept in the cockpit whilst awaiting the arrival of the patrol. It took two days before the patrol reached me, and I can only take my hat off to the native guides...'

A month later, another Fox Moth was purchased, VH-GAS, as a replacement for UTY, whose bones are still corroding at Lake Myola. It, too, had a short life. On 23 March, 1954, Henk Van Santen was taking off from Abau in GAS with a single passenger, ADO Campbell Fleahy. Just as the Moth became airborne, long *kunai* grass at the end of the short strip became entangled in the undercarriage, flipping the aeroplane onto its back. Pilot and passenger escaped injury, but GAS was badly damaged, and was out of action for a long time. It was eventually sold south, in June, 1955.

As was the case with so many early post-war operators, the Avro Anson was initially the backbone of the Patair fleet. The company was to employ a total of ten Ansons at different times, the last purchased in March, 1958. Two, BMD and BMN, joined the fleet during 1954. Both had originally been purchased from Disposals by James Loneragan of Mudgee, and both were once owned by Eric McIlree.

The Patair partners were well aware of the shortcomings of the Anson, but suitable alternatives at an acceptable price were not easy to obtain. But from the beginning, Patair was always ready to consider new and original aircraft designs.

The most promising Anson replacement when Patair was getting started was the Percival Prince, a British high-wing monoplane of all-metal construction, with twin engines and a tricycle retractable undercarriage. The Prince was an advanced design. It had an all-up weight of 4990 kilograms and cruised at 151 mph. A Prince which made a demonstration tour of PNG in 1954 so impressed Cliff Jackson that a provisional order for two was placed. At the same time, Patair applied for an airline licence, to operate the Princes on a regular service from Port Moresby to Popondetta, Kokoda, Bereina, Tapini and Embi.

Sadly, the Percival Prince failed to live up to its promise, particularly in regard to engine performance, and the order was cancelled. Nor was the airline application successful. The company had little alternative but to build up its Anson fleet. Says Jackson: 'Each of our Ansons carried fourteen passengers on utility seats, or about 2,000 pounds [907 kilograms] of freight. With the extended load door, we were able to carry sheets of building materials and by loading through the front storm windows we were able to carry timber up to fourteen feet [four metres] in length.

'The Anson was capable of operating in and out of most aerodromes at that time, but each aircraft could only be used for about a year, as during that time the high humidity and dampness deteriorated the timber wingspar and the glued joints to

134. Frank Goossens, World War II Mustang fighter pilot. *Photo: Frank Goossens*

134

a point where repairs were uneconomical, so we had to replace them every year...'

The Patair Ansons thus mostly had short lives. The first, BMP, was the longest-lived, being struck off the register in August, 1957. The second two, BMD and BMN, were struck off in September, 1955, but actually expired in December, 1954, after accidents. In both cases, Frank Goossens was the pilot. On 18 December, Goossens came in for a landing at Embi, with fourteen passengers on board. Unknown to Goossens, one of the tyres was flat, and on touchdown the Anson veered off the runway into thick *kunai* grass, and the undercarriage collapsed. The Anson remained in the *kunai*, washed by heavy night rain and baked by the searing, humid heat until glue deterioration rendered it unrepairable.

Just four days later, Goossens was on a flight back to Port Moresby when foul weather forced him to land at Bereina. The Anson skidded on the wet strip, and a wing was torn off. Goossens and his passenger were not hurt, but the Anson was eventually completely submerged by floodwaters and was declared a total loss. The Command radio equipment — all Patair Ansons had radio — and instruments were salvaged.

This was Frank Goossens' third accident in a little over twelve months, and he was required to undergo DCA flight-tests before he could continue flying. He was passed as fully fit for command.

In July, 1955, the company became Papuan Air Transport Limited. Although the airline application had failed, Patair was granted an exemption under the Air Navigation Regulations to operate a number of scheduled services on its charter licence, and as time went by the network was progressively extended. A tentative order was placed for a Scottish Aviation Twin Pioneer, valued at £50,000, but this was not proceeded with. The Ansons soldiered on. Between 1955 and 1958 another seven were placed in service: BBG, BFM, MMC, BLL, EFR, BKU, AOA. All expired through deterioration of the structures under tropical conditions.

This was the first phase of Patair's existence. Until 1958, the Anson was the company workhorse, but in that year the introduction of new and superior aircraft types began, a story that will be told in a later chapter.

Jackson is today proud of his involvement in another aspect of civil aviation in PNG that had its beginnings when Patair started: the Aero Club of Papua.

The aero club movement has a long and honourable history in Australian aviation, and has always enjoyed the active support of government. In the early years of aviation, the only practicable avenue for aspirants to train as pilots was through an aero club or flying training school, and governments have from the first recognised the value of the movement, particularly as a source of trained pilots in time of war.

The outbreak of the Second World War forced a virtual end to private flying, but on the cessation of hostilities, government assistance to the clubs was resumed. Direct financial grants began in 1945/46. A subsidy of £25 was paid for each fifty hours flown by a club trainer, with the maximum amount payable in one year set at £100. £50 was paid for each club member trained *ab initio* and qualifying for a private licence, and £10 for each member renewing his private licence at his own club. By 1947/48, some £40,000 was being paid to Australian aero clubs each year by DCA, to enable them to provide cheap training facilities for young people wishing to become pilots.

Many of the pilots who flew in PNG after the war received their initial instruction as members of aero clubs, and the movement has always been supported by concerned professional aviators. During 1951, informal discussions began in Port Moresby between persons interested in forming an aero club. The Minister for Civil Aviation, H.L. Anthony, visited PNG in July, with the Director-General, Air Marshall Williams, and met with a deputation of enthusiasts. The minister later announced that DCA would give 'every aid that is open to us in the formation of this and similar clubs. We think they are a very good thing'. Air Marshall Williams said that if the proposed club met requirements for a licence as a training school — it would have to have a qualified instructor, and ground instruction facilities — the usual subsidies would be paid.

On 13 August, 1951, Lloyd Nesbit — who was a DCA officer — wrote as 'Honorary Organiser' to the minister, who had returned to Australia, seeking his active assistance: 'For some time, a relatively large number of keenly- interested people have been trying to establish an Aero Club in New Guinea, with headquarters at Port Moresby. They have been frustrated by, one, a lack of hangarage, and, two, an inability to obtain, at reasonable cost, suitable training aircraft...

'There is a high demand for DH82 aircraft owing to the RAAF Civil Training Scheme...£600 to £800 for ex-Disposals machines with full C of A is outside the resources of the proposed Club, as a minimum of five will be needed...'

Nesbit asked the minister to make an approach to the RAAF for the release of a number of Tiger Moths. This Anthony did, but unfortunately only Airspeed Oxfords were available, and these were not suitable for *ab initio* training.

Matters rested. During 1952, a steering committee was formed, and a number of meetings were held, chaired by DCA officers. On 30 January, 1953, John Arthur announced that a meeting would be held at the Aviat Mess on 3 February, to form a club committee, which would study the establishment of an aero club. The meeting would be chaired by Tom Drury, Regional Safety Officer, who held an A grade flight instructor rating, and had extensive experience with the Newcastle Aero Club. Drury remembers: 'There was a great attendance — I didn't expect the result. I was made manager and chief flying instructor of the club, and Rube Murray was elected president. Vice-presidents were Vince Sanders and Gil Bugg, Peter Grimshaw was secretary and Peter Rudder, treasurer. Committee members were Ken Montagu, Des Sullivan, Don Hodson, Bill Bruce and Tony Parsons. We had thirty-six members by late April, including ten women. Everything went swimmingly...'

At first, the aero club was a social organisation. It was of course recognised that suitable trainer aircraft were the first priority, and the committee began to look around. Negotiations began in February, 1954, with the owners of two Tiger Moths, at Lae and Kavieng, but were ultimately unsuccessful. Accommodation was short then in Port Moresby, but DCA came to the rescue with an old Quonsett hut that had once served as married quarters. It was built of iron and concrete, but it served. The aero club was formally opened by the Administrator, Brigadier D.M. Cleland (who had been appointed in 1953, following the dismissal of Colonel Murray) on 10 October, 1953. The club had been incorporated as a company the week before.

In 1954, the club acquired its first aircraft, a pair of ex-RAAF Tiger Moths. A number of these trainers had been released by the Department of Supply to flying training groups, and were available at Tocumwal, New South Wales. They had to be paid for, but the price — £250 each — was low, and the aeroplanes in excellent condition. An appeal by Tom Drury to various companies in Port Moresby for financial support for the aero club had failed to realise the £5,000 sought, and Vince Sanders, who did more than any other person to foster the aero club movement in Papua, provided a loan to cover the cost of one of the Tigers.

The DH82's were flown to Sydney by RAAF pilots, dismantled, and shipped to Port Moresby on the M.V. *Sinkiang* in June, at a special concessional freight rate. They were loaded onto two of Vince Sanders' largest semi-trailers (Sanders operated a transport and salvage business) and transported through the streets of the town to Jacksons Airport, to the huge delight of the populace. Vince Sanders shepherded the trucks and their precious loads along like a sheepdog, in an old jeep.

Club members and well-wishers then commenced the task of assembling the Moths, which were registered as VH-ERM (Rube Murray's initials) and EVS (for Vince Sanders). While this work was in progress, Patrick Berg applied on behalf of the Club for flying school and aerial work licences. The applications were supported by F.G. Shanahan, Acting Regional Director in the absence of John Arthur, on leave: 'Flying instruction is to be done by E. Reid, who was the holder of an A2 instructor rating and is about to undergo a re-rating test, and A.H. Nisbet, holder of a current B1 rating...Mr Ray Harris, a former A2 instructor, who is currently joining DCA, will assist in the supervision of flying instruction...Mr C.R. Jackson has given the club full use of the facilities, workshop and equipment of Papuan Air Transport at Jacksons...'

Cliff Jackson was the first honorary engineer of the club, and gave a lot of his time to its early development. Two other club members were licenced aircraft maintenance engineers: L.R. Weeks and R.E. Slotter.

Flying School Licence No. 36 was granted on 8 October, 1954, together with an aerial work licence, which allowed the club to charter its Tiger Moths to earn additional revenue.

The first Tiger, VH-ERM, was test-flown in September by Ken Montagu of

135. Vince Sanders with his Tiger Moth. *Photo: Ray Thurecht*

135

Qantas, and Ray Harris flew EVS soon after. It was his first flight since joining DCA. By the end of that month, membership of the club stood at seventy, and flying training was actively in progress. Trainees were charged £3/15/- per hour solo, and £3/18/- dual. The normal DCA subsidies were paid.

The club was nothing if not ambitious. The Tigers were worked hard, for there were many eager to learn to fly, and there was plenty of room at Jacksons then, when so few aircraft were based there. In a memoir on the early days of the club, the late Vince Sanders recalled: 'The Tigers had no radio, and the only means of signalling by ATC was with an old Aldis lamp from the tower. Most flying was carried out in the early part of the day as the Tiger Moths did not have brakes, and a strong breeze made it difficult to taxi without two helpers holding the wingtips. Many is the time the fire crews used to race out to help a pilot 'turn the corner'. Generally, aircraft movements were infrequent then, and usually by midday the only aircraft left on the airfield were our two Tigers and the original Patair fleet — the Waco, a Fox Moth and two Ansons, and an old Gibbes Sepik Airways Norseman. All other aircraft had returned to Lae, where they were based, by early afternoon.

'Some of the early honorary instructors were Bob Wilson, Bob Larder and Frank Minjoy of World Wide Helicopters*; Ernie Reid of Goilala Air Services; Bob Gordon of Adastra, and Ray Harris and John Herrick of DCA...'

So enthusiastic were the trainee pilots that an anonymous correspondent in *Pacific Islands Monthly* was moved to remark: 'The noise of low-flying Tiger Moths is becoming a nuisance during the weekends...'

The aero club held its first Field Day on 17 April, 1955, at the DCA emergency landing ground at Fisherman's Island (which had proved to be an absolute white elephant. Commercial pilots refused to use it). The Tigers were in the air from dawn to dusk, and the day was voted an outstanding success. It was a trial run for a bold experiment: the first air pageant ever to be held in PNG. The president of the club was now Vince Sanders, and he announced that aircraft from all parts of the Territory would be invited to take part in a spectacular fly-past. Club pilots would engage in flying competitions, and club instructors would compete in aerobatics. Sanders also announced that a World War Two paratrooper, Peter Goodwin, would make a swallow-dive from a Tiger Moth from 1524 metres, and would fall for 305 metres before opening his parachute. DCA, however, refused to allow Goodwin to make the jump.

The air pageant was held on 14 May, and was another great success. Over 2,000 people attended, and aircraft from Qantas, Patair, Goilala Air Services, MAL and GSA took part in the fly-past, as did World Wide Helicopters, and the two club Tiger Moths. Vince Sanders later wrote an account of the day: 'The fly- past of aircraft, comprising the two lead machines, ERM and EVS, DC3's, Ansons, helicopters and a Norseman had all rendezvoused at Fisherman's Island, presented a startling spectacle.

'The two Tiger Moths led, but their pilots, worried about their low speed, left Fisherman's too early, and arrived over Jacksons far in advance of the main body. Consequently, as they turned on to a reciprocal over the Eight Mile, we were all startled to note that they were flying back towards the main body, which was still coming up from Bootless Inlet. Air Traffic Control nearly had a fit, as did the committee members, as the Tigers had no radios. However, the pilots used their brains, and passed under the main body. Our guest of honour, Judge Gore (our patron, the Administrator, was away) thought that this was quite spectacular!

'At this stage we had to get ready for the formal christening of our Tiger Moths by His Honour. I asked one of our helpers to get the champagne, but to our consternation he said that the clubhouse was locked up, and he did not have a key. "Who has the key?" says I. 'It's up there,' replies he, pointing skywards, and, sure enough, one of the pilots in the fly-past had the key in his pocket.

'A quick discussion, and our worthy helper departed at high speed to a local trade store, where, for one shilling, a bottle of creaming soda was produced, and this is what Judge Gore used to christen the two aircraft.

'After the ceremony, one Tiger Moth wouldn't start — they had to be handswung on the prop. It was the old story, the impulse magneto needed a tap. No tools were handy, so our duty engineer used the heel of his shoe. What all the spectators thought about this I never did find out...'

The Aero Club of Papua was the first in PNG, but in 1953 an effort to form a club

*See Chapter 5, The Helicopter Appears

at Lae was made. It is perhaps surprising that Lae was late into the aero club field, for it was still indisputably the centre of Territory aviation. However, the question of the future of the Lae airstrip had been under discussion and argument for many years, which may have been one reason for the late start.

As early as 1946, an alternative to the Lae strip (which had originally been constructed in 1927 to take the small biplanes that had pioneered aviation in New Guinea) had been considered essential. During the Second World War, the Japanese built a strip at Malahang, some three kilometres from Lae, on land owned by the Lutheran mission. It was capable of development to a high standard, and in 1946, only weeks after his arrival in PNG, John Arthur was instructed to investigate the possibility of abandoning Lae in favour of a new strip at Malahang. The matter was considered to be urgent. Lae was reckoned to be unsafe for anything larger than a DC3, and could not easily be lengthened.

On 27 November, Arthur advised the Director-General that it was going to be difficult to investigate Malahang, because it was still sewn with Japanese mines and booby-traps. 'Natives are being blown up weekly,' he reported. Nevertheless, he was ordered to proceed. Army personnel still at Lae de-loused Malahang, and in February, 1947, Arthur, surveyor Jack Ecclestone and a Department of Works team roughly pegged a strip to a length of 1828 metres which could readily be extended to 3048 metres if required. Most of the strip was on Lutheran mission land, the remainder on land owned by the people of Yanga and Wagan villages (who had been told by ANGAU officers that the Malahang land would be returned to them after the war).

In June, 1947, DCA airstrip surveyors moved onto Malahang and began to lay out the design of the new strip. Bishop John Kuder of the Lutheran mission protested, saying that building Malahang airport would 'destroy the great work of the mission'. The Yanga and Wagan people stated, firmly, that they would not sell their land.

Much discussion and negotiation followed. By the end of 1947 the single gravelled runway at Lae was taking an average of 1,000 aircraft movements a month, and maintenance costs were rising. The Lutheran mission and the villagers gave in and in March, 1948, it was announced that a 'great international airport' would be built at a cost of £3 million. The first stage of the project was costed at £120,000, and a requisition was raised in April, 1949, for construction of a sealed gravel runway at Malahang, 1737 metres long by 46 metres wide.

Work on the strip had still not begun a year later, although a workshop building was erected, and stumps sunk for a passenger lounge for Qantas. The cost estimate had risen to £132,700 with a decision to include concrete aprons twenty-eight centimetres thick instead of the twenty centimetres originally specified.

The uncertainty about the future of the Lae strip naturally had an effect on the post-war reconstruction of Lae, which had originally developed, in typical New Guinea fashion, around the strip. There was a good deal of public concern about the delay, and in February, 1951, the Director of Works, D.J. Rooney, was moved to write a letter to the *South Pacific Post* defending his department. While it was true that funds had been made available for Stage One, no instructions to commence had been received. In fact, it was not until July, 1950, that DCA had given details of the type of pavement that Works was to design, and in November had decided to replace the concrete aprons with aprons of sealed gravel, to cut costs. Work had to be suspended while fresh designs were prepared.

Early in 1951, tenders for the Malahang work were finally called. Two were received, both totalling over £200,000 which was considered excessive. So fresh tenders were called, closing on 20 September, 1951.

In August, the Minister for Civil Aviation, H.L. Anthony, announced that the existing Lae strip would definitely be scrapped. The modern aerodrome to be built at Malahang would be developed as the 'nerve centre of PNG's airline network'. When Lae was out of commission through bad weather, all other centres were adversely affected. Lae could not be developed to international standard, because the range of mountains in line with the runway made radio navigational aids 'hazardous'. DCA would press on with the development of Malahang 'as rapidly as possible', the minister promised.

What an enormous amount of bother would have been avoided had Arthur Affleck's 1947 recommendation to swing the direction of Lae strip by fifteen per cent to allow for future extensions been accepted!

Despite Anthony's promise of action, nothing happened. In April, 1952, the Administration announced its intention to resume the land required for the Malahang airstrip, and also land for an extension to the south-eastern end of Jacksons. The Director of Works was anxious to get going with the Malahang work, so that expensive plant purchased for the job could be put to use. John Arthur wrote to the Director-General: 'The whole aviation economy in New Guinea is bound up with the provision of an all-weather instrument airport at Lae...urgent action is required.'

However, the question of the construction of the airport at Nadzab rather than at Malahang had again been raised, and Arthur was advised that 'a close study of all factors' would be made. It was estimated that a Nadzab development would cost £278,300 as against £288,260 for Malahang — but the Nadzab figure did not include the cost of a road from Lae.

The constructing authority, the Department of Works, was for Nadzab, and DCA was firmly in favour of Malahang. John Arthur wrote again: 'There is little or no likelihood of civil development at Nadzab, at least for a very long time...Nadzab would serve no useful purpose to the government or the community at present...the best interests of all would be served by the development of Malahang with appropriate aids, even to the inclusion of an instrument landing system or a localiser.'

The argument continued, and was still in progress in mid-1953. Other airports in PNG were being steadily improved, and new ones built. Tari in the Southern Highlands was opened, and a fine new strip at Bulolo completed, which avoided the notorious Pimple. This was built by Bulolo Gold Dredging Ltd to a DCA design, and was suitable for DC3's. Fisherman's was finally completed, and a search for a replacement strip for Rabaul was under way. A strip was opened at Buka, greatly improving air services to Bougainville. Major improvements to Jacksons Airport were announced, at a cost of £250,000. A new strip at Boram, eight kilometres from Wewak, was under construction. When it was completed, Wewak strip was to be abandoned (to the intense fury of Bobby Gibbes, whose facilities — such as they were — were at Wewak strip). But Lae remained in a state of limbo.

The citizens of Lae were becoming quite fed up with the Nadzab versus Malahang argument. Most wanted Lae strip to remain, and the Morobe District Advisory Council passed a motion calling on the Administration to reconsider the whole question. But once again, no positive action followed. The unsatisfactory state of the Lae strip was given unwelcome emphasis in September, 1954. As mentioned in Chapter 4, Success And Failure, rising public dissatisfaction with the frequency of services on the Bird of Paradise route had prompted Qantas to put an additional Skymaster on the run, which terminated at Lae. In late August, Lae was hit by prolonged heavy rains, and on 2 September, Lae was declared closed to Skymasters, following complaints from pilots that grass growing on the surface was too slippery for the safe operation of heavy aircraft. DCA disagreed, pointing out that Qantas had been sending Skymasters into Lae for years past. Qantas wanted all the grass removed from the strip, but DCA said that this would be a major project, costing many thousands of pounds, and would take a year to complete.

Lae was re-opened for Skymasters five weeks later, but Qantas refused to allow them to land there. John Arthur announced that DCA 'would not touch' Lae strip until Qantas had declared its intention to use it for Skymasters. He said: 'Reconditioning the airstrip until it is suitable for DC4's will be very costly. The department hopes to start work on the new 2072-metre airstrip at Malahang in the 1955/56 financial year, and this is another factor influencing the department's attitude on Lae renovations...'

There was thus a stalemate. Lae was the fastest-growing town in PNG, rapidly developing a light industrial base, and still the centre of PNG aviation, and yet without an adequate airport. Clearly, it was a situation that could not be allowed to continue. All sections of the community were calling for action. DCA had to give way. A small allocation of funds — £6,500 — was made, and the Department of Works removed all grass from the main runway surfaces, and completed re-gravelling. The project was started in mid-May, 1955, and two shifts were operated, night work being carried out under flood-lighting. Skymasters began landing again at Lae on 2 August eleven months after operations had been suspended. But the problem of a permanent aerodrome for Lae remained — and a strip never was

136. The remains of Fox Moth VH-UTY piloted by Frank Goossens which crashed near Kokoda in 1953. *Photo: Frank Goossens*
137. Patair's Fox Moth VH-GAS at Port Moresby. *Photo: Sten-Erik Molker*

136

137

built at Malahang.

It was against this background of uncertainty that the Lae Aero Club was born. In September, 1953, a group of about thirty enthusiasts met and decided to form a club, with the aim of 'encouraging young people to learn to fly without asking them to lay out considerable sums for lessons'. It was hoped to raise funds to purchase a pair of Tiger Moths. The Lae Airport manager, Bert Hayes, was elected president, and Alan Smith of Amalgamated Wireless (Australasia) Ltd, honorary secretary. John Simpson of DCA and Laurie Crowley were elected vice-presidents.

There was initially plenty of support promised for the club, and Jack Gray offered the use of his Tiger Moth, AQS, until trainers could be acquired. However, there was little activity for a considerable time. The name of the club was changed to New Guinea Aero Club, and in July, 1955, a drive was launched to raise funds for the purchase of the Tiger Moth, VH-CAH, owned by DCA, which had been replaced by a Cessna. The price was £800.

The funds drive had limited success. In October, vice-president Jeff Martin said that the club could not hope that DCA would hold open the option to purchase CAH indefinitely. Of the entire population of Lae, only twenty people had joined the club. Many of these had contributed funds, but not sufficient to cover the purchase of the Tiger. 'With an aeroplane available for the asking,' said Martin, 'the club had a "golden opportunity" to become airborne.' Training aircraft were becoming increasingly hard to obtain, and those that did become available were usually snapped up by Australian clubs. He appealed to aspiring pilots to join the club, and support its fundraising efforts.

The New Guinea Aero Club slowly grew. Newly-arrived pilots, flying for Qantas, MAL and the smaller companies, took an interest in its development, and finally VH-CAH was purchased, and flying training began.

On 19 May, 1957, the club was officially opened, and the event was celebrated by Lae's first air pageant. Aircraft of Qantas, MAL, Crowley Airways, TAL, Madang Air Charters, and Adastra Aerial Surveys took part in the fly-past. Some were modern aeroplanes — Beavers, and several Cessnas. The day of the Dragon and the Anson was almost over, but the DC3 still soldiered on.

138. Sir Donald Cleland (right) opens the Papua Aero Club. *Photo: Ray Thurecht*
139. Guests at the opening of the Papua Aero Club. *Photo: Ray Thurecht*
140. Kokoda airstrip. *Photo: Cliff Jackson*
141. Woitape airstrip. *Photo: Cliff Jackson*

138

139

140

141

142

144

145

142. Waco VH-VYD at Daugo Island near Port Moresby. *Photo: Cliff Jackson*
143. Patair Piaggio VH-PAP (left) and Patair Waco VH-UYD, October, 1959. *Photo: Cliff Jackson*
144 & 145. The Aero Club of Papua's first Tiger Moth in original livery, and as it appeared later as VH-EVS. *Photos: Ray Thurecht*
146. Patair Anson VH-BMP at Tapini, mid-1950s. *Photo: Cliff Jackson*
147. Tapini airstrip. *Photo: Cliff Jackson*

146

147

Exit The Dragon

In 1953, a small company was formed at Madang by MAL pilot Dave Robertson and Qantas engineer, Wally Ferguson. It was called Madang Air Charters.

Dave Robbie had been MAL manager at Madang for about eighteen months, and being a man of notably independent mind, had been considering 'having a go' at charter work on his own behalf for some time. As it happened, Qantas had another Dragon for sale: VH-AQW. Qantas had been operating this Dragon out of Lae since January, 1949. Dave Robbie remembers: 'At that stage, Wally Ferguson and I got our heads together, and we bought the Dragon. Father Joe Walachy flew me from Madang to Lae in his Tiger Moth to have a look at the aeroplane — quite a hair-raising trip, I tell you. So we bought the aeroplane for £1,400, Wally and I, in April, 1953...'

AQW was in good condition — Qantas invariably maintained their aircraft in top condition — but the C of A had only four months to run. The partners naturally wanted to put the Dragon to work without delay. On 8 May, Robertson applied for a charter licence.

Before a licence could be granted, however, maintenance, loading and operations manuals satisfactory to DCA had to be produced, and this would take some time. In the interim, Dave Robbie asked for temporary approval to operate charter services pending the completion of the manuals. Permission was refused. John Arthur took the view that Madang was already sufficiently served by W.E. Passlow's charter operation, Riverlea Market Gardens.

Bill Passlow had for years past been supplying Madang with fresh vegetables out of the Highlands, particularly Chimbu. In June, 1951, he had been granted an aerial work (and later a charter) licence allowing 'carriage for the purpose of trade of goods being the property of the pilot, the owner, or the hirer of the aircraft'. The aircraft in question was a Fairchild Argus, VH-AKY, a small, single-engined high-wing monoplane with a capacity of three passengers that had been constructed by the Fairchild Airplane Manufacturing Company in 1942, and purchased by Passlow in March, 1951.

Bill Passlow had purchased the Fairchild on terms, and by November, 1952, the aeroplane was in an advanced stage of disrepair, and was in the hands of the receivers. The C of A expired in February, 1953, and was not renewed. In March, shareholders of Highland Traders, a firm based at Kundiawa, joined with Passlow to form Riverlea Market Gardens Limited. A Dragon, VH-AOT, was purchased from Qantas, and the aerial work and charter licences were transferred to the new company. In 1954, the shareholding in Riverlea Market Gardens was changed to Passlow and D.E. Kelaart. Dick Kelaart was an ex-patrol officer who had left the Administration to go into business at Kundiawa.

At the time that Dave Robertson made his application for interim permission to operate the Dragon VH-AQW, the Riverlea Dragon was in full operation, and Bill Passlow was a most competent and experienced pilot. Robertson and Ferguson appeared to be stymied. However, the Catholic mission came to the rescue.

As mentioned in Chapter 3, Gibbes Sepik Airways, the Catholic Mission of the Holy Ghost was quick to re-enter the aviation field after the war. The flying service of the mission was known as Divine Word Airways, operated by the Society of the Divine Word as a non-profit organisation. Divine Word Airways was principally concerned with servicing the string of mission stations in the hinterland of the Madang district, and in the Highlands. The first aeroplanes to be acquired were a Tiger Moth, VH-BXA, and a Dragon, VH-AMO. Another Dragon, AOE, was purchased in December, 1947, and a third, URD, in August, 1948. As previously recorded, AMO crashed at Mingende in December, 1948, and the pilot, Father

148. Dave Robertson inside the cockpit of a Cessna 310. *Photo: Dave Robertson*

148

John Glover, was killed. URD was sold to Qantas in May, 1950.

On 22 April, 1953, the Tiger Moth crashed into heavily-timbered ridges while on a flight to the tiny mission airstrip of Kegsugl, which at a height of 2590 metres was reputed to be the second-highest strip in the world. The pilot, Father Walachy, and his passenger, a seventy-five-year-old priest, were unharmed, but the Tiger was a total loss.

The story goes that following this crash, Father Joe was having a quiet beer in the Madang Hotel as was his wont when one of the locals came up and said to the universally popular priest/pilot, 'Well, Joe, "J.C." was certainly on your side this time, wasn't he?' Walachy turned around, looked the local up and down, and said, "Well, he ought to be, I work for the guy, don't I?" Great roars of laughter from the crowd at the bar.

The crash placed the mission in an embarrassing position. Father Walachy was not authorised to fly Dragons into the Highlands, and the only qualified pilot, Father Henry Hoff, was away from New Guinea.

In May, 1953, John Arthur visited Madang on departmental business, and was approached by Dave Robertson and Father Walachy. The priest told Arthur that the mission was having grave difficulty in supplying the inland stations, and asked that Dave Robertson be permitted to operate VH-AQW on behalf of the mission for the time being. John Arthur wrote to the Director-General: 'I told Father Walachy that I would agree to this request as a result of an approach from the mission, and told both he and Robertson together that I would confirm this on receipt of a proper application. The approach was received from Robertson, but not from the mission, and I therefore decided on 26 May to authorise Robertson to operate VH-AQW on mission work, in accordance with the provisions of the mission's aerial work and charter licences, and subject to compliance with the Air Navigation Regulations, Air Navigation Orders, and the provisions of the mission's maintenance manual, operations manual, and loading system...

'While flying an Avro Anson aircraft in the Highlands with the chief airport engineer during the first week in July, I noticed that the mission's Dragon aircraft, VH-AOE, was operating in the Highlands with cargo, and it was being flown by Walachy. My subsequent enquiries disclosed that Father Walachy has been operating the mission Dragon in a private capacity in the Highlands for some time...

'Further enquiries on the activities of D.I. Robertson and VH-AQW disclosed that he had set up a bi-weekly regular service operating to a schedule over the route Madang-Goroka-Madang, and that VH-AQW had, in fact, been utilised on very few occasions on mission work. The matter of Father Walachy's operations has been investigated by the Superintendent of Air Navigation. Immediately on discovery of the activities of Madang Air Charters mentioned above, I stopped their air charter operations, the regular scheduled service, and informed them quite clearly that they had only been authorised to operate on behalf of the mission.'

It was unfortunate for Madang Air Charters that they had fallen foul of the Regional Director even before they had a charter licence! John Arthur usually supported small operators, but he could be perverse when crossed. Robertson and Ferguson worked hard to produce the operations and maintenance manuals, and again submitted an application for charter and aerial work licences. There was, however, a problem. Dave Robbie had left MAL, but Wally Ferguson was branch engineer for Qantas at Madang, and he was not yet ready to resign. So the applications were submitted in the names of the wives of the two men; June Robertson and Hilda Ferguson. The application specified the use of Dragons equipped with Command SRC274NX radio. Maintenance would be done by Ferguson — who held B, C and D tickets — and Robertson, who it will be remembered was a pilot/engineer.

In an earlier letter, Robertson and Ferguson had advised John Arthur that Madang Air Charters intended to establish a regular link between Madang and the Highlands, perhaps employing a Cessna 170B. As soon as their licences were granted, they would go ahead with the erection of hangars, stores and a workshop, and would obtain essential spare parts.

Forwarding the licence applications in September, 1953, Arthur noted that although Mrs Robertson and Mrs Ferguson were registered as the nominal owners of Madang Air Charters, the business of the company was being conducted by their husbands. He recommended that the applications be refused. Qantas, MAL and Bill Passlow were meeting all of Madang's needs. John Arthur continued: 'In the light of the foregoing, and as Madang Air Charters is not properly set up, and has only undertaken to set itself up properly if we grant them permission to operate...I consider that a charter licence should not be issued, irrespective of the fact that the operations manual, maintenance manual, loading system and company registration are in order.'

The Director-General could see 'no technical or operational objection' to the granting of licences to Madang Air Charters, despite John Arthur's recommendations. Charter and Aerial Work Licence No. 195 was therefore issued, in the names of Mrs Robertson and Mrs Ferguson. The way was now clear for the company to legally operate, although Dave Robbie and Wally Ferguson kept a wary eye out for John Arthur, and stuck strictly to the regulations.

During 1954, Dave Robertson flew VH-AQW into most of the airstrips of the hinterland, and even into the remote Southern Highlands — a long trip in a Dragon.

DCA was now trying to discourage the use of the venerable DH84. It was obsolete, vulnerable to tropical conditions, lacking in high-altitude performance, and increasingly costly to repair and maintain. Modern light aircraft were at last becoming available at reasonable cost, and the department frankly wanted to see the end of the elderly British biplane that had served PNG so well for more than twenty years. From 1954 on, 'DCA began bluffing about radio for the DH84', in Ray Harris's words, and eventually new performance charts were introduced that virtually drove the Dragon out of the skies. A spate of Dragon accidents in 1954/55 hastened the inevitable end. On 20 January, 1954, an MAL Dragon ran off the end of Wewak strip. Pilots Norman Lather and William Englebeck and several native passengers were bruised and shaken, and the Dragon was badly damaged. Following this incident, DCA officer, W.J. Mort, said that DCA had 'tightened flying regulations in New Guinea to reduce small plane accidents...all small planes, including Austers, now had to carry radio...'

Carrying radio was not yet mandatory, although it was on the way. Another Dragon crash occurred in August. Pilot Jim Davis was at the controls of MAL Dragon VH-AOP when the brakes failed while taking off from Hayfield. The Dragon swung off the strip into a deep ditch, and burst into flames. Davis and his seven passengers were uninjured, but the Dragon was a total loss.

These Dragon accidents were bad luck for MAL. It will be recalled that following the loss of the Dragon VH-BDB piloted by John Jenkins in April, 1951, the Director-General ordered a full investigation of MAL's operations to be carried out. For a variety of reasons, this was not immediately done and in the meantime the management and operations procedures of MAL were thoroughly overhauled. In June, 1951, Doug Elphinstone was given exactly seven days to set up an approved training and checking organisation for his pilots — and did it. In a letter to the Director-General, John Arthur wrote: 'Mandated Airlines Ltd have displayed a marked reluctance to incur any expenditure whatsoever on the proper training and thereafter periodical checking of air crews. It has been necessary to force this company to contribute a percentage of its revenue to the effective training and checking of its air crews. Company pilots have demanded this, and it can only be done with the constructive assistance of the Department...

'Overall "bush pilot" days in New Guinea are rapidly drawing to a complete end, except in the uncontrolled areas where development is taking place. As Mandated Airlines Ltd is in the final stages of transition from a charter organisation to an extensive regular public transport organisation, the whole of the company's organisation and administration must be developed to meet the higher efficiency necessary for the regular public transport category...'

The chief MAL check pilot was Les Davis, and he began to put the company pilots through a series of instrument rating and night flying tests. Davis himself, and senior captain Jim Perry, were about to go to east Africa to fly back a pair of DH104 Doves, which the company hoped would enable the phasing- out of their Dragons. Before they left, both Davis and Perry were tested by a DCA examiner of airmen.

The de Havilland Dove was a delightful little aeroplane, designed in 1943 specifically as a replacement for the DH89 Rapide, which was itself a development of the Dragon. It was one of the most successful light civil transports ever

149. Bill Passlow's Tiger Moth at Vagau airstrip. *Photo: Jim Glindemann*

149

produced in Great Britain. Of all-metal stressed-skin construction, the Dove was a low-winged monoplane with retractable tricycle undercarriage, equipped with two 330 hp Gipsy Queen engines and with seating for eight passengers. It was in every respect a vast improvement on the old Dragon.

It will be recalled that in 1951, Trans Oceanic Airways entered into an agreement with MAL to provide feeder services for its short-lived flying-boat operation. The Dragon was clearly not the right aeroplane for this work, and Doug Elphinstone looked around for better ones. He remembers: 'Eventually, I got onto Shackleton in London, and we were able to buy two Doves from East African Airlines for £15,000. Remarkable little aeroplanes. What decided us on them was that Western Australian Airlines had them, and had had fantastic service out of them. They arrived, and we had just started to operate them when a wing came off one in W.A. — they apparently had a fault in the spar — and everyone on board was killed. So that put a mocka on the Dove, and DCA wouldn't let us fly them until the spars were modified. Well, you can imagine, brand new aeroplanes and not being able to fly them, and the kicks I was getting from Sydney for buying something that couldn't be used...but they were beautiful little aeroplanes.'

Les Davis and Jim Perry travelled to Nairobi, East Africa, in June, 1951, to ferry the 'brand new' second-hand Doves back to Lae. Jim Perry recalls: 'A pleasant interlude in Nairobi, a quick endorsement on the Dove, and we were off on a solo journey, taking in many airports and countries over twenty days and seventy-seven flying hours, from 7 to 26 July. Although the Dove was the first of the mini-airliners and an absolute pleasure to fly, the performance figures as claimed did not hold true for our tropical climate. It was not, therefore, the magic replacement for the DH84, but for approximately two years flitted about the Territory skies, much admired by passengers and pilots alike...'

The two Doves, VH-MAL and MAB, were formally christened with the names Rai and Talio at a ceremony at the MAL hangar in Lae. Barely a month after the arrival of the Doves, VH-MAL was the centre of a drama that aroused world interest. Jack Stammer tells the story: 'The Dove was very modern for New Guinea. In August, Les Davis checked me out on MAL, and we took it up to Wewak, trying it out on the different strips in that area available to it. Went into Hayfield, Angoram, Tadji, Maprik — was a bit dicey on Maprik. After about three days we started back to Lae. We landed at Madang to refuel. Les Davis, you might remember, was always the most immaculate person, his dress never seemed to get ruffled or anything. My wife, Lillian — she had been a secretary with Guinea Air Traders — was with us as well and before leaving Wewak she ironed a clean uniform for me and I changed into it at Madang, because we knew half of Lae would be out to see the Dove land. A scrupulously pressed, beautifully clean uniform...

'After leaving Madang, Les discovered that the pneumatic system controlling the undercarriage wasn't working, although it had functioned perfectly in the Wewak area. Over Lae, we managed to get the main wheels down using emergency methods, but the nose-wheel only went a third of the way down, and wouldn't lock. So we couldn't land. Couldn't get the main gear up for a belly-landing either, so it looked like a crash-landing was inevitable. We circled Lae for over two hours, to burn up as much fuel as possible...'

150. Captain Jim Perry. *Photo: Jim Perry*

150

Meanwhile, on the ground there was 'a bit of a panic' in the words of the MAL chief pilot, Tom Deegan. Engineers and pilots examined the other Dove, VH-MAB, to see if there was any way of forcing the nose-wheel into the locked position while in the air. There did not appear to be. Then Jim Perry and Tom Deegan took off in MAB with a load similar to that carried in MAL, and made a landing. There was no way that the nose-wheel could be kept from making contact with the strip at a safe speed. Says Deegan: 'We estimated that the nose-wheel would hit the deck at 60 mph, which wouldn't do the aeroplane much good. Then I remembered that when I was in Noumea during the war, a Liberator bomber came in with a stuck nosewheel. Apparently they'd cut a hole in the floor somehow, and dangled a fellow down by his arms, and he pushed the nose-wheel down with his feet, and locked it. So I opened the front baggage compartment of the Dove, in the nose, and there was the little box that the nose-wheel recessed into when it came up. I looked up, and I could see the floor where the pilot sat. I knew they had a crash axe in the Dove, and I wondered if they could cut a hole in the floor...'

Overhead, Les Davis had almost exhausted the possibilities open to him. The emergency air bottle had failed to lock the wheel, and several risky 'bumps' on the strip with the main gear to try and jolt it down had not worked. Deegan's solution looked to be the only way to avoid what could be a very dangerous crash-landing. While Les Davis remained at the controls, Jack Stammer grabbed the axe and set to work. Deegan and the others on the ground were wondering: even if Stammer could cut through to the nose-wheel, how was he to push it down?

In Wewak, Lillian Stammer had seen a length of conduit lying around that would be perfect for a curtain-rod for her home in Lae. In 1951, housewives had to use what they could find. She had thrown the rod into the back of the Dove just before takeoff. It might have saved the lives of all on board MAL on that August day. Stammer continues: 'I chopped a hole under the co-pilot's seat which let me get into the nose baggage compartment, and then I chopped a hole through there into the nose-wheel, and I got Lillian's lump of conduit onto the axle of the wheel, and finally managed to push it down into place by which time, having cut my way through two layers of aluminium with jagged edges my arms were covered in blood, I was sweating like a pig, and my immaculate uniform had suffered greatly. But on the ground, Les as usual got out of the aircraft, ran his comb through his hair and put his hat on, and looked as though he was just starting a day's work instead of finishing a rather tortuous one...'

Despite this experience, Jack Stammer — like most pilots who flew this little aeroplane — loved the Dove. Just after Qantas took delivery of the first Drover, Stammer was flying a Dragon from Lae to Slate Creek: 'I was climbing to go through the pass into the Snake valley — bit quicker that way than going through the Mumeng gap — when Bill Forgan-Smith in his Drover came and shot past me, illustrating the high performance of this aeroplane. I finally caught up with Bill when I was flying a Dove. He was going somewhere in his Drover, and I got a little bit above him and then dived down alongside him with one engine feathered, and shot past him to show him how good the Dove was in comparison with the Drover...'

Youthful high spirits in an aeroplane had earlier landed Jack Stammer in court. He had flown a load of passengers in a Dragon from Lae to Finschhafen: 'I dropped them off at the strip, and took off for the return to Lae. Couldn't see any sign of the jeep coming for them from the station, a mile or two away, so I decided to entertain them with a bit of low flying. I made a pass down the strip just off one side. It was quite clear there, but I hadn't noticed a telephone line, the only one in the area, and the rudder clipped and cut it. There was a great fuss, both DCA and the District Officer, Horrie Niall, were rather upset. So I had to go in front of Horrie, and he fined me £55 for the effort, and at that time I was only getting about £13 a week. Couldn't afford to pay the fine, so I borrowed the money from the senior DCA air traffic controller, Bert Garland. Had to pay it back at £5 per week for ten weeks...'

Many a pilot in pride and exhuberance 'beat-up' airstrips and settlements in those care-free days, and it was Stammer's misfortune that his escapade coincided with a decision by John Arthur to clamp down on such activity. Jack Stammer joined Qantas in 1952 and continued to fly in PNG until 1958, when he started on the road that was to end with his retirement in 1978 as a well- respected Boeing 747 captain, to whom the 'beat-up' of an isolated airstrip would be unthinkable.

The pretty little Doves did not last long in MAL service. They proved unsuitable for New Guinea conditions, and were sold to Airlines of Western Australia; MAL in December, 1953, and MAB in April, 1955.

By this time, Doug Elphinstone was a coffee planter at Goroka, having had more than enough of airline management. Harold Hindwood replaced him as MAL manager in June, 1952. The long-threatened investigation of MAL operations finally took place in September, 1953, and was conducted by Tom Drury. He found that Harold Hindwood was exercising 'very thorough supervision of sectional heads and other staff', and that company operations were now safe. But management was 'severely handicapped' by 'antiquated and dilapidated hangar facilities at Lae', which were MAL, not DCA, property, and well below DCA standards. In fact, hangar and workshops were frequently flooded in the wet season.

At this time, Tom Deegan was MAL Flight Superintendent, responsible for 'safe and efficient flying operations'. Deegan was in charge of eleven pilots. Les Davis was senior check captain, and Jim Perry the DHl04 check pilot. Drury found that Deegan policed air navigation regulations 'very thoroughly' and carried out periodical flight checks on all pilots. Deegan, with a total of 8,000 flying hours, mostly on DC3's, was one of the most experienced pilots in PNG.

Drury found that the MAL training organisation, nevertheless, still needed to be improved. No regular night-flying training was being done. 'The fault is not with Deegan, but with the directors of the company, who are reluctant to release funds'.

The MAL engineering section was skilled and efficient, headed by chief engineer Bill Humphreys, who had eight engineers at Lae, one at Madang where a DC3 and crew was based, and two at Wewak, where two Dragons and three pilots were stationed, under the supervision of agent Alan Corrigan. The MAL traffic manager, Doug Treeby, was capable and experienced, but Drury found the Lae passenger lounge 'a disgrace'.

Drury concluded his report with words of praise and blame: 'The company has a sound and well-organised system of maintenance, operations and traffic handling, but is primarily handicapped by a complete lack of reasonable facilities...it is considered that the reluctance of management to renovate hangar, storage, cargo and engineering facilities is brought about by indecision as to the proposed move to Malahang, if, and when, this project takes place.'

Clearly, MAL was living up to its nickname of 'Mean and Lousy', but official dithering on Malahang versus Nadzab was plainly having an adverse effect on aviation development at Lae. Despite poor facilities, MAL was a happy company with an excellent safety record. Prior to the two Dragon incidents referred to earlier in this chapter, MAL had been almost accident-free for two years. In May, 1952, Tom Deegan had the bad luck to stand a DC3 (VH-MAC) on its nose at Kaiapit when he landed there with an unsuspected flat tyre, but nobody was hurt and the aeroplane was not damaged.

Deegan had a narrow escape in a Douglas in February, 1953. He was coming in to land at Madang in VH-MAE with six passengers when he saw right ahead of him a Qantas DC3. He recalls: 'The bloke in the tower let Qantas taxi out in front of me! I'm just about to touch down. The bloke in front was talking to the tower, so the tower couldn't warn him. I touched down, and started to catch up on him at a great rate of knots. Instead of running up his backside, I decided I'd try to make a turn-off into the taxi-way. I pulled the wing out, but there was green algae slime on the strip, and as soon as we hit it the kite started to go sideways, and down into the drainage ditch we went.

'Stopped right in front of the tower. They could hear me swearing in the tower with no trouble at all. The bloke who'd given me the landing clearance ran out of the tower — he thought I was going to kill him.

'We got Tom Huxley to drag the Three out of the ditch — he charged us about two thousand quid. An undercarriage oleo leg had gone, so Bill Humphreys went across to the old DC3, a wartime prang, that was behind Father Joe's mission hangar. We pulled an undercarriage leg off that, and used part of it to fly our aeroplane back to Lae. Should have been changed at Lae, but they left it on, just painted it silver and left it. And that was the cause of a prang years later at Wapenamanda...They were great aeroplanes, the Threes...best of the lot. There were no fatal DC3 accidents in PNG, just incidents...'

151. Laurie Crowley pilots his original 170 (CSD) taken from Territory Airlines' original 170 (GKA). *Photo: Jasper Maskelyne*

151

The DC3 was the DC3, and there never was another aeroplane like it. But the Dragon was another matter. It was almost at the end of its time. In January, 1955, John Arthur suspended all Dragon operations into the Western and Southern Highlands. Dave Robertson of Madang Air Charters immediately protested, claiming that the move was designed to protect a Gibbes Sepik Airways monopoly over these routes. A lively verbal battle in the pages of *South Pacific Post* followed between John Arthur and Madang Air Charters, but the ban on Dragons, except in emergency circumstances, remained.

Dave Robbie was anxious to obtain permanent work for his little company in the Highlands. In October, 1954, the MAC DH84, VH-AQW, became the first Dragon in PNG to be equipped with radio, and Robertson hoped that this would persuade DCA to allow him to continue to fly into the Southern Highlands. John Arthur, however, had long since concluded that the Dragon was unsafe to operate over long distances into high altitude strips in remote locations, and gradually the viability of the Dragon was reduced by revised performance charts and the introduction of increased reserve fuel requirements. The Southern Highlands had been pioneered by GSA Austers and Qantas Dragons and Catalinas, but Qantas had withdrawn the DH84's from the route for safety and operational reasons, and the Gibbes Norsemans had filled the gap. There was no room for Dave Robbie — the Dragon, in any case, could never have competed with the Norseman.

Dave Robbie perhaps had reason to feel aggrieved. In October, 1954, the new airstrip at Erave in the Southern Highlands was ready for an initial landing. GSA did not have an aeroplane available, so DCA asked Madang Air Charters to do the job. Dave Robbie made the landing, and then improved the shining hour by ferrying cargo between Minj and Erave on behalf of the Administration for some weeks. Naturally, he hoped to continue and develop this business, but the decision to suspend Dragon operations forced him to stop. It was a bitter pill to swallow.

Dragons continued to have accidents. A remarkable incident in April, 1955, however, proved how tough these little aeroplanes really were. The Dragon involved was TAL's VH-AIA; the pilot, A.V. Chadim.

Tony Chadim was a Czechoslovakian, who started flying gliders and sailplanes in his native land just after the Second World War, when he was sixteen. Before he was eighteen, Chadim held the international Silver C certificate, and was an expert glider pilot — experience he was to find invaluable years later, when flying under-powered Dragons in New Guinea.

In 1947, Tony Chadim joined the Czechoslovak Air Force, as an officer cadet. The Air Force was at that time directed by patriots who had fled Czechoslovakia when Hitler's armies marched in, men who thereafter fought the Luftwaffe from the ranks of the Royal Air Force. In February, 1948, the Communist coup in Czechoslovakia altered the course of history in that unhappy land.

During the purges that followed, most of the ex-RAF senior officers of the Air Force were sacked, arrested, or again forced to leave Czechoslovakia. Chadim and his fellow cadets found their flying training was continually interrupted by petrol shortages and political interference. By mid-1948, only fifty of the original group of eighty-five officer cadets remained. They were under severe pressure to join the Communist Party; those who refused were dismissed. Just before Christmas, 1948, sixteen officer cadets were sacked, including Tony Chadim and his best friend, Jaroslav (Fred) Tosner. They, and other cadets, were transferred into various army units to complete their compulsory military training.

Chadim and Tosner began to plan to escape from Czechoslovakia. Over a period of time, they acquired a compass, automatic pistol, civilian clothing, and documents. Chadim remembers: 'My mother bought fifty Austrian schillings for an exhorbitant price on the black market, somebody else managed to steal a map from a government office, and Fred pinched a teargas grenade from his unit's ammunition store. He also got somewhere a book by a Doctor Alfonso, *English in Useful Phrases*, which we tried to memorise...'

Their preparations complete, Chadim and Tosner applied for weekend leave, made certain alterations to their leave passes, and after many adventures succeeded in reaching Vienna, and the West. Both eventually ended up in Australia.

Both young Czechs wanted to continue with their interrupted flying careers, but the RAAF was not accepting foreign-born applicants in the early 1950s. Commercial aviation seemed the only possibility. The language difficulty was enormous, but both mastered enough English to pass the written examinations for commercial pilot's licences. Flying jobs were not easy to find for brand-new foreign-born commercial pilots, and the paths of the two young friends now diverged. Tosner became a flying instructor with Kingsford Smith Flying Service, and a few years later was killed in a mid-air collision. Chadim went to New Guinea.

In Melbourne in 1952, Chadim had met a fellow-countryman, pilot Mat Matus, who had also escaped under dramatic circumstances from Czechoslovakia. Matus had just obtained a position with Gibbes Sepik Airways. Bobby Gibbes was being pressed by DCA to install radio equipment in his Norsemans, and Matus told him that he knew just the right person to do the job: Tony Chadim. So Gibbes offered Chadim a position as radio maintenance engineer, with the chance of a job as pilot later. Says Chadim: 'I had to brush up on my knowledge of electronics quickly in the Melbourne Public Library, because I had to pass more exams to get the various radio licences. I could not get the Aircraft Radio Maintenance Engineer's Licence until I got the Aircraft Radio Operator's Licence, which I could not get until I had the P.M.G. Certificate of Proficiency, and for that I had to be able to receive Morse code at twelve words per minute...'

Chadim passed his exams, installed the Lear radio equipment in the GSA Norsemans, and serviced them (until AWA set up a workshop at Goroka some years later, and took over). He did a little Norseman flying after Mat Matus and John Wells left GSA. As recorded in Chapter 4, Territory Airlines, Matus joined Territory Airlines late in 1954. Noel Peacock had departed, and Les Taylor was about to go. In February, 1955, Mat Matus decided he was not interested in a permanent position with TAL, but would rather devote most of his time to his business interests.

The TAL pilot situation was becoming serious. Jack Gray looked around for another pilot, and offered the job to Tony Chadim. Chadim wanted to fly, but he did not want to leave GSA in the lurch, so he came to an agreement with Vic Cox to continue to service and sign out GSA radios while flying for TAL. Chadim joined TAL on 7 March, 1955. He was given a quick DH84 conversion and route familiarisation by Mat Matus.

On 24 April, Tony Chadim took off from Lae in VH-AIA for Boana, a small airstrip on a flat-topped hill flanked by a deep, narrow river gorge and surrounded by high ranges. It was a Lutheran mission stronghold, less than twenty minutes flying time from Lae, the centre of a considerable population of mountain tribesmen, who had been taught by the industrious Lutherans to grow superb European fresh vegetables. Laurie Crowley had got his start buying vegetables at Boana and selling them in Lae, and TAL, too, earned much-needed revenue in this way in the early days. But Boana was a dangerous strip for the inexperienced. Says Chadim: 'Dragons, if kept to Air Force standards of serviceability, would not have been such a bad aircraft. Their handling characteristics were quite pleasant when not overloaded, typically English or European, quite unlike the small American aircraft like a Cessna. They were hopelessly under-powered, at least with the worn-out engines we had. In the Highlands, this meant that full power was applied for take-off, and the only adjustment for a climb — if you could call it that — was to synchronise the engines, no throttling back. The nut of the throttle levers was tightened, and that was that.

'The slow speed of the Dragon was an advantage in bad weather, one could turn in a small space. Because the windscreen was made of two pieces of perspex joined in the middle at a relatively sharp angle, not much rain water accumulated on it, and this, together with the triangular side windows which could be opened, gave the pilot good visibility.

'The fact that one was never too far away from treetops in a loaded Dragon was also of some use. Later, on my regular routes, I came to know every tree that looked different from the others, and that was very handy in emergencies. I had one hair-raising experience when I had to go to Kerowagi one day on a medical emergency. Near Mount Elimbari, the ridge was in thick cloud, and after some probing I found a tiny sliver of light. I managed to enter the Chuave area between two tall trees whose tops were in the cloud. If I had not known exactly what lay behind them I could not have done that, neither would I have succeeded in any other aircraft. I had to dodge the treetops between Chuave and Chimbu, but I arrived in Kerowagi alive, to my surprise...

'Flying the Dragon from Lae to Goroka in the afternoon was sometimes a nerve-wracking experience. The weather in the Markham valley was usually quite good and visibility excellent, with huge banks of beautiful white clouds at the entrance to the Highlands at Karanka. It took three quarters of an hour to reach Karanka in the painfully slow Dragon, and a lot of cloud buildup can take place in that time.

'So it meant sitting in the cockpit, and wondering whether there will be any gap between the clouds, and watching them for forty-five minutes whether they grow faster than the Dragon could climb. Very often, I could have made it to Goroka in full sunshine, if only I could have climbed two or three hundred feet [sixty to ninety metres] higher. Instead, I had to dive under and struggle through the turbulent Arona gap or other narrow entrances. The prospect of having to go back to Lae when the clouds were too high and the mountain ridges covered was even worse than negotiating narrow valleys in bad weather: BP's freezer would have already closed, and 1,200 pounds [544 kilograms] of meat would be melting in the Dragon overnight...

'At the time of my Boana episode, I had only thirty-nine hours on Dragons. The accident was a simple combination of inexperience and misadjusted engine idling speed. Because of the poor compression and resulting problems with the spark plugs oiling up, our engines were set to idle rather high, and it was necessary to touch down very close to the end of short airstrips like Boana — I had the impression that one must land the Dragon within a few feet of the threshold, otherwise one would fall off the abyss at the other end.

'This was only my second landing at Boana. I brushed the top of a large clump of *kunai* grass that was growing on the airstrip's threshold with my left wingtip. That turned me a little, and I did not have time to straighten the Dragon before touchdown. As a result, I ground-looped, and the tail of AIA broke off...'

When the dust had settled, Chadim was unscathed, but AIA was in a mess. The entire tail section of the Dragon had been torn off, and other damage had been sustained. It was the kind of accident that would require a complete reconstruction with a modern aircraft.

It will be remembered that MAL sold TAL the dismantled Dragon, VH-AAC, which had been badly damaged at Slate Creek in January, 1951. The various bits of AAC were in storage in Lae. The tail section was in excellent condition, and Jack Gray decided on a radical solution: he would attempt to splice the tail of AAC onto the fuselage of AIA.

The tail section was carried overland for three days into Boana by a group of enthusiastic Boana schoolboys, accompanied by Tony Chadim. Jack Gray then flew in to Boana in the Tiger Moth, AQS, and in less than a week succeeded in fitting the tail section onto AIA. Laurie Crowley then flew Mat Matus in, in his Curtiss Robin, and Matus piloted AIA to Lae, escorted by the Tiger Moth and the Robin. In Lae, AIA was very carefully inspected by a DCA aircraft surveyor before being allowed to depart for TAL headquarters at Goroka. Says Chadim: 'The rigging must have been improved by the repair, because AIA flew a few knots faster after that. That was not the first time a tail broke off a Dragon. Apparently, one of my predecessors did something similar at Wantoat or somewhere, except that DCA never found out about it. He had a bundle of *sarifs** on board, drilled holes in them and spliced the tail on with the self-tapping screws he pulled out of the Dragon's internal linings. He flew to Goroka very carefully in the last light, and by morning all traces had been covered — the whole company worked all night. That was before my time...'

One of the saddest of all Dragon accidents took place soon after Chadim's experience at Boana. On 2 August, 1955, Bill Passlow was flying VH-AOT from Madang to the Seventh-Day Adventist Hansenide colony at Togoba, in the Western Highlands. The Dragon was carrying a full load of freight and passengers — SDA pastors Martin, Grieve and Gander. AOT was not fitted with radio, and nobody will ever know just what happened, but the aeroplane crashed into the side of a high mountain gorge. Bill Passlow died instantly. Pastor Martin was seriously injured; Grieve and Gander were also hurt.

It was the first PNG air fatality since the death of Dr John McInerney in February, 1952. During 1954, there had been 66,700 aircraft movements throughout the Territory, all but a minute percentage trouble-free. Critics always seem to overlook such figures when discussing air crashes in PNG.

From his bed in a Sydney hospital, Pastor S.H. Gander made a public statement three weeks after the crash: 'Bill Passlow definitely sacrificed his life so that we, his passengers, might live. He could see the crash coming, and planned his actions to the last degree...

'He flew the plane on the port side of the gorge, thus giving security to the passengers, strapped on the starboard side. He tried everything he knew. He at last said, "Gentlemen, hang on tightly. We are not going to make it"

'I saw the port wing go through the grass on the wall of the gorge, then a small bush, and, finally, I was unconscious. At the critical moment he had switched the ignition off, thus saving us from certain incineration. At that moment, Bill met instant death...'

Bill Passlow left a widow and four children. A public collection of funds for their benefit was well supported throughout PNG. With Passlow died his little charter company, Riverlea Market Gardens.

It is less than fair to missions and missionaries, but from the earliest days of aviation in PNG, pilots and air companies have often suspected air charter loadings of certain missions. Mission organisations in general were invariably short of funds, and air charters were invariably expensive. To the missionary,

*Long knives of hoop-iron, used for cutting grass.

every pound saved in air charter expenses was a pound that could be better applied to the welfare of the people of PNG. Pilots sometimes harboured unworthy suspicions that some missionaries deliberately under-stated the weight of their loadings. Certainly, missionaries in general usually insisted on using every bit of available aircraft capacity.

After Bill Passlow died, a whisper spread through the ranks of the pilots: the Dragon had been overloaded with SDA cargo. It was always fervently denied by the mission.

One pilot recently wrote to the author: 'Bill Passlow got killed because he was overloaded, and could not outclimb the rising terrain. After this, I and several other pilots insisted on personally checking all SDA cargo for weight. Once I checked a load in Lae, and it added up to 1,197 pounds [542.9 kilograms] when the limit was 1,200 pounds [544.3 kilograms]. The SDA pastor said, "Oh, excuse me, I forgot something", jumped into his jeep and before I knew it, he was off. Here I was, sitting in the stinking heat, and he came back from Chinatown with three loaves of bread!' The last DH84 Dragon to be lost in PNG crashed in November, 1955.* The pilot was Dave Robertson. Madang Air Charters was managing to keep going, despite the loss of the Southern Highlands work to GSA. A third principal had joined Dave Robbie and Wally Ferguson. Clive Sanderson was a young bank officer who bought in to the company, and who took over the bookkeeping. Later, the trio invested in a number of other concerns, which gave Madang Air Charters (which became Madang Air Services Ltd in September, 1956) a firmer financial base.

On 8 November, Dave Robbie took off from Madang in the Dragon with two passengers, education officer Tom Stanley and patrol officer Barry Griffin, and a load of building timber. Just after becoming airborne, an engine spluttered and cut. Robertson tried to turn back. He remembers: 'With the experience I have today, I'd have gotten away with it. Because of the wind direction and speed, if I'd turned into the dead engine, kept my speed up, and been carried by the wind, it would have carried me straight round, and onto the strip. I reckon I'd have made it. But it was inexperience on my part. I'd been taught never to turn into a dead engine and things like that. As it was, I almost got back, going in the opposite direction...sort of carried me out in a larger arc, and I got tangled up in some of the growth at the bottom of the garden so to speak, and ended upside down in the harbour.

'Nobody was hurt, although I got a punch in the chest getting out of it. I'd been in and out of a Dragon so much that even though I was upside down, I knew just how to get out. When I got out into the cabin section I could see a bit of light in the front window, and I smashed the perspex out and went through there. Got scraped on a jagged bit of perspex. Got out, but I couldn't see the other fellows, so I ducked under the water and went round to the other side of the aeroplane, looking for them. At the same time, they were doing the same thing, looking for me. Eventually, we all popped up together. We got onto the wing, and boats were by this time on the way. I was covered in blood by then.

'My thoughts were that the Dragon could have been salvaged, but Wally said "It's been in the water, wood and rag"...so they got a boat alongside it and hoisted it up, got a sling around it, and as they lifted it, it crumbled. Not designed to be lifted. So it never flew again..,

'I guess I flew the Dragon everywhere, and I only ever had one mechanical problem apart from the Madang one. I was coming back from Hagen to Madang, fortunately very very light. Always made it a habit when crossing those areas to get as much height as possible, and I was at 9,000 feet [2743 metres] or so. One engine just went, Bang! Threw a conrod out the side, broke the engine mounts. I reckoned I was nearer to Banz strip than anywhere. I tried to stop the engine from windmilling, I'd stalled it a few times, but it kept turning, windmilled all the way back to Banz. I got over the ridge into the Wahgi valley and could see Banz...I kept saying to myself, Will I make it, or won't I make it. Slowly going down, airstrip getting closer and closer, couple of Dragons sitting on the top end of it...

'I got close to the strip, and knew I was going to make it. I thought, Be my luck for somebody to take off, and that's exactly what happened. I'm on the way in and couldn't do anything about it, no radio, and he took off. Fortunately, he spotted me in time, and veered off just after he was airborne. If I'd had a load on, I'd have been in trouble...'

152. Mat and Lidia Matus with Joan and Jack Gray. *Photo: Joan Gray*
153. Mandated Airlines DC3 on its nose at Kaiapit. Pilot Tom Deegan looks out from the hatch. *Photo: Tom Deegan*
154. Madang Air Charters' first aircraft, a DH84, VH-AQW, at Faita airstrip in the Ramu valley. *Photo: Dave Robertson*
155. Salvaging Dave Robertson's Dragon from Madang Harbour. *Photo: Isobel Pert*
156. DH83 Fox Moth similar to the aircraft acquired by Territory Airlines (VH-URI) in the mid-1950s; C of A was never completed. *Photo: Talair*
157. Territory Airlines Dragon VH-AON ready for departure to Australia, December, 1959. Pilot, Dick Cresswell. *Photo: Talair*

*I have been unable to discover details of this Dragon.

known as Missionary Aviation Fellowship, and Hartwig decided that he would form a fellowship in Australia. At a meeting in Melbourne on 30 June, 1947, it was resolved that an autonomous body called the Australian Missionary Aviation Fellowship be formed, which would request affiliation with the fellowships in Britain and America. At the second meeting, on 18 August, Mr (now Sir John) Nimmo was elected president, Harry Hartwig secretary, and B. Morton, treasurer. Headquarters of the MAF in Australia were — and remain — located in Melbourne.

The first fields of Australian MAF endeavour were to be in northern Australia and New Guinea. Harry Hartwig was appointed full-time field worker. A survey of northern Australia followed, and then a survey of New Guinea. As a result, it was decided to establish MAF flying programmes supporting mission work in both regions. An air service to support the Lutherans in New Guinea would begin the work.

The MAF in PNG was to eventually provide flying services to many Protestant mission societies, including Anglican, Apostolic, Assembly of God, Brethren, Lutheran and Methodist missions, and interdenominational groups such as the Christian Radio Missionary Fellowship, East and West Indies Bible Mission, Evangelical Bible Mission, South Sea Evangelical Mission, Unevangelised Fields Mission, and others. While the various Catholic missions tended to develop their own small air operations, the Protestant missions were almost all served by the MAF, which was virtually an inter-mission co-operative organisation, which freed individual missions from having to operate and maintain expensive aircraft that would perhaps be only used on occasion.

The question of a suitable aircraft for New Guinea conditions was finally resolved by a decision to purchase an Auster. By the very nature of its operations, MAF throughout its existence has almost always used small, versatile aeroplanes, able to get in and out of short, rough strips. The Cessna 170 series, which with the C180 and C185 were to become the favoured MAF aircraft in Mexico, Peru, Ecuador, Ethiopia, Sudan, and Dutch New Guinea as well as PNG, was not introduced until 1948, and the model did not really catch on until 1953, when the C170B became available, fitted with slotted Fowler-type flaps which dramatically improved low-speed performance. In 1950, when the decision to enter New Guinea was made, the Auster seemed to be the most suitable available aeroplane. It had a simple but rugged airframe, it had good short take-off and landing performance, and it was fitted with the reliable Gipsy Major engine.

A J5 Autocar was ordered for delivery in 1950. It was agreed that the aircraft would be provided by the Lutheran mission, while MAF would contribute a pilot, and a pilot/engineer, and operate the Auster. At this time, Australian MAF was still in its infancy, and had but one pilot available, Harry Hartwig. The American MAF was asked to send a pilot/engineer. Bob Hutchins was sent, thus beginning an association between the Lutheran mission and American MAF that was to last for thirty years and beyond, and significantly affect the development of the Australian MAF.

The Auster, registered as VH-KAN, arrived in Melbourne in February, 1951. In common with most other Austers then, it had no radio, starter generator or battery, and the instrumentation was basic. Hartwig and Hutchins flew KAN to the MAF base at Madang in April, and the MAF-Lutheran mission aviation programme formally began on 7 May.

From the beginning, the programme was a great success. During May, the little Auster with its capacity of 272 kilograms lifted thirty passengers and 6804 kilograms of freight. During June, the Auster flew ninety-five hours and carried twenty-six passengers and 10,506 kilograms of cargo. So successful was the operation that Harry Hartwig began to make tentative plans to extend it to Dutch New Guinea, where the Unevangelised Fields Mission was working under pioneering conditions.

It was not to be. New Guinea was about to claim another victim, an experienced pilot — but not experienced in New Guinea conditions.

On Monday, 6 August, 1951, Harry Hartwig took off from Madang in KAN for the Lutheran mission station at Asaroka, in the Eastern Highlands. He delivered his three passengers, and at 3.15 p.m. left on the return flight to Madang, via the Goroka (Asaroka) Gap. He was the sole occupant. The flight normally took about forty minutes. When the Auster failed to arrive, a search began. Aircraft went out,

158. Vic and Joan Ambrose and children, Banz, 1956. This Cessna 170, VH-BUX, was MAF's second aircraft. It is still flying for MAF at their training base at Ballarat. *Photo: MAF*

158

and Highlands radio operators were asked to be on the lookout. Madang strip was lit for a possible night landing.

On Tuesday, the crew of a searching Qantas DC3 noticed two natives in the branches of a tall tree in the cloud-shrouded Goroka Gap, flashing a mirror to attract their attention. Nearby, a thin column of smoke was rising. But the weather was bad, and the search had to be broken off.

In the meantime, a mission teacher at an isolated outpost saw an aeroplane circling in the gap, and then heard the sound of an impact. He sent word to Goroka, and a ground party was hastily organised and left for the scene. Lutheran missionaries Ralph Goldhart and Russell Shearer set off independently on horseback, and after camping overnight in the bush, met the ground party next morning. At about 9 a.m. villagers called to them from across a deep mountain gorge: the Auster had crashed; the pilot was dead.

The wreckage was found near the top of a high ridge by the gap. Hartwig had obviously been killed instantly. What happened was impossible to establish, but the probability was that Hartwig had been caught by the notoriously changeable weather pattern in the gap, had been battered by turbulence common in the mountain heights, had lost control — and crashed.

This was a sad blow to the MAF, and the mission. Hartwig, more than any other individual, had been the guiding light of the Australian MAF. The loss of one man, regardless of his standing, could not justify the abandonment of the grand design, but MAF was still a tiny organisation. There was no pilot available to fill Hartwig's place, and for long months, the MAF programme was suspended.

The Lutheran mission turned to the American MAF to re-establish the aviation programme. The Americans, with characteristic generosity, agreed to take the responsibility as agents and representatives of Australian MAF. It was decided to abandon the Auster in favour of the Cessna 170. A C170A was ordered, and Charles J. Mellis, secretary/treasurer of American MAF, was chosen as its pilot. It was further agreed that if possible, Lutheran Mission New Guinea would assume responsibility for its own aviation service at the end of two years.

Mellish was an ex-US Air Force pilot, who had spent most of his time since the war behind a desk. A tall, intense man and an excellent pilot, Mellish was also an aircraft engineer. In October, 1952, Mellish flew the Lutheran Cessna, VH- AMF, to Madang, and resumed the flying programme.

In January, 1953, the president of American MAF, J. Grady Parrot, visited Australia and New Guinea, and at the request of the Australian fellowship made a six-week survey of the then largely unexplored and partly evangelised interior of Dutch New Guinea. Both the Dutch Government and missionary societies welcomed MAF with open arms, for reliable air transport was desperately required. The largest mission in Dutch New Guinea was the Christian and Missionary Alliance, with its main base located on Wissel Lakes. The mission had its own Sealand amphibian, but with twelve stations to supply, and more on the way, a single aircraft was not enough. The Royal Dutch Navy had been providing a Catalina service, which it was anxious to discontinue. Parrot had arrived at an appropriate time.

159. Vic Ambrose farewells Charles Mellis, President of MAF in the United States, after a survey they carried out together in the Dutch New Guinea-registered MAF Piper Pacer, JZ-PTB in 1957. *Photo: MAF*

159

It was agreed that the Dutch New Guinea work would be a joint programme by the American and Australian fellowships. Vic Ambrose was now general secretary of Australian MAF, and at his suggestion another base was established at Wewak, to serve the Sepik missions, and also support the base the Americans were to set up at Sentani, Hollandia.

In the meantime, the MAF-Lutheran mission flying programme continued as a separate project from the Madang base. So successful was the first Cessna that the Lutherans ordered another, a C170B, in 1953. This aeroplane, VH-AMM, was registered on 13 August. It was flown by a Lutheran pilot, Ray Jaensch, who with another Lutheran, Tom Johnson, was to form the nucleus of Lutheran Mission Aviation Department.

Meanwhile, construction of the Australian MAF base at Wewak proceeded rapidly. Funding was always a problem for the MAF. The organisation depended largely on donations, interest-free loans from supporters, and from commercial charter operations whenever the demands of missionary work allowed. Australian MAF did not have the funds to purchase an aircraft, and again the Americans came to the rescue, with a Piper Pacer, VH-MFA. This aeroplane arrived in January, 1954, and was soon flying into the interior, even as far as the new Methodist Overseas

Mission and Unevangelised Fields Mission stations at remote Tari, in the Southern Highlands District.

With two Cessnas and the Piper, more pilots were needed by MAF, and Vic Ambrose was asked to leave his desk and take over the controls of an aeroplane once again. Ambrose had scarcely flown since his war days, but he passed the necessary tests and exams for his commercial licence, and moved with his family to Wewak. Anglican minister/pilot Reverend Doug McCraw became secretary at Melbourne. Charles Mellish transferred to the new Wewak base. Another MAF pilot arrived, Max Flavel. Flavel held maintenance engineer's licences covering airframes and engines, and obtained his commercial licence in 1953. He moved to Madang with his wife, Bessie, a trained nurse, in 1954.

The little aeroplanes of MAF soon became the life-supports of Protestant mission stations throughout the interior of mainland New Guinea, and in Dutch New Guinea. Many flights were made between Wewak and Sentani by MAF aircraft. A second Piper Pacer, on floats, was purchased and assembled in Madang. It was flown to Sentani and given the Dutch registration, JZ-PTB. Charles Mellish and Bob Hutchins made the first float landing in the interior of Dutch New Guinea in March, 1955, when they landed the Piper on Lake Archbold.

MAF pilots were working Christians, and prepared to go almost anywhere to serve the missions. Their aeroplanes were small enough to land on the shortest, steepest airstrips, and they went into landing grounds commercial pilots rarely visited. In July, 1955, another MAF base was set up at Banz. The Lutheran mission's Missouri Synod provided another Cessna for Banz. This was the fourth Lutheran Cessna, a C170B, VH-BUX. Their third, C170B VH-AMO, had been registered in December, 1954. Lutheran Mission New Guinea provided the land for the Banz base, and the Baptist mission, timber for housing. By this time, the Dutch New Guinea work was mainly the province of American MAF, although the Australians contributed pilots when they were required.

One of the most remote airstrips serviced by MAF aeroplanes was Telefomin, in the Sepik headwaters and close to the Dutch New Guinea border. This was in the very heart of mainland New Guinea, and it was a long and often dangerous flight from Wewak in the little Pacer. A Baptist mission station was opened at Telefomin in 1953, and it was the Baptist missionary, Reverend R.W. Draper, who first advised the outside world that the only other whites in the area, Patrol Officer Gerald Szarka and cadet Geoffrey Harris, had been killed by wild tribesmen, together with two of their police, constables Purari and Buratori.*

Pilots and aircraft of Gibbes Sepik Airways were involved in the aftermath of the Telefomin massacre. It was probably the unscheduled landing of a GSA Norseman that upset the plans of the tribesmen — from Eliptamin — to wipe out Telefomin. Following the killings, there was an unprecedented concentration of Administration strength at Telefomin. One hundred police were flown in, with great quantities of equipment and supplies, and a number of experienced assistant district officers and patrol officers led police detachments on long patrols to arrest the killers. These men and supplies were flown in in GSA Norsemans, and MAL Doves.

The GSA manager at Wewak was Robin Gray, a D.F.C. winner and ex-Wing Commander, who had flown with Bobby Gibbes in the Desert. At the time of the massacre — 16 November, 1953 — GSA was pulling out of its financial and operational problems, and was able to provide four Norsemans — VH-BNE, BNT, ASN and ASS — for the emergency. Pilots Mat Matus, John Wells and Tony Chadim made so many flights between Wewak and Telefomin that they ran out of hours. GSA was then short of experienced pilots. Roy Shaw had not been cleared into Telefomin, and Robin Gray decided he would do this. Bobby Gibbes was in Australia, and Rinus Zuydam, the GSA chief pilot, had just resigned to fly for Laurie Crowley in Lae.

Gray had done little flying in recent months, and a few teething problems were still being ironed out with the Norsemans, newly-fitted with the geared SIHI-G Pratt and Whitney motors. These engines, with their three-bladed propellers, were appreciably heavier than those they replaced, with the result that the aeroplanes were slightly nose-heavy, and had to be carefully handled. The problem was cured soon afterwards, by re-positioning the heavy starter battery from the nose to the tail, but not soon enough to save Robin Gray. He and Shaw flew to Telefomin in VH-BNE, on Friday, 13 November — an ill-omened date — with a full

*For details see *Sepik Pilot*, op. cit.

cargo of bagged rice, and one passenger, the personal servant of one of the officers out in the bush chasing the Telefomin killers. Gray and Shaw, of course, had seatbelts in the cockpit, but the native passengers sat, belt-less, on a bag of rice. It was against DCA regulations, but in view of the emergency situation Gray agreed to let him travel.

On landing at Telefomin, BNE hit a soft spot and overturned. The passenger suffered a broken leg, and BNE a shattered wing and strained airframe. The aeroplane had to be repaired in situ by a team of engineers under the direction of the then chief engineer, Ray Pugh, and it was more than a year before BNE was ready to be flown out.

A far more serious consequence than the temporary loss of an aeroplane was the suspension of all GSA operations, on 17 December. The usual DCA investigation followed the accident to BNE, and the fact that the passenger had been carried without a seat and seatbelt was discovered. But this was not the only breach of regulations uncovered by the investigators. Writing to the Director-General, John Arthur listed 'a number of dangerous practices on the part of Gibbes Sepik Airways Limited' that had been disclosed. Inspection periods for the engines and airframes of certain Norsemans had been exceeded; the airworthiness of aircraft had been certified without the relevant logs being available to the engineers; load sheets had been improperly compiled; incorrect tare weights of aircraft had been used; components had been removed from Norsemans to reduce tare weight without advice to DCA, and unauthorised overhaul and repair of aircraft engines had been carried out 'in circumstances extremely unsuitable' — a reference to the occasional field emergency repairs that most operators were forced to make from time to time.

It was a daunting indictment, and when Tom Drury, Superintendent of Air Navigation, served Bobby Gibbes with the notice of suspension, Bobby was beside himself with rage. He was in a state of nervous and physical exhaustion from his efforts to save his air company, and the suspension was to him the last straw. He tore the notice into shreds, and flung them into Drury's face, together with much explicit language. He 'would go to Canberra and lodge a protest', yelled Gibbes. Tom Drury recalls: 'My word, Bobby was cross — I thought he was going to explode. Some of the things he called me made me cross, too, but Bobby and I have always been mates, and I didn't take them to heart. I think that John Arthur only wanted to slap Bobby's wrist, because the suspension was soon lifted and the Norsemans got back into the air again. Bobby was a reasonably good boy for quite some time after this...'

In fact, the suspension was lifted after only five days, but it was a sharp lesson for Bobby Gibbes, and one that he took to heart.

After the Telefomin massacre, the district was closed to missionaries and other whites by the Administration, except for official patrols. Early in 1955, the region was again opened to missionary work, and Reverend Draper went back. The MAF once more began to service the Baptist mission. The following year, the Piper Pacer was damaged while landing at Telefomin, piloted by Bruce Lindsay, a highly-qualified engineer/pilot who was later to become the MAF chief engineer.

Although as mentioned earlier in this chapter the Lutheran mission had decided on the Cessna for New Guinea work, the MAF initially preferred the Piper Pacer — and in 1956 the American MAF purchased and modified on behalf of the mission, a Piper Super Cub, VH-MFB, with STOL (short take-off and landing) performance for use on short strips like Wanuma and Begesin. The Piper was much cheaper than the Cessna — a significant consideration — and it was of tube-and-fabric construction. The theory was that in isolated locations, the Piper could more easily be repaired in situ after minor accidents by the pilot — MAF pilots invariably had engineering qualifications.

But the Piper Pacer was not the easiest of light aeroplanes to fly. Once in the air there was no problem.

Pipers were light and the engines fitted, powerful — the Super Cub had a 150 hp engine and was a notable performer but on take-off and landing, the aeroplane could be a handful. These early Pipers were tail-wheel aircraft, with a very sensitive rudder. When full power was applied, the slightest rudder movement resulted in a very rapid change of direction. The brake pedal was set high up on the rudder pedal, and the pilot had to act very quickly with the brake if he was to avoid the swing on landing that invariably developed if there was a cross or

160. MAF pilot Max Meyers and family, at Wewak in 1962. *Photo: MAF*

160

quartering tail wind.

Lindsay did not act quickly enough. The Pacer ground-looped, turned 180 degrees, ran into the side of the strip, wiped off the undercarriage, and damaged the tail. It was repaired on the spot — vindicating the MAF theory — although the work occupied six weeks.

In January, 1956, an American MAF pilot named Nate Saint was killed by jungle tribesmen in Ecuador, together with four missionaries. For months, the team had been trying to contact primitive Auca Indians; Saint had evolved an extraordinary technique of flying his aeroplane in tight circles above a fixed point, while a bucket filled with gifts was lowered by rope to waiting Auca below. The killings attracted world-wide attention and acted as a challenge to the MAF and missionary societies endeavouring to bring the word of God to the tribes of inland New Guinea. The MAF had begun to use the Cessna 180 in the jungles of South America, and in 1956 the first C180 was purchased for the work in Dutch New Guinea. In this year the American MAF pilot on loan to Australian MAF, Bob Hutchins, returned to his homeland with his family on leave, and soon after his return to PNG word came to Hutchins that his home church, Lake Avenue Congregational, had raised £15,000 and presented a C180 for use in the Highlands. The Cessna, a C180A, was registered in August, 1957, as VH-BVJ. Another C180A, VH-BVL, was registered in November, 1958. The new Cessnas rapidly proved their superiority over the Piper Pacer.

By this time, the Cessna was being used increasingly in PNG. DCA sent a C170A, VH-CAS, to the PNG region as official aircraft in January, 1955. Divine Word Airways — the Catholic mission aviation arm — put a C170B, VH-SVD, into service in January, 1954, and a second, BVE, in May, 1956. SVD was sold in December, 1957, and the registration was transferred to a C180A.

On 25 March, 1949, the flying service of the Diocese of Wewak, Society of the Divine Word, was inaugurated by the 'Flying Bishop', Bishop Leo Arkfeld. It was called Wirui Air Service, and operated under aerial work and charter licences, and with the assistance of Administration mail contracts.

Prior to the formation of Wirui, the Mission had very briefly owned a DH84, VH-URD. This had been registered on 18 August, 1948, and a week later was passed on to Divine Word Airways (who disposed of it to Qantas in May, 1950. This was the Dragon lost at Menyamya in December, 1951, piloted by Dick Davis — see Chapter 3, The Woes Of Qantas). When Wirui was started, its only aircraft was Bishop Arkfeld's J5 Auster, VH-KSR. Another Auster, VH-KAR, was registered in June, 1951. The first Wirui Cessna was a C180, VH-AVM, registered in September, 1955. A second C180, AVV, was acquired in December, 1956, and a third, MIV, in August, 1960. A C182A, VH-KLJ, joined the Wirui fleet in January, 1959.

Cessnas were employed by other missions. The Capuchin mission at Mendi — an American order — registered a C180, VH-BVG, in December, 1956. In May of that year, the New Guinea Anglican Mission registered a C170B, VH-ABM. One private owner, Alan Ferguson of Kinjibi Plantation, owned two Cessnas at different times during the 1950s; a C170B, VH-BVA, registered in April, 1956, and a C180A, VH-KIM, registered in January, 1958.

The main commercial user of Cessnas during the 1950s was Madang Air Services. Dave Robertson had a high opinion of the Cessna, and in later years held the PNG franchise for this manufacturer. The first MAS Cessna was a C170B, VH-BVB registered in March, 1956. In February, 1957, Robertson registered the only C190 to fly in PNG.

The Cessna company built the C190 and C195 in comparatively small numbers — 233 C190's and 890 C195's — between 1947 and 1954. They were unique among post-war single-engined Cessnas in being powered by radial engines — a 240 hp Continental in the case of the C190, and Jacobs radials of 245 and 300 hp in the case of the two models of the C195. Apart from the engines, the C190 and C195 were virtually identical. They were fast high-winged monoplanes without wing bracing, and introduced the celebrated Cessna spring-steel main landing gear legs. Dave Robertson remembers: 'The 190 was very nice. I bought it in the States, second-hand, shipped it out in a crate, and assembled it in Madang. It had a Continental radial, and a very good one too, just as trouble-free as the horizontally-opposed cylinders. It had about the speed of a Cessna 185. We had quite a few tussles coming out of the hills with Father Joe and Henry Hoff, side by side you know, and while the 190 was turning over at about 1,950 rpm, they would be doing about

2,450-2,500 in their Mission 180...

'When we had just the Dragon, I gave a fellow called Frank Smith a job as engineer in our workshop. He had a commercial licence, so I checked him out on the Dragon, and let him loose. Turned out to be a crackerjack pilot...another fellow we had about then, an Englishman by the name of John Green, loved the 190, because it had plenty of instruments, and he'd been used to flying Vampires or something.

'We got a good contract from the Administration by having the 190. There were some houses to be transported to Bundi and places, and things like refrigerators. I never said "No" to anything, so we tendered for the job, and got it. When they brought us these refrigerators, we couldn't get them into the DH84 — too big. So I said to Clive and Wally, "Let's take the door off the 190." Stuffed the refrigerator in through the door and down on an angle behind the pilot's seat. Got it in with about two feet [over half a metre] of it sticking out the door.

'I said to Clive and Wally, "I'll take it in when the wind is blowing down the runway, and take off with the pilot's side towards the tower, so DCA can't see what we are up to." Off I went towards the seaward end of Madang, turned round and took off. I didn't notice anything different with the aeroplane. Flew quite well, even with all that stuck out there. Only one thing worried me. What would happen when I got my speed down, going in to Bundi? No chance of practising at the airport, DCA would have got onto it. I never got special approval for the load. They would have shuddered. But apart from a little bit of buffeting at low speed on the tail end it performed quite well. Went in without trouble. Pulled out the fridge, and went home.

'We transported all the refrigerators that way. That left us with things like galvanised roof iron for the houses, and timber, which was cut into lengths that would fit into the DH84...'

161. MAF Cessna taking off from Telefomin for Wewak. *Photo: MAF*

162. Reverend Doug McCraw, an Anglican clergyman, who served as a pilot with MAF in the early 1960s. *Photo: MAF*

161

162

In December, 1957, MAS registered another Cessna, a C170B, VH-BVH. In the following year, Dave Robbie introduced the first tricycle undercarriage single-engined aircraft to fly in PNG: a C182A, VH-BVC. Says Robertson: 'I introduced the type. Father Joe said to me, "No, Robbie, you are looking for trouble with tricycle gear," and I said to him, 'You have to be a bit modern with this now, Joe. Got to keep up with the times.' So we got this 182, and not long after, Max Flavel of MAF had a forced landing on the new strip at Yellow River, in the Piper Pacer. The drain plug in the engine sump had cracked, and he lost all his oil. Ray Harris asked me would I go in and get Max out, and I said 'Yes.' So I went in the 182, and flew over Yellow River and spoke to Max, who was on the radio in the Piper. I said, 'Max, it doesn't look too good to me, pretty wet and boggy.' He said, "No, it's OK, bit slippery on the top, but quite firm..."

'So I made a couple of passes and decided where I was going to land. I ran about fifty yards [about forty-six metres], and it was so thick with black mud that down she went, to her belly. The nose-wheel went in, and she flipped straight on her back. I wasn't hurt, but the182...as it flipped over, it sort of raised up in the air, and then came straight down. Drove the tail fin right into the ground, and we had to fit two new wings. Of course, everyone then said to me, "Tricycle gear is no good".'

Max Flavel had managed to repair the Piper Pacer's drain plug, and the two pilots drained the oil from the 182's engine, and filled the sump of the Pacer. They then flew out, leaving the 182, forlorn, on its back in the sticky mud, like a crushed fly on a giant strip of flypaper. It was months before BVC flew again. In November, 1959, MAS registered another C182A, VH-BVE.

In fact, the introduction of tricycle gear had given the sale of the 170 series (170, 172, 175, 182 and variants) an enormous boost. Tail-dragger Cessnas were much harder to handle on the ground than tricycles, and a generation of later pilots have never flown tail-wheel aircraft. But the real work-horse Cessna, the 180-185 series, introduced in 1953 as a more powerful partner for the 170 series, were tail-draggers, and remain so to the present day. The original 180 used the same wing and flap system as the 170B, with a new fuselage and tail unit, and a 225 hp Continental flat-six engine instead of the 145 hp of the 170. The 180 was faster than the 170, and carried six passengers as against four in the smaller aircraft. TheC180was strong and rugged, and the series won wide acceptance in PNG, where it was ideal for rough, short bush strips.

Laurie Crowley registered his first Cessna, a C170A, VH-CSD, in October, 1955, and in August, 1956, Territory Airlines registered a C170, a model B. This was VH-GKA. TAL was to become the major user of Cessnas of all types, from the 170 to the Citation jet, in PNG, but the major buildup of the fleet took place during the 1960s.

That initial Cessna 170B, GKA, was a remarkable little aeroplane. It kept the struggling company afloat during the most critical days of TAL's existence, following the tragic deaths of Jack Gray and Jim Smith in the crash of Tiger Moth VH-AQS, on 1 June, 1957.

164

165

166

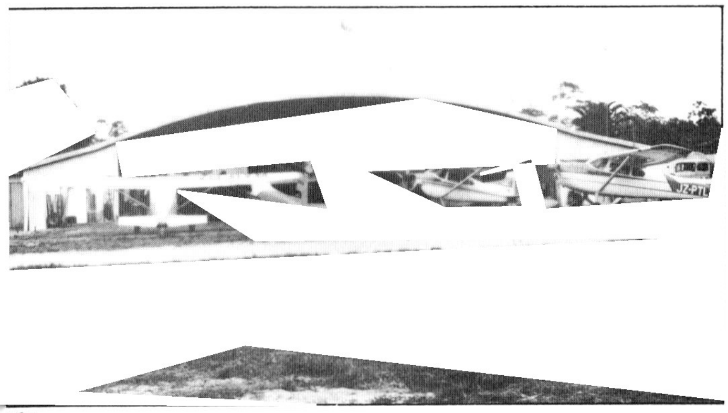

167

168

163. MAF's first Auster at Bankstown on the day it was dedicated in April, 1951. Harry Hartwig is on the extreme right. *Photo: MAF*
164. MAF headquarters hangar at Mt Hagen. *Photo: MAF*
165. MAF Cessna 180 is refuelled at Telefomin for the return trip to Wewak. *Photo: MAF*
166. MAF Cessna 180A, VH-BVL, at Wasua, 1958. *Photo: MAF*
167. The original MAF hangar at Wirui airstrip, Wewak. *Photo: MAF*
168. An MAF aircraft arrives at April River in the East Sepik District. *Photo: MAF*

169

170

5

Towards The Future

Territory Airlines Changes Hands

After the departure of Ray Harris to join DCA, the fortunes of Territory Airlines gradually began to decline. The company's Dragons were increasingly costly to operate and maintain. Performance restrictions and the mandatory fitting of radio equipment cut down on payloads, and made it harder to earn a profit with the DH84. The company was faced with the same problem that was besetting other small operators: aircraft replacement. It was no longer possible to pick up ex-RAAF aircraft cheaply, work them into the ground (sometimes literally!) and then repeat the process. Modern replacement aircraft cost real money, and TAL had not set up a replacement fund.

It probably would have been better for the long-term future of TAL if Sibil Plantation had not been purchased. The company could have had a Cessna flying and earning revenue a full eighteen months earlier than it did, had scarce money not been sunk into a plantation that was in the early stages of development, and which soaked up more money as time went by. Too great a percentage of TAL's financial resources went into Sibil, to the detriment of the flying operations that had been the reason for the company's formation.

It was understandable. Coffee was the magic word in the Highlands in the early 1950s. Everybody who could do so went out in their spare time and located land, made a deal with the native owners, and generally before purchase was officially finalised by the Administration began planting coffee, with crops of sweet potatoes and peanuts on the side for quick sale to help cover development costs. Most of those who stuck at it eventually made money, but it was impossible to work full-time at a job and develop a coffee plantation at the same time. Something had to give.

Sibil was purchased from Dick Kelaart and Brian Heagney, another ex-patrol officer. It had an area of approximately thirty-nine hectares, and at the time of purchase about three hectares were planted in coffee, with a further hectare cleared and ready for planting. There was also a good seedling nursery.

As noted in the preceeding chapter, the purchase of a Cessna proposed in February, 1955, was postponed following the purchase of Sibil. In May, the manager of Cottees Passiona (NG) Ltd approached Jack Gray, and expressed a possible interest in buying into TAL. Cottees had pioneered commercial passion-fruit production in PNG, and the question of speedy transport of the fresh concentrate was a vital one. The TAL directors decided to offer Cottees 645 shares at par, conditional on an advance of £5,000 for three years at two- and-a-half per cent interest. The cash would be used to purchase a second-hand Cessna, to transport passionfruit between Mount Hagen and Goroka.

The plan fell through. The Samarai trading firm of A.H. Bunting Limited was offered an interest in TAL, but again, nothing eventuated, and once more the purchase of a Cessna was left in abeyance. In an effort to generate more funds, Jim Smith was sent to Rabaul, to explore the possibility of opening a TAL branch there. Upon receiving Smith's report, the directors appointed him manager at Rabaul and offered Shirley Smith, his wife, the post of TAL representative. Vic Smith was asked to recruit a suitable pilot, at a salary of £120 per month. It was planned to commence the Rabaul operation by 30 July, 1955.

Again, nothing came of this plan. In August, Jim Smith was appointed manager of Sibil. Arrangements were made to transfer £150 per month from TAL funds to a Sibil account each month, to cover operating and development costs. Later, Sibil labour costs were met by TAL.

These were significant amounts of cash to a small company, and in December, 1955, a Bank of New South Wales loan of £3,000 was negotiated, for the final payment due on Sibil. A proposed dividend payment to shareholders had to be

169. Sinake Giregire, Captain Bill Johns and Paul Bolger with VH-AIA at TAL, Goroka in May, 1956. *Photo: Talair*
170. Captain Bill Johns and tribal leaders at Kerowagi with Dragon, VH-AON, June, 1956. *Photo: Talair*

held off until the loan was paid out. While progress at Sibil was judged by Lloyd Hurrell — who had left the Administration and was himself farming at Wau — to be satisfactory, there was no denying that the plantation was costing money, and producing no revenue. Its long-term prospects, though, were good.

Despite the drain on TAL's resources that Sibil represented, financial results continued to be quite good. Earnings fell, but operating costs fell also. However, no allowance was made for depreciation or replacement of aircraft. The company relied on a substantial bank overdraft for most of its operating cash. On 31 March, 1956, a special director's meeting was called to 'organise funds for the purchase of a Cessna aircraft'. It was agreed to call for another 3,000 shares at par. The original shareholders were to be allowed to each take up a further 100 shares, making the total issue 3,700 shares. Proceeds would be used for the purchase of the Cessna.

Jim Smith wrote from Sibil to his brother Vic, in April, 1956: 'We are forcing ourselves to buy a Cessna, because if we wait until we can afford it, we'll never get one, and although the DH84's are making a quid, they are eating it up in maintenance. The work for Dragons is not what it used to be, and they have not the range to compete in places that are opening up, so the Cessna is necessary.

'You will notice also that we have decided to sell some shares to outsiders, which seems to be the only way of seeing that the banks don't take us over if things go bad...'

One thousand new shares were eventually taken, by three pilots: Ronald Horlock and David Campbell of MAL, and John Downie, who was flying for TAL. The cash received, £1,500, was not sufficient to cover the cost of a Cessna, and was diverted to Sibil. The Bank of New South Wales again came to the rescue, as Jack Gray disclosed in a letter to Vic Smith: 'You will be pleased to hear that the Bank has approved a loan of £2,000 for the purchase of a Cessna, and we have been offered what I consider would be a good buy for approximately £4,000, landed Lae.

'With regards to the disposal of the Dragons, would you keep an ear open as to whether anyone wishes to purchase them down there. It was decided at a meeting to let them go for as low as £1,000, but naturally we would prefer them to go for between £1,250 and £1,500...'

Jack Gray had been working tremendously hard, driving himself to keep Dragons flying, managing the company, and even on occasion acting as traffic officer and secretary, when nobody else was available. It will be recalled that in July, 1952, Noel Peacock was flying one of the original TAL Dragons, VH-BDC, from Minj to Goroka when a propeller shattered, badly damaging the fuselage. In March, 1955, Gray was still making repairs to the Dragon, as he told Vic Smith in another letter: 'BDC is progressing slowly, the reason being that I have only just completed building a freight store and new office, the previous ones having been sold. The majority of component parts are ready for covering — the two main planes are already covered. There is still some more woodwork being carried out on the fuselage at present.

'You must realise that continuous work on BDC is out of the question. We are not employing a traffic officer, and I have to do this work along with other things.

'AON is due for C of A...I have been doing progressive work on it to try and cut down the C of A to approximately three weeks. I am also trying to meet DCA requirements regarding an approved workshop...'

Of the seven Dragons owned at one time or another by TAL, only AON and AIA were still capable of earning revenue. APL had been lost at Chimbu in September, 1952; AAC was just a collection of bits and pieces, and AGJ had never been more than a dismembered corpse, suitable for spares — it never flew in PNG. ASX, the 'Reluctant Dragon', had been sold in September, 1954. The overhaul of BDC never was completed, and the aeroplane was struck from the register in July, 1957.

After the first TAL Cessna, VH-GKA, was delivered in July, 1956, VH-AIA was sold to Mitchell Aerial Services, Cairns. TAL now had just two serviceable aeroplanes: The Cessna 170B, and the DH84, VH-AON (although Jack Gray's Tiger Moth, VH-AQS, was sometimes used for light work).

Pilots had come and gone during these early years. Tony Chadim was at times the sole pilot. After Mat Matus finally resigned, he still did a little flying for TAL, but only as his own business commitments allowed. In 1955, MAL pilot Wally Pawlawicz considered joining TAL, but backed out after Harold Hindwood asked him to stay on.

Efforts were made to recruit pilots in Australia. One young man arrived in Goroka, flew nervously through the formidable mountains of the Highlands for a week or so, then departed for the safer climes of flat Australia. Other possibilities for one reason or another did not work out.

During 1955, Bobby Gibbes purchased three Junkers JU52/3m aircraft in Sweden*. He flew one back to New Guinea late that year. After his return to Goroka, Gibbes asked Tony Chadim if he would go to Sweden to fit Lear radio equipment to the two remaining Junkers, so that they could also be ferried out. Relations between GSA and TAL were as usual cordial, and Jack Gray agreed to give Chadim leave so that he could travel to Sweden, fit the gear, and return as co-pilot in one of the Junkers. Chadim left by sea with his wife and daughter, arriving in Stockholm in December, 1956.

It would have been impossible for Gray to grant leave to Chadim had not Harold Hindwood of MAL lent a helping hand. MAL had a temporary surplus of young first officer pilots, all of whom were desperately anxious to build up in-command flying hours. To a young commercial pilot, *nothing* was more important than in-command hours, for this was the road to promotion to the command of a DC3, the essential step for the pilot ambitious to go on to bigger and better things. The only way a first officer could gain in-command hours was at the controls of a light aircraft, and as mentioned earlier, MAL was rapidly disposing of its Dragons. In 1956, the MAL fleet consisted of just five operational aircraft: two Dragons, VH-AOQ and API, and three Douglas C47's, VH- MAE, MAC and MAN (MAE was formerly registered as BFV, and MAC as BFW). Both Doves had been sold, and the great expansion of MAL operations was still to come. Hindwood was only too pleased to lend TAL first officers for short periods — pilots like Jim Rhodin, and Bill Johns.

William Joseph Johns was still flying in PNG in 1985, the longest-serving Air Niugini pilot, with over 22,000 hours in his logbook. When he came to PNG, Bill Johns had accumulated just 365 hours, mostly on Tiger Moths and Austers.

Bill Johns was born in north Queensland in 1928. Both his mother and father were musicians, and from his earliest years, Bill Johns' ambition was to follow in their footsteps. He became a professional musician in 1943 with the American Entertainment Services, and went on to play percussion instruments for J.C. Williamson, assorted dance bands, and finally with the Tasmanian Symphony Orchestra.

Despite his preoccupation with music, Johns had a boyhood interest in flying, and in 1950 while playing at the Theatre Royal in Brisbane he obtained his private licence. In 1953 he studied for, and obtained, his commercial licence. Says Johns: 'In 1955, I decided that I didn't want to be under the conductor's baton for the rest of my working life, so wrote to all airlines in Australia and all English-speaking airlines abroad. I received many negative answers, but Mandated Airlines offered me a position as a DC3 co-pilot. I accepted, and then engaged in a hassle with the ABC over termination, as they required two months' notice from a principal player, and MAL wanted me to start in a month. Honour was satisfied all round when I stayed six weeks, and broke the new bloke in.

'Arrived Lae on 22 July, 1955. Tom Deegan, the chief pilot, had me fully fledged within a few days, and I flew as F/O on DC3's for nine months...'

Tom Deegan well remembers the arrival of Bill Johns, who was short in stature, mild in manner, and quiet in speech: 'I'll never forget Billy Johns when I gave him his job. In his application, he said his profession was a tympanist. I went to Harold Hindwood and said, 'What's a tympanist?' Harold said, "Buggered if I know...is it a poofter or something?" Bill was a good pilot...'

When Tony Chadim applied for leave to go to Sweden for Bob Gibbes, Jack Gray asked Hindwood if he could help out with the loan of a pilot. Bill Johns was about to be endorsed on the MAL Dragon, VH-AOQ, and he was selected to go to TAL. Over the next four months, Johns flew 326 hours on TAL Dragons, and twenty-five hours on the Tiger Moth. He shared a house with the then accountant/traffic officer of TAL, Paul Bolger, who had been engaged following the resignation of Jeff Joynton-Smith in March, 1955. Bill Johns recalls: 'Tony Chadim endorsed me on VH-AON, on 27/28 February, 1956. As it was a one-place cockpit, this consisted of a demonstration flight with a change of pilots during flight — no mean feat — then some circuits. I did a few flights under Tony's supervision to Chimbu, Kerowagi,

171. Bill Johns and VH-AON at Kerowagi, June, 1956. *Photo: Talair*

171

*See Chapter 5, The End Of Gibbes Sepik Airways

Kainantu etc.

'I then returned to MAL and the DC3 to await Tony's departure for Sweden. I began my 'loan' flying with TAL on 30 April, '56, on VH-AIA (AON was in the hangar for C of A). I was 'chief pilot' — the only pilot! Often I would fly the DH84 and the Tiger Moth, AQS, during the one day. One would be loaded while one was flying. The Tiger was loaded mainly with freezer goods for Chimbu, in the front cockpit and rear locker. It was rarely flown in the afternoons, due to the windy conditions.

'I made a couple of flights with Jack Gray in the Tiger. One to Nambaiyufa when it first opened, and another to Faita, when Frank Smith went up on his nose after landing the Madang Air Services Dragon there. Jack Gray repaired it and Frank flew it out a few hours later. Shortly afterwards, Jack ran off the strip at Gusap whilst trying to take AQS off in very windy conditions. I then flew 'One-Eye', the TAL engine *boi*, down in AON complete with tool box so that Jack could repair same.

'By this time another pilot, Rudolf Teichmann, had arrived, and I began to instruct him in the DH82. I also familiarised Allan Woodward of DCA in the DH82.

'Teichmann then flew as supernumary with me on most flights with a view to being checked out on the DH84. On one of these flights — 6 August, '56 — I took off from Kainantu for Lae with Teichmann, and Harold Hindwood, the MAL manager, as passenger. Over the Markham the port engine of AON began to vibrate and backfire. I shut it down and headed for Kaiapit. Made it, and we spent the night there with the patrol officer. Neat rum and poker was the entertainment for the night. We managed to drink all the PO's rum, and win all his money. Next morning, engineer Ernie Shearer arrived in Crowley's Cessna, piloted by Keith Mullan. Harold went with Keith in the Cessna to Lae, while Ernie replaced the six failed spark plugs in AON's engine, and then returned to Lae with me.

'Mat Matus then took over the endorsement and training of Teichmann, as I was then posted back to Wewak to fly VH-AOQ. I was only there a short time when I began command training on the DC3, back in Lae. Brian Badger took over from me at Wewak, and he was followed by Jim Rhodin...'

Soon afterwards, another pilot joined TAL. John Downie was born and bred on his parents' farm, on the south-west slopes of New South Wales. He left school at the age of fourteen, to work on the farm, became interested in flying, and joined the local flight of the Air Training Corps at its inception in 1942. A bid to join the RAAF for aircrew training was unsuccessful, because of his lack of formal education, so Downie served for four-and-a-half years as a wireless maintenance mechanic, taking his discharge in 1949. He then went back to school, to improve his education, and in mid-1950 returned to the Air Force for his long-delayed pilot training. He obtained his wings in December, 1951, and later went to Korea with 77 Squadron. However, ill-health forced John Downie to retire from the RAAF. He recalls: 'I did a bit of farm work and some truck driving for two years, felt fine, presented myself for civil flying licences, was successful, then looked for a job — not too plentiful in mid-1955. After some correspondence with a chap called Peter Manser, I boarded a DC4 in Sydney bound for Goroka — wherever that was — and a flying job with Gibbes Sepik Airways.

'I settled quickly into the way of life and to the flying, although chaffed under the necessity to spend 200 flying hours in the right seat of a Norseman, learning the country. Once I had achieved command, and was on my own over some of the wildest country, I understood the advantages, even the necessity, of such an introduction, and believe there would have been fewer tragic accidents if all PNG operators had adopted the principle...'

John Downie joined GSA at a time when the volatile fortunes of the company were temporarily on the ascendant. The Gibbes Norsemans were then almost entirely responsible for keeping the isolated Southern Highlands stations and outposts alive. It was pioneer bush flying at its purest, and involved a considerable amount of airdropping, to officers constructing new airstrips, and to patrols probing into new country. It is hardly surprising that there were a number of accidents, although there were no fatalities. Downie himself had a close shave in VH-GSA in October, 1956. He was returning to Minj after a flight to Tari when his engine failed, over Mount Hagen. Downie was forced to land, just failing to reach the Hagen strip. The nose of the Norseman ploughed into the soft ground, and the aeroplane overturned. It was damaged, but repairable, and Downie was unhurt.

Peter Manser, Downie's correspondent, had earlier overturned another Norseman, VH-BLM, at Vanimo. Coming in to land with two passengers, the Norseman hit a log that had mysteriously appeared in the middle of the strip. Pilot and passengers received slight injuries, but BLM was so smashed up that it had to be sent to Australia for overhaul. Barely a fortnight later — in June, 1953 — pilot Patricia Toole (nee Graham: she had married the GSA Wewak branch manager, Col Toole) forced-landed in Auster VH-KSK in wild, mountainous country about 130 kilometres from Wewak. She had been trapped by weather, and spent so much time circling a thick gray blanket of cloud that by the time she found a break, and landed in a dry river-course, there were only four-and-a-half litres of fuel left in the tanks. It was a masterly landing: to avoid collision with a stand of tall trees, Pat Toole ground-looped the Auster. She spent two nights in the cockpit before an Administration patrol walked in to her rescue. KSK was only slightly damaged, and Peter Manser flew the aeroplane out to Wewak a few days later.

Shortly before John Downie arrived to fly for GSA, the pilot of another Norseman had a close brush with death, in the Southern Highlands.

Adrian Nisbet had joined GSA late in 1953. He was born in Brisbane in 1923, son of a radiologist father and a mother who was the fourth woman pilot in Australia. When Nisbet was six, his parents divorced. His mother then married the Australian vice-president of Chrysler Corporation, who was keenly interested in aviation. He imported Taylor Cub aircraft for the Australian market, which were test-flown by a pilot who lived next door, called Rod Julius. Julius taught the young Nisbet to fly. He actually soloed, quite illegally, at the tender age of fourteen, at the old Mascot aerodrome.

Nisbet's stepfather was a constant world traveller, often accompanied by his family. Adrian spent eighteen months at school in England. When the war came, it was inevitable that he would join the RAAF. Not surprisingly, Nisbet gained his wings with ease, and with his trainee group was sent to England.

Nisbet's war career was distinguished. He did one-and-a-half tours as a Lancaster bomber pilot of 5 Group, engaged in specialist precision bombing of selected targets over Germany — oil dumps, refineries, gun emplacements, canals. In one low-level raid, only two of the ten Lancasters that took part returned. Nisbet was awarded an immediate DFC for this and other dangerous raids.

Following the war, Adrian Nisbet joined Qantas. He remembers: 'For almost a year, I was on the American DC3 charter run, between Brisbane and Manila, six or seven days up and down, six or seven days off, then we did it again. I became fed up with being a glorified bus driver, so much to my stepfather's chagrin and the annoyance of the Operations Manager, Lester Brain — an old family friend: he taught my mother to fly — I resigned from Qantas, and went to Canada. I joined a small show called Sawage Air Services — bought into it. This was early '46, and I stayed in Canada until '49.

'We flew mainly Fleet Canucks — high wing, Cessna type four-six seat monoplanes — on floats and skis, and we also had two Stinson Voyagers. We had seven aircraft, all singles, and flew mainly on passenger charter through Ontario and into Quebec. Then we got into a very handy business, flying American fishermen out to the bush lakes. Strap canoes to the struts of the floats — very pleasant and lucrative. Getting into and out of some of those lakes was a bit tricky...kept the adrenalin flowing...

'Well, we decided to keep going in our big-headed way through one of the worst winters on record, 1949/50. We got a contract from one of the Hudson Bay companies to supply them during this winter, and we lost pretty well all the aircraft — no casualties. They were lost in the snow, ditching, by damage, and wear and tear. Very difficult to warm up engines sufficiently in those conditions, so they cracked up...'

Following the collapse of the little company, Nisbet worked for six months as a nickel miner to raise some funds, then went to England and applied for a short-term commission in the Royal Air Force, in Bomber Command. But he was an experienced bomber pilot, and was made an instructor, much against his will. He recalls: 'I applied again and again to go back to Bomber Command, but they kept me on instructing. I was a bit of a naughty boy, and finally they allowed me to resign my commission, in 1953. I'd gotten very friendly with a squadron-leader who was a friend of Bobby Gibbes, and when it was time to get out he said, "What are you going to do?" I was thinking of going to Rhodesia to grow groundnuts. I'm glad I didn't go. The squadron-leader told me about Bobby Gibbes and New

Guinea, and said that GSA wanted a pilot "rather desperately". So I flew back to Australia, and got the job.'

Nisbet flew to New Guinea as co-pilot to Bobby Gibbes in Norseman VH-ASN, which had just had a major overhaul at Bankstown. He remembers: 'There was everything, *including* the kitchen sink, in that aeroplane, things for the new house Bobby and Jean were building in Goroka. David Bennett, the GSA agent at Minj, was on board, Arthur the corgi and Paddy Gibbes, the sheepdog, Jean with her new baby, Julie...everything.

'An hour out of Bankstown we had a runaway prop, couldn't control it, and it looked as though we'd have to put down on a road. Finally, we urged ASN back to Archerfield, and there we worked on her for three or four days — took us about nine days to get to PNG.

'The first sight of the Highlands was marvellous. We flew direct to the Asaro valley from Horn Island, a wonderful experience. I stayed in Goroka the first year, in a little pit-sawn hut. All the GSA buildings were native materials, and they were just starting to build the big hangar. I was studying for my senior commercial at that stage, and every time I had a navigational problem, I'd shoot up to the district office and see Ian Downs, the DC. He used to be a naval officer, and he was a great deal of help.

'I got my senior commercial. We were still flying mainly out of Goroka. We had a base at Banz, but Minj base hadn't been established. Then when Bobby and Jean went to Europe to look for JU52's, I sneaked a base into Minj, with the connivance of Vic Cox. Shifted all the stuff over from Banz, Peter Manser and I and two Norsemans, for days on end, five minute flight each way across the Wahgi.

'The DC3's started coming in from Madang to Minj, and the big push into the Southern Highlands began...

'I started tramping into the bush in my spare time, and about six or eight miles [ten to thirteen kilometres] west of Minj I found about 420 acres [170 hectares] of good coffee land, negotiated with the local people, and was granted a lease in February, 1955. Bobby had Tremearne at the other end of the valley, and Col Toole had Kimel, the company plantation. There was Kinjibi, owned by Alan Ferguson and Jim Matthew, and Rinus Zuydam had a place in the Asaro. At one stage, we reckoned that GSA just about populated the Wahgi valley. David Bennett had the plantation next to me. Arthur Corby, an engineer with GSA, was further down the line...'

Adrian Nisbet did a great deal of pioneer flying into Southern Highlands airstrips in Norsemans, and airdropped supplies to a number of exploratory patrols*. His handling of the Norseman was beautiful to behold, and his airdropping the most precise of all the Gibbes pilots. But even the most skilled aviator could come to grief on raw new airstrips, and the one at Ialibu, situated at a height of 1920 metres on the bitterly cold flanks of the great mountain of the same name, was one of the rawest of them.

In April, 1955, Ialibu strip was inspected by John Arthur in the DCA Cessna, and was opened for commercial operations, under stringent rainfall restrictions. Adrian Nisbet made the first Norseman landing. He will not forget the experience: 'I hit a soft patch, had no cargo in the back, and over we went. All the locals standing at the top to see the first Norseman landing came tearing down in great glee, with a policeman in front. He leapt over the shattered wing as I hauled myself out of the cockpit, and said, in Pidgin, "It's all right, *masta*, don't worry, there are plenty of people here, we'll turn the *balus* over, and you can try again!"

'That was the second time I'd gone into Ialibu. First time I forced-landed, when there was no strip at all — they were still building it. Landed in the *kunai*, and was there for seven days, building a strip out of stone brought in by the people from a stream about a mile away [about one-and-a-half kilometres]. We bedded them down, and rolled the Norseman over the stones, to and fro, and got about 800 feet [244 metres] laid. I put white sticks down the edge to mark the places where I had to be going at twenty knots, thirty knots, forty knots, then off the ground sort of thing. John McDermott was with me. He stayed behind, and I flew the thing out, in one piece. It was a hairy take-off, but the local people had done a magnificent job of work, laid those stones like a crazy paving...'

John McDermott had a truly spectacular crash on 23 August, 1955. He took off from Minj in Norseman VH-BNT on a flight to Mendi. The Norseman climbed to a

*Including many led by the author, who was a field officer in the Southern Highlands from November, 1954, to January, 1959.

height of eighteen metres and then the engine failed, for want of fuel. The aeroplane went straight in. Says Adrian Nisbet: 'He was extremely lucky to get away with that one, because we found him sitting in his seat still. The bits of the aeroplane were scattered all around him, and there he was, strapped in his seat, right side up and sitting on a tuft of *kunai* grass...'

BNT was totally destroyed in this crash, but, by some miracle, McDermott escaped with minor injuries.

John Downie did his full share of flying into the Southern Highlands and by late 1956, was feeling the need for a holiday. Before he could leave, he was invited by Jim Smith, manager of the TAL plantation, Sibil, to invest in, and fly for, the company. Downie left GSA, purchased 250 shares in TAL, and took over the controls of the recently-arrived Cessna, VH-GKA. Bill Johns had returned to MAL, and Tony Chadim was the only other TAL pilot. German pilot, Rudi Teichmann, had left after a few months. Chadim continued to fly the Dragon, VH-AON. He remembers: 'John Downie avoided flying the Dragon for five months, but I did not mind too much. The Cessna had a nice quiet engine, but flying a Cessna is like milking a mouse. It might sound unjust to say that the worst part of our DH84's were the engines, because they never actually stopped on me, but they were certainly the most frustrating part. It was all a vicious circle of operating old aircraft which were cheap to buy but expensive to maintain, and diverting the profits away to develop coffee plantations. The engines should have been thrown away, but they had to be sent for overhaul, which was expensive, and every last hour had to be squeezed out of them...'

In November, 1956, auditor E.B. Davis presented the directors of TAL with audited balance sheets and accounts for the year ended 30 June. Although the Cessna, VH-GKA, was operating successfully, too high a price had been paid for it, and the radio equipment installed was obsolete. The value of the aeroplane had to be written down to market value. Considerable stocks of DH84 spares were held, and were now of very limited value. Referring to Sibil Plantation, Dick Davis noted: 'Development of Sibil has only been made possible by the deferring, to a certain extent, of payment of creditors together with financial assistance from the Bank of New South Wales. It is obvious that to further develop the investment, without overloading aircraft operations, additional capital is required.'

TAL was still operating profitably, but by a slim margin. Clearly, there could be no major expansion without a lot more capital. Before any moves in that direction could be made, however, Jack Gray and Jim Smith were killed.

On Saturday morning, 1 June, 1957, the two men left Goroka in Gray's Tiger Moth, AQS, for Lae, to attend an auction sale. The aeroplane failed to arrive on schedule, and thirty minutes later, five aircraft were in the air, searching. All airstrips between Lae and Goroka, and all clear areas of ground, were examined, to no avail. The most likely place along the route for a Tiger Moth to find trouble was the rough mountain country around the Karanka and Arona gaps, and the search was concentrated in this region. No sign of the missing Tiger had been seen when the aircraft had to break off, in order to reach Lae before night fell.

At dawn on Sunday the search was resumed. At 1.30 p.m., a band of natives signalled to a searching aircraft, and pointed up at the Arona Gap. A ground party was approaching that region and at 3 p.m. the natives made contact, and told the party that the Tiger had crashed in heavily-timbered country at the foot of the gap. They led the ground party for four hours, to the scene. Gray and Smith were dead. The bodies were carried to Goroka, and buried on Monday. Joan Gray was still in Goroka Hospital, recovering from the birth of her son.

A tragedy of this dimension would have been hard for any small company to absorb. It was particularly difficult for TAL, at such a critical stage of its existence. Jack Gray had been manager, chief (and only) engineer, secretary, office boy, traffic officer, aircraft cleaner and general factotum of TAL. Jim Smith had been manager and developer of Sibil, the principal asset owned by TAL. They, with John Downie and Tony Chadim, were the sole white staff of TAL at the time of the accident, although Mat Matus still flew on occasion for the company, as did Allan Woodward, a commercial pilot then employed by DCA in the Goroka tower. Sam 'One-Eye' had developed into a very competent engineering hand, and a few other Papua New Guineans were doing good work, but the loss of the entire management of TAL at one blow was a tremendous shock, and the initial reaction of the shareholders was to cease operations.

On the Saturday following the crash, a special meeting of the directors was held at Goroka to decide about the future. Auditor Dick Davis told the meeting that in his opinion the position of the company was 'quite sound', and that 'any move to discontinue business would be a retrograde step'. His advice was heeded, and it was decided to carry on. John Downie was offered, and accepted, the post of temporary manager, and Jim Smith's widow, Shirley, took over the secretarial side of the business. A manager, Allan Booth, was appointed to take over Sibil Plantation.

Once again, the close relations that had always existed between GSA and TAL provided a solution to the aircraft maintenance problem. The GSA chief engineer, Eric Cross, had just resigned, and his place had been taken by Ray Pugh. A month after the death of Gray and Smith, another engineer joined Pugh's staff: Frank Newell. Newell looked after the TAL Cessna and Dragon for some months, signing the aircraft out so that they could continue to fly and earn income.

Frank Newell came to Goroka to work for two years — and stayed for twenty. Like Frank King — back working for GSA — Newell was an engineer of the old school. Born in 1917, Newell had always been interested in aircraft, and after leaving school he was apprenticed to the Kingsford Smith Flying School. Frank Newell remembers: 'This was in the days of Smithy and Tommy Pethybridge. I worked on the *Southern Cross* and the *Faith in Australia*. I was the boy around the hangar — cleaning the benches down at 5 o'clock, getting the men's lunches, pedalling all the way up to the Mascot shopping centre and back again, and woe betide me if I got the order wrong. Lot different to apprenticeships these days. Apprentices are a chosen race today.

'I gradually came up through the ranks, and got my licence in 1941. Aircraft Engineer's Licence — I still have it. No. 1148. After a long time I got endorsements — Klemm Swallow, Gipsy Moth, Avro Ten, Avro Trainer, Short Scion, all that old fabric and wood stuff. I've got a whole list of endorsements...

'That was in the days when they were doing joyrides in the Avro Ten, across Sydney Harbour Bridge and back for ten bob. My job was to wind up the inertia starters on the old Armstrong Siddeley Lynx's. When you got them wound up, you'd pull the string, the clutch would hit and the prop would turn over and you'd pray it would start, because if it didn't you'd have to wind it up again, and it took a long time. I'd work in the hangar and workshop during the week, then at weekends go back there and do tarmac duty. Put the dual controls in the old Gipsy Moths for instruction, and pull them out for solo flights, wind up the Avro, all part of the job. I was living at Manly, and I'd catch the first ferry to Sydney in the morning, 6 a.m. — sometimes I wouldn't get home until nine or ten at night, buggered. KS Flying School was nearly broke. They were battling to make a living off joyrides and the bit of flying training they were running. Hard days. No union rules then. I stuck it out — I had to, in the Depression...

'Then the war came, and our aircraft were taken over by the Air Force. The flying school became part of the Empire Training Scheme — No. 4 EFTS — Elementary Flying Training School. I tried to join the Air Force and got as far as the interview and screening part of it, but when they found out that I was an aircraft engineer at Mascot they sent me straight back there. It was a reserved occupation, and they said, "Don't waste our time, or your time." So I stayed at Mascot until the war was finished.'

After the war, Frank Newell worked in succession for the Royal Aero Club, Aerial Agriculture and as a partner in a small company called Somerset Airways, which operated a number of Austers. After several years with Somerset, he left and set up in business for himself as a freelance engineer at Longreach, almost alongside the original Qantas hangar. Newell recalls: 'I started to look after a lot of graziers' aircraft, touring around in a Standard Vanguard utility with all my toolboxes in the back, from homestead to homestead, fixing up elevators that'd been chewed by goats — terrific experience. One chap had a Percival Gull, another a Piper Tri-Pacer, others had Austers. Quite a variety, but all single-engined things, no problem to look after them. Never a dull moment, but didn't make much money! Some of these blokes had me on ninety days' credit. Pretty rough on a one-man show. I had to buy petrol for the truck, materials, fabric, dope, all the things they needed on aircraft in those days. Kept that up for about five years.

'Finally, I got so fed up with travelling around in a hot climate, living in old, broken-down houses, that I left Longreach, and went to work for Western Air, in Charleville, just down the line. Broke with them after less than twelve months. So I decided to write to Bobby Gibbes in New Guinea. We had been schoolboys together, at Manly. Bob sent me a telegram, telling me to get on the next boat. This was July, '57. So I caught the *Malaita* with my wife and two boys, and went to New Guinea...'

Frank Newell's rich experience fitted him exactly to work on the Norseman, which was fabric-covered. While he worked on the airframe, Ray Pugh's daughter, Jan, 'bagged-up' the fabric, in engineer's parlance. Although Newell signed-out the two TAL aircraft, Sam 'One-Eye' did most of the routine work on them.

The struggle to keep TAL afloat proved too much for Shirley Smith, Joan Gray and the other shareholders. John Downie resigned as manager/pilot on 6 October, 1957, leaving Tony Chadim as the sole pilot. Naturally, aircraft utilisation hours fell sharply, GSA took up the slack, and TAL revenue plummetted. Most of the shareholders felt that the company could no longer survive, and in November a director's meeting was called, attended by Les Gray, Shirley Smith, Joan Gray, Lloyd Hurrell and Ray Harris (who still retained his TAL shares) to discuss the disposal of the assets of the company.

An attempt had been made soon after the fatal crash to have Ansett Airways take over the TAL charter licence, but this had come to nothing. Dave Robertson of Madang Air Services had expressed an interest in taking over TAL, but again nothing eventuated. The directors had hoped to sell TAL, and apply the proceeds to the development of Sibil Plantation, but Lloyd Hurrell pointed out that this would be a long-term proposition, involving the expenditure of upwards of £20,000 to bring the plantation to full production. Finally, it was agreed that both TAL and Sibil should be placed on the market, the airline for £14,000 reducing to £12,000 if necessary to clinch a sale, and the plantation for £20,000, reducing to £15,000.

The figure placed on the value of TAL was realistic. The only assets apart from the charter licence itself were the Cessna and one spare Continental engine; a well-worn Dragon (plus the partly-overhauled BDC) with one overhauled and nine time-expired Gipsy Major engines, so worn as to be practically valueless; sundry tools, spares, office equipment, and a two-bedroomed house on a town block in Goroka. In truth, TAL was so run-down that it could hardly be regarded as a prize. Fourteen hectares of coffee had been planted at Sibil, of which ten were bearing. Another five hectares were ready for planting. There was an old jeep, a sawbench, lighting plant, house, tractor and trade-store, but little else.

On 19 January, 1958, a special meeting of TAL shareholders was held at the office of E.B. Davis in Lae, to receive a report on action taken for the sale of the company's assets. An offer of £17,000 had been accepted from Mr Hilton Jarrett for Sibil. The money, after the discharge of creditors and the collection of outstanding debts, would be distributed to shareholders by way of dividend. Dick Davis, in a letter to Vic Smith in Sydney, outlined the position with Territory Airlines: 'The Company has entered into an Agreement with Mr Dennis Buchanan whereby they will sell him shares of TAL for the sum of £12,000 after the distribution of a dividend resultant in the sale of Sibil. The sale to Buchanan is 7,052 shares for the sum of £12,000...the payment being made by Buchanan is £6,000 on completion of the Agreement, and £6,000 in four equal quarterly payments...'

An Era Begins

Dennis Buchanan was only a young man — twenty-five — at the time he purchased TAL. His years with Gibbes Sepik Airways had matured him and given him an intimate knowledge of light aircraft operation, and some idea of the problems of running a small air company, but he was still fresh-faced, tousle-headed, boyish in appearance and apparently without any driving ambition. Appearances were deceptive. Buchanan had married, and his life-style had changed a good deal over the previous two years. Despite this, he was still generally known as 'Junior' to his friends and acquaintances. A typical reaction to Buchanan in those early years was that of Adrian Nisbet. 'Junior was just the kid about the place then — very energetic, very willing, but just a bagsnatcher.' Mal Shannon remembers the opinion of a certain well-known Qantas captain after watching Buchanan directing the unloading of a GSA aircraft, with much calling of orders and waving of arms: "Who is that bloody idiot? I wouldn't feed him". Says Shannon: 'I've often thought about that, and I'd like to laugh about it with him today, and ask him: "Do you know what that 'bloody idiot' went on to do, and what he has built up?" '

It will be remembered that after Buchanan began to work for Bob Gibbes, he was sent to Wewak to learn the ropes, first as an office boy/rouseabout, and later as a cargo handler and aircraft loader. Strangely, for a young man constantly around aeroplanes and pilots, Buchanan had not the slightest desire to learn to fly.

In June, 1951, Buchanan was sent to Popondetta after the eruption of Mount Lamington volcano, to handle the unloading of the leased Lockheed Lodestar, VH-FAC, which was at that time piloted by Frank Wiza, and First Officer Ron Pilling. The Lodestar made many flights from Port Moresby with emergency supplies to Inonda, which were then ferried to Popondetta in a Norseman piloted by John Wells or Bob Bell. After the emergency, Buchanan was transferred to Port Moresby, where he remained for over two years. It was during this period that Bob Gibbes briefly established the headquarters of his company at Port Moresby. Buchanan was by this time working mainly as a traffic officer. He remembers those days: 'Peter Howse was manager in Moresby. I'd get up at four in the morning, pick up the freezer from BP's, load the aeroplanes in the dark and make sure they took off at first light. We had two Norsemans and the Lodestar at Moresby then — the Austers were practically finished.

172. Dennis Buchanan, 1949. *Photo: Sten-Erik Molker*

172

'I liked Moresby. I played hockey against North Queensland. I was nineteen when I went there, and I went downhill and got on the booze a bit — I hadn't been allowed to drink in Wewak. I went to Sydney on leave and had my twenty- first birthday there. Peter Manser was on leave, too, and we whooped it up together. I had saved about a thousand quid in those first four years in New Guinea. I did the thousand, and had to borrow money to get back! Borrowed about five hundred, and when I got back to Wewak — Bob Gibbes had transferred me back there — I thought, It's going to take me twelve months to pay that back! I was only getting £14 or £15 a week, and I owed five hundred! I shook my head in disbelief. Stupid. I promised myself I'd never do that again.

'But it was a great holiday. Stayed with my father a lot — he was in a Housing Commission flat. Plenty of birds around. My brother David and I used to have parties in the flat when my father was away — big booze-ups. Eventually, my father got a bit short-tempered. People complained to him about the parties, and keeping the neighbours awake all night, and I think he was pretty pleased to see me go back to New Guinea. I didn't have another leave for about seven years. I went back to Wewak with Peter, February '54, on the *Bulolo*. Got drunk every day — nothing else to do.

'I was under Robin Gray at Wewak, on traffic. I was happy, but I got run-down in health, and in January '55 I got malaria, pretty bad. Later that year, I got blackwater fever, and the doctor rang Robin and recommended a transfer to the Highlands. So Robin sent me to Goroka. Bobby was there, and he said, "Nothing wrong with you, son," and sent me back. Robin sent me straight back to Goroka again! This went on for about a week, and then Bob realised that I was sick, and then I stayed in Goroka. I lived with Frank King, in a little house he had built, and after twelve months I was virtually out of debt. I saved about £350. Bobby took it out of my wages, for shares in Kimel, the GSA co- operative plantation.

'Frank had a plantation, in the Bena — half a block, the other half was owned by Jack Gray and Bruce Blackley. He asked me to go into partnership with him, so I did. Cost me about £700. I raised the money, sold my motorbike and my shares in Kimel. Then Frank decided to go back south, so I looked after the plantation to the end of '56.

'We had seventy-five acres [thirty hectares], and planted over thirty [over twelve hectares] in coffee. I used to walk out there, six or seven hours, seventeen miles [twenty-seven kilometres]. No proper road then. I'd leave at four in the morning on a Monday, and be back for work on Tuesday. It kept me very fit. Bob Gibbes didn't mind — he liked to see people up and doing. Most of the farmers in the Wahgi were ex-GSA. Old Jim Brugh, he lived on the side of the airstrip, he got a bit sorry for me, and he gave me a little BSA motorbike which made it a lot easier. I hammered that Besa...

'At the end of '56, a Yank called Andy Anderson arrived, and offered to buy the place. So Frank and I sold it to him for £8,000. So I had £4,000, a lot of money in '56. I was only getting £25 a week. So I thought I'd get married...'

While living at Wewak, Buchanan had become friends with one of the authentic characters of the mighty Sepik River, 'Shanghai' Brown, a labour recruiter and trader. 'Shanghai's' wife, Emily Pelayo, was Shanghai-born, of Spanish-Chinese extraction. They were married in Shanghai in the 1920s. When the Second World War came, Brown left Shanghai, and went to the Mandated Territory of New Guinea. Emily and her daughters, Della and Paula, remained in China until 'Shanghai' could arrange for their passage.

The Japanese onrush that followed the outbreak of war in the Pacific, trapped Emily and her children in Shanghai. They were British subjects, and were interned, spending two years in a prisoner-of-war camp. They were well treated, and after the war stayed on, waiting for 'Shanghai' Brown to send for them. But then the Chinese revolution occurred, and again Emily and her daughters were trapped. Paula married a Swiss engineer, John Comtesse, manager of the Shanghai Aluminium works, and remained in Communist China. Emily and Della were finally allowed to leave in 1951, although they were not permitted to take out anything but their personal effects.

For two years, Emily and Della — who was then eleven — lived in Macao and Hong Kong, and in 1953 were at last re-united with 'Shanghai', who was living at Angoram. Dennis Buchanan frequently visited 'Shanghai' Brown, and of course he got to know young Della.

Buchanan has had a life-long eye for a pretty girl, and Della, though hardly more than a child, was certainly that. She was petite, slim, with the oval face, flawless creamy skin, high cheekbones and lustrous dark eyes of her ancestry. 'Shanghai' Brown was a rough diamond, but he knew the value of a good education, and he sent Della to school in Brisbane. Buchanan kept in touch. Then 'Shanghai' died, and Emily was left in poor circumstances. She went to work in Madang and then Lae to support herself and Della. Later, Buchanan found Della a job in Goroka, and in 1956 they decided to marry. The civil ceremony was performed in Lae by the District Commissioner of the Morobe District, Horrie Niall, on 27 December.

Dennis and Della settled down happily to life in Goroka. Buchanan's work with GSA kept him busy, and Della soon became pregnant. He certainly entertained no thought at this time of entering the airline business for himself.

After the death of Jack Gray and Jim Smith, it was not long before rumours swept Goroka: Territory Airlines was in trouble. The little company was obviously going downhill, and few were surprised when it was announced that it was on the market. Buchanan decided to make an offer. He had but £5,000 and he invited the GSA engineer, Frank Newell, and a GSA pilot, H.O. Tschuchnigg, to come in as partners. Both said they were interested.

Heliodor (Helly) Tschuchnigg, an Austrian, first arrived in PNG in January, 1952, as a flight traffic officer on a Qantas Sandringham flying-boat. He was a foundation

173. Territory Airlines engine staff. *Photo: Gerd Franck*

173

member of the Aero Club of Papua, and served for a time as honorary secretary. In 1955, he joined Gibbes Sepik Airways as a pilot, and made many flights into the remote Southern Highlands District, servicing outstations and airdropping to patrols.

In June, 1957, Tschuchnigg was the pilot of Norseman VH-BNE, engaged in dropping supplies to an exploratory patrol led by the author* when an engine failure forced him to land in a swamp near tiny Lake Kopiago. The Norseman overturned, and Tschuchnigg and his drop-crew (Barney Madden, District Education Officer, Mendi; Dave McLaughlin, Public Works Department, and Rick Hill, cadet patrol officer, Minj) were stuck, unarmed, in new country, in the midst of tribal groups who were constantly warring, and who had seen little of the white man. After a harrowing night in BNE in the middle of the swamp, they were led by Kopiago people to a base camp we had earlier constructed at the lake. By this time, the author and medical assistant Bert Speer, had reached the lake from the Strickland Gorge. Contact was made with Mendi by portable teleradio, and on 1 July, Ken Montagu flew in to the lake in Qantas Beaver floatplane, VH-EAU, from Middletown, the Australasian Petroleum Company base near Kikori. He landed, and took Tschuchnigg and his crew on board. We watched the take-off with anxious eyes; the lake was so tiny, the Beaver so vulnerable against the backdrop of heavily-timbered mountains in this wild country. At the last possible moment the floatplane lifted off, and Montagu climbed up and over the ranges, for the long flight back. It was a remarkable operation.

There was an extraordinary sequel to this copybook rescue. Bob Gibbes badly wanted to salvage BNE if he could, but it was on its back in a swamp in uncontrolled territory, scores of kilometres from the nearest airstrip. He made a characteristically bold decision: he would send engineers in, on foot, dismantle the Norseman, and carry it out to Tari.

We did what we could to help by building a raft of logs in the swamp in front of BNE, and then pulling the big monoplane over onto its wheels, using heavy ropes dropped by Bobby Gibbes and the combined labour of 100 men. We erected a barrier of tree-trunks around the Norseman, and left it in the care of friendly tribesmen while we carried on with our patrol.

Gibbes flew in sixty Mount Hagen carriers to Tari, together with engineer Albert Buzzi, carpenter Noel Robertson, ex-patrol officer Laurie Nolan, and ample supplies and equipment. Since BNE had come down in restricted territory, it was necessary for the Administration to provide an official escort: patrol officer Roger Claridge, and twelve members of the Royal Papua and New Guinea Constabulary. Another sixty carriers were recruited at Tari, and the great expedition left for the lake on 17 July, 1957. They reached Kopiago in six days, dismantled BNE (discarding the heavy engine: years later it was recovered, and installed in a place of honour at the new Lake Kopiago patrol post) and carried the aeroplane back, through swamps, over mountains, across rivers, for 113 terrible kilometres. A track, four-and-a-half metres wide, was cut through the forest for a distance of sixty-four kilometres, and hundreds of tribesmen — thrilled at this fantastic sight — helped the carriers as they staged the dismantled Norseman from camp to camp, day after day. The great wings, with their massive spruce main-spars, caused the most problems. As this strange cavalcade inched along, Norsemans dropped supplies at regular intervals to feed the exhausted carriers. They left Lake Kopiago on 28 July, and arrived at Tari on 15 August.

Gibbes flew the pieces of BNE to Goroka, and there it was rebuilt. In the words of a DCA spokesman, it was one of the most difficult salvage operations ever undertaken by an airline company anywhere in the world. The new regional director, V.A. Rule, went further: the salvage was an 'amazing feat, unique in the history of Commonwealth aviation'.

....

Vern Rule was appointed Regional Director of civil aviation for PNG in January, 1956. He had served in New Guinea as a Liberator pilot with the RAAF during the war, and was an experienced and competent officer. He arrived in Port Moresby in mid-February, accompanied by R.W. Green, former senior route captain with TAA. Bob Green had been appointed Superintendent of Flying Operations, replacing Tom Drury.

174. From left: Shirly Smith, Allan Booth and John Downie. *Photo: John Downie*

174

*For details, see *Sepik Pilot*, op. cit.

These appointments were a result of the new order in the department, following on the retirement of the Director-General, Air Marshall Sir Richard Williams. He had been replaced by Donald (later, Sir Donald) Anderson, who took office on 1 January, 1956, and who was to prove a brilliant Director-General. Anderson made a number of staff stages at senior level in several regions, including PNG.

John Arthur and Tom Drury were recognised 'Dicky Williams men'. As Drury says: 'John was appointed to New Guinea by Sir Richard Williams. Dicky was one of Nature's gentlemen. Never made a mistake in his selection of men. Dicky, John and I got on like blood brothers. He agreed to any of our reasonable requests. So John had a more-or-less direct line to the DG. This by-passed a whole heap of superintendents, assistant DG's etc., and upset a lot of people in DCA who thought our budget and financial requests, for example, should have gone through proper channels. We'd go direct to Dicky, and solve the whole problem in five minutes, without fuss or bother...Then Dicky retired.

'Don Anderson was a pretty straight shooter, and he didn't go for this sort of thing. I remember John ringing me one Saturday morning at about 7.30. "Have you heard the ABC news? I'll be at the office in half an hour," he said.

'I went down. There was John; Stan Coles, Chief Communications Officer; Frank Roberts, Personnel Superintendent; Bob Whitecross, Chief Air Traffic Controller. And then I heard that Don Anderson was the new DG. I pulled out ten quid, put it on the table, and said, "I'll be the first to go," and John covered it. 'He hates my guts,' John said, 'I'll be the first to go.' But I went first, so I picked up the twenty quid. That was at the end of '55.

'John and I played into Anderson's hands. We had a brawl in the Pacific Islands Regiment Mess, at the official opening. Not the place to put on a brawl, with the Administrator, colonels, various directors and brass watching us. John and I were honorary members. There were three Royal Air Force Mosquitos in Moresby, sent up to photograph the Dutch New Guinea border. We'd all been drinking, of course, and we got into an argument about how long the job was going to take. I told one of the English pilots, 'You'll be here for years! You're going to photograph from 25,000 feet [7620 metres] — you'll have to wait for completely cloud-free conditions to 25,000 feet." Then John Arthur said it would be a simple matter. He told me to pull my head in, in front of the Administrator, and I told him to go to buggery, and belt the bloody job...one word led to another...one of us threw a fist, and we finished up wrestling on the floor.

'That was the end of the line for our marvellous friendship, but it didn't really cease. John and I remained mates right up to the time of his death, in 1979.

'John was transferred to a new position of Co-ordinator of Operational Planning at head office in Melbourne. I reverted back to my permanent status of Examiner of Airmen. John did a marvellous job, but he never really saw eye to eye with Don Anderson, and he left DCA in 1961...'

....

Before making a bid for Territory Airlines, Buchanan asked Bobby Gibbes if he was interested in the company. Gibbes said, 'Not really, but I'm putting in an offer of £9,000.' Buchanan said he was going to offer £10,000. He remembers: 'Bob said, "That's all right, I don't really want it." He didn't mind me going ahead, even though I was still working for GSA. So I sat down with Dick Davis in Lae, and made him the offer. We wheeled and dealed. He said, 'We want £12,000 for TAL,' and I said, "I haven't got £12,000, I'm only offering you ten, and I've only five. I've got partners who'll come in." Old Dick said, 'Give me six now, and six over two years, and we'll charge you six-and-a-half per cent interest.' That seemed all right, so he drew up an agreement, and I signed it. I paid a thousand down, and I had until 31 March to pay the other five. Dick said, 'Right, you can have TAL on 1 January, 1958.' My son, Ranald, had just been born...

'I had to raise another thousand, quick. I wasn't worried about the second six thousand — I still had Helly and Frank up my sleeve. I was covered. When I got back to Goroka, I wrote to my father and my brother, David, and told them I wanted a thousand pounds. There was plenty of time...I had until 31 March.

'Then I talked to Frank Newell. He was a bit worried for personal security reasons, said, "If I get involved with TAL, I might lose my job with Gibbes...I've got a wife and two kids. But I'll maintain your aeroplanes, no worries, if Bob Gibbes agrees."

'Frank didn't want to risk his job. TAL wasn't making any money, and didn't look like a good deal. Then, on 6 January, Helly Tschuchnigg sent me a telegram, from down south. He'd just gone back to Qantas, and he was out, too. Crikey! Now I had no partners, and no pilot. TAL had a pilot, Tony Chadim, but I wanted another, one of my own. There was a conflict of personalities between Tony and me. Anyhow, I still had some time before I had to pay the five thousand, so I took over and ran TAL, with young Jan Gane doing the office work...'

Janice Gane, the daughter of pioneer Seventh-Day Adventist missionaries, was just seventeen years old when she started to work for Dennis Buchanan. Shirley Smith stayed on for a month to help her learn the ropes, and then Janice was on her own. She kept the minimal books, issued passenger tickets, weighed cargo, and supervised loading. By this time, TAL had only one scheduled run, a weekly flight to Lae and back every Friday, in the Dragon, VH-AON. Cash receipts were miserably small. Chadim was the only pilot, and Sam 'One-Eye' the only competent workshop hand. The problem of aircraft maintenance was overcome with the assistance of Bob Gibbes. He agreed to allow his engineers to continue to sign-out the TAL aeroplanes, although Sam and a few other trusted locals did the routine work on them.

Buchanan remained in the employ of GSA, a company operating in competition with his own. It was a curious situation, as he remembers: 'I still worked for Gibbes, and Jan Gane looked after TAL. She did it all. March 31 was getting nearer, and it was looking serious. I only had four thousand but my father and my brother came good with £500 each, and I paid off on time. We operated TAL on a day-to-day cash basis, helped by the carry-on from the old TAL. We got through it somehow. I just traded, and paid another £3,000 the first year, and the last £3,000 the second year.

'The old Cessna 170, GKA, carried the company...'

To fly the Cessna, Buchanan recruited another pilot, G.U. Franck. Gerd Franck, German-born, arrived in Australia with £2 in his pocket, in 1954. A tailor and textile engineer by profession, he wanted to fly, and knew that on his salary he could never hope to afford flying lessons in Germany. He was an immense young man, 188 centimetres tall and weighing more than ninety kilograms, and he was determined to gain his commercial pilot's licence. This he did, in July, 1956, with an instructor's rating, to boot. In April, 1958, he joined TAL. Gerd Franck recalls: 'My starting salary was £24 a week, with free accommodation — a shack at the back of Dennis's house. A little primitive, but adequate for a bachelor. My first flight was with Tony Chadim in the old DH84, Goroka- Kainantu-Lae-Goroka. This was my first 'big' aeroplane, two engines!

'I remember being impressed by the scenery, and startled by the Dragon's lack of power. We had to keep full throttle all the way from Lae to Goroka, to make it over the hills. This lack of performance wasn't helped by the fact that after putting on the 'normal' load for Dennis, Tony had added some timber for the house he was building in Goroka. I learnt that many of the pilots who had trade stores in the Highlands did the same.

175. Jan Gane. *Photo: Gerd Franck*

175

'I flew the Dragon on only a few occasions. I can remember one day after taking off from Goroka being told to return. There was a very strong cross tail-wind blowing, and many turned out to watch what they thought might be a spectacular landing. I did manage to keep it straight, but it prompted Dennis to ask me if I smoked. When I said 'No,' he gave me a box of matches and said, "Well, if you ever prang it, climb out and throw a match back in, it's worth more in insurance than actual value!"

'I got my first taste of "airline operations" flying the Dragon. Whilst I was in the Goroka tower, flight-planning, Sam 'One-Eye' would climb into the cockpit, and start up and warm the engines for me. I can remember how pleased he was with himself...'

Buchanan well knew that the day of the Dragon was over. Initial experience with the Cessna quickly convinced him that this was the aeroplane for the future. He decided to get rid of VH-AON. Tony Chadim was still flying the old Dragon and bringing in revenue, but the annual C of A inspection was due towards the end of 1958, and this always involved an expensive, major overhaul. It would mean, too, that the aeroplane would be out of service for a month or more. Thus Tony

Chadim would have no aeroplane to fly while AON was in the workshop.

In September, 1958, Buchanan found a buyer for AON, and sold her. He dismissed Chadim — he could not afford to keep him on, idle. At the very beginning of his career as an airline operator, Buchanan demonstrated that he was prepared to take the tough decisions that all successful entrepreneurs have to take in the course of building up large enterprises. Chadim was, and remains, bitter at the sacking. He remained in Goroka until 1965, working as a regional hydrographer for the Commonwealth Department of Works, and then went to Canberra as a senior computer programmer with the Bureau of Agricultural Economics. He was certainly a versatile man.

Buchanan was still working for Bobby Gibbes throughout this first year of his ownership of TAL, but GSA was in a bad way, and Gibbes was almost at the end of his rope. It is ironic that Gibbes Sepik Airways was about to finish as Dennis Buchanan was beginning his career in airline operation, for without Bob Gibbes, Buchanan would probably never have come to PNG; without the encouragement and assistance of Gibbes, he would probably not have been able to purchase and run TAL in the vital early stages. Buchanan acknowledges today: 'The salary I got from GSA looked after Della and Ranald, enabled me to pay Jan Gane for looking after TAL, and left me with a bit of cash for TAL if need be. So there was really no draw on TAL. That's how it happened. Della was working, too, until she had the baby, and we were living in the GSA flats...I was able to buy my first aeroplane, a Cessna 180, with a bank loan, and cash flow paid it off...'

From the very beginning, Gibbes Sepik Airways was under-capitalised, and life for Bob Gibbes during the ten years of the existence of the company was one long fight to survive, to obtain the aeroplanes and the routes that were essential to his company's existence. By 1958, Bobby and Jean had almost had enough. Then came a blow when GSA was at its nadir, and which helped to decide them to sell the company, and get out of commercial aviation.

Over the years, GSA had suffered many aircraft accidents, as this account has shown, but never a fatality. This is all the more surprising when it is remembered that GSA aircraft operated over some of the roughest country in PNG, and into some of the worst of the airstrips. One fine morning in July, 1958, a Gibbes Norseman landed at Goroka with the mangled body of a pilot as cargo. It was a ghastly shock to young Jan Gane. The dead pilot, Ronald Alan de Forest, had been flying for GSA for twelve months. He was a likeable young man, with pleasant, easy manners, and only the day before had been talking to Jan Gane about his forthcoming marriage.

On 8 July, Ron de Forest left Minj in Norseman VH-GSA on a routine flight to Mendi, with a full load of bagged cement. It was late in the afternoon, and the weather in the Wahgi was building up. It was a lot worse in the Mendi valley. The head of the valley was blocked with boiling black cloud, and at 5.20 p.m., de Forest called the Goroka tower to say that he could not get through, and was turning back. Nothing further was heard from him. Minutes after making his last call, Ron de Forest was dead. He lost control, no doubt blinded by rain, lighting and cloud, and spun in. He was killed instantly.

Four months later, Gibbes Sepik Airways was sold to Mandated Airlines.

177

178

176

176. VH-BNE down at Lake Kopiago, 1957. *Photo: James Sinclair*
177. Salvaging Norseman VH-BNE at Lake Kopiago, 1957. *Photo: Roger Claridge*
178. TAL Dragon, VH-AON undergoing maintenance. *Photo: Gerd Franck*

179. Captain Adrian Nisbet flying Norseman VH-GSA, drops supplies to the patrol led by the author through the Strickland River area in 1957. *Photo: James Sinclair*
180. JU52, VH-BUU, of Gibbes Sepik Airways, takes part in the airdrop. *Photo: James Sinclair*
181. Dragon VH-AON, ex-Territory Airlines. *Photo: Talair*
182. Beaver floatplane evacuates survivors of the crashed Norseman from Lake Kopiago. *Photo: James Sinclair*
183. GSA Norseman airdrops to patrol led by the author, Lake Kopiago, 1956. *Photo: James Sinclair*
184. Territory Airlines Cessna 170, VH-GKB. *Photo: Bill Sherwood*

17

180

18

182

18

184

The End Of Gibbes Sepik Airways

Gibbes Sepik Airways folded for a number of reasons, but principally because of a lack of sufficient development capital. Bobby Gibbes did not have the money to replace his obsolete Norsemans with the modern light all-metal aircraft which would have allowed him to compete with rival companies, and he could not afford DC3's, which were now very expensive on the open market.

At the beginning of 1958, Gibbes was operating six Norsemans and three JU52's: VH-BNE, BNL, BLM, BHG, ASN and GSA, and VH-BUU, BUW and BUV. Of these aircraft, the Norseman, BNE, was in the workshops being rebuilt after the Lake Kopiago crash, and the Junkers, BUW, was likewise under rebuild after another crash, later described. Only by operating the remainder of the fleet to high capacity could Gibbes have hoped to survive. The Norsemans carried GSA, but with the coming of the Junkers, Norseman utilisation dropped alarmingly. The Norseman, at a charter of £44 per hour, could lift 1043 kilograms on average. For £75 per hour, the customer could charter a JU52, lifting up to 3630 kilograms on short hauls, and the Junkers could get into any strip that the Norseman could land on. Obviously, most customers wanted the Junkers. Bob Gibbes had unwittingly cut his own throat with the introduction of the JU52.

185. From left: Robin Gray, Bobby Gibbes and Peter Manser. *Photo: Sten-Erik Molker*
186. From left: Brian Walker, Paul Raasch and Robin Gray. *Photo: Sten-Erik Molker*

185

186

The JU52 was the last aeroplane on Gibbes' mind when he set about acquiring heavy aircraft to extend his airline operations. Like every other airline executive, Gibbes wanted DC3's. As previously recounted, the supply of these magnificent aircraft at the end of the Pacific war was strictly regulated by the Australian Government, which allocated them to the major airlines. Shortly after the war, used DC3's were selling in Australia for as little as £5,000 each, but by the time GSA got started the second-hand price was starting to climb. By 1955, airlines were paying £20,000 for DC3's with many thousands of hours behind them. Demand was falling, however. Companies in various parts of the world were developing what they all said would be DC3 replacements — a myth, as it turned out. The Douglas company itself failed to develop a modern substitute for the DC3, although they tried hard with the Super DC3. No aeroplane offering the blend of price, performance, reliability, longevity and operating economics of the DC3 appeared. As a result, demand for used DC3's again soared. By 1957, Australian companies were paying £50,000 and more for average specimens. In April of that year, MAL paid £65,000 for a DC3 (actually, a C47-35DL, VH-MAB) owned by the Commonwealth Bank. It was, admittedly, a superb example, luxuriously equipped and with very low hours. A DCA spokesman said that the aeroplane was a bargain at that price, probably the highest ever paid in Australia for a DC3.

Expenditures of this order were quite beyond Gibbes' financial reach. He did have a possible chance to acquire a DC3 in excellent condition at a favourable price in March, 1952.

On 29 March, an RAAF Dakota took off from Lae, bound for Goroka, on a combined training and inspection flight. On board was a party headed by the Australian Minister for Works and Housing, W.S. Kent-Hughes. The pilot, an RAAF flight- lieutenant, had never before flown in the Highlands. In time of war or national emergency, service pilots are required to land anywhere the demands of the situation dictates, and so the Air Force has never followed the civil aviation practice of restricting pilots to routes and airstrips into which they have been checked.

The weather in the Highlands was poor, with a low ceiling. The DC3 entered the Asaro valley, and the pilot was confronted with a massive wall of racing dark cloud. When he was over what he thought was Goroka — blanketed with cloud — he prepared to land. He contacted the tower, and asked the controller if he could hear his engines. The controller left the tower, and heard, high above, the sound of aircraft engines. He cleared the DC3 to land. The pilot replied that he had Goroka in sight, and was coming straight in.

However, the engine heard by the controller was not that of the DC3, but of a GSA Norseman, en route from Madang to Chimbu, and piloted by John Wells. The airstrip on which the RAAF pilot landed was not Goroka, but the tiny emergency strip, Finintegu, thirty-two kilometres away. Putting a DC3 safely down on a light aircraft emergency strip was a considerable feat of airmanship, but the pilot managed it.

Landing a DC3 at Finintegu was one thing; getting it out another. The ministerial party was ferried out, one person at a time, to Goroka late that afternoon by Laurie Crowley in his venerable Curtiss Robin, and George Larum in his Tiger Moth. An attempt by the flight-lieutenant to fly the DC3 off from Finintegu ended with the aeroplane bogging, and standing on its nose.

Bob Gibbes saw a chance to pick up a good DC3, cheap. George Larum flew Gibbes and his then chief engineer, Murray White, to Finintegu. White found the DC3 very little damaged. Gibbes was sure he could extend Finintegu, and fly the aeroplane out. He made an offer of £10,000 to the RAAF for the DC3, as it was. (MAL only offered £1,000). The offer was refused. The RAAF was red-faced over this incident, which was the cause of many ribald comments among Territorians generally, and civil pilots in particular. A ground party walked in, stripped the DC3 of all excess weight, extended the strip, and flew the DC3 out. It was an expensive exercise, but it vindicated the honour of the RAAF.

Finally, Bob Gibbes had to forget his dream of operating DC3's, and turn to something cheaper. The DCA Senior Aircraft Surveyor, Rube Murray, had for some time been extolling the virtues of the JU52/3m to Gibbes. They were produced by the thousands during the war — the JU52 was to the Luftwaffe what the DC3 was to the Allied Air Forces — and they were robust, simple aeroplanes with excellent short-field performance, and should be easy to maintain and repair.

The JU52/3m was by any standard a remarkable aeroplane. It was virtually a modernised version of the tri-motor G31 freighter that had broken world records during the opening of the Morobe Goldfield in the 1930s. 'Iron Annie', as the JU52 was affectionately known to the men of the Luftwaffe, was a large, low- winged, fixed-undercarriage monoplane, skinned with corrugated light alloy and equipped with the characteristic Junkers 'double-wing' flaps that gave it extraordinary low-speed performance. Most production JU52's were powered by three BMW 132T nine-cylinder air-cooled radial engines each of 830 hp, turning fixed-pitch twin-bladed airscrews. After the war, converted JU52's served in European airlines in much the same role as did converted military DC3's.

It so happened that Gibbes had a considerable sum in Dutch currency to his credit in Holland, returns from flying that he, Peter Manser and Rinus Zuydam performed for the Dutch New Guinea Government in 1951 and early 1952. This allowed Bobby and Jean to take an overseas trip in 1952 to look for suitable JU52's. It was during their absence that the crisis referred to in an earlier chapter developed — and they did not locate any suitable aeroplanes.

In 1955, Gibbes was told of the availability of three JU52/3m's, in Sweden. They had been operated by the famous Captain Albin Ahrenberg, whose company was in liquidation, and were said to be in flyaway condition. The asking-price — £15,000 sterling each — was too steep for Gibbes, and the Swedes refused to budge. On impulse, Gibbes flew to Stockholm to negotiate, and after a series of manoeuvres*, succeeded in purchasing them for only £1,500 each, including spares.

At that price, the aeroplanes were bargains, but they were in badly-neglected condition. Two were without engines, and only one of the three BMW 132Z-3 engines in the third machine was operational. A number of BMW engines included in the deal were in such worn condition that they were beyond repair.

It had been Gibbes' intention to dismantle the aircraft and ship them back to PNG, but he soon found that the cost would be so formidable that it was not a practicable proposition. The Junkers would have to make the long journey across the world under their own steam — or not at all. He engaged an experienced Swedish engineer, who, after careful inspection assured Gibbes that the best of the three, which bore the Swedish registration, SH-BUE, was fully capable of the flight, given efficient engines.

BUE had only one good engine, with some 300 hours still on it. Through the

187. Captain Albin Ahrenberg with VH-BUU. *Photo: Sten-Erik Molker*
188. Paul Raasch (left) prepares JU52 for flight to New Guinea, Stockholm, December, 1956. *Photo: Paul Raasch*

187

188

* See *Sepik Pilot*, op. cit.

good offices of Captain Ahrenberg, Gibbes managed to purchase two fully-overhauled BMW 132Z-3 engines in Finland, for the bargain price of £200 sterling each. Whilst his engineer methodically went about the task of installing these into BUE, Gibbes travelled Europe, looking for bits and pieces for his old aeroplanes. His financial position was by now exceedingly precarious — no new experience for Bobby Gibbes — and he was in a fever to depart. His plan was to fly BUE home himself, and arrange for the delivery of the other two later.

One big problem remained. Gibbes had acted on impulse when he flew to Sweden to purchase the Junkers, and before he had obtained permission from DCA to use these German ex-military aeroplanes in an Australian territory. Should the department refuse to issue Certificates of Airworthiness, the Junkers would be so much expensive junk. Rube Murray had privately recommended JU52's to Gibbes, but the department had given him no undertakings.

Gibbes telephoned the Director-General, Sir Richard Williams, in Australia. Williams, an old RAAF acquaintance, did not mince his words: DCA would require the fullest technical data and overhaul and maintenance manuals before any consideration would be given to the question of C's of A. Fortunately, Gibbes was able to obtain a number of engineering publications from British-European Airways, who had operated one of the largest post-war fleets of JU52's. They were a long way from meeting the requirements of DCA, but they were a start.

189. Robin Gray. *Photo: Sten-Erik Molker*
190. Robin Gray and Captain Ahrenberg check their map before departing for New Guinea, December, 1956. *Photo: Sten-Erik Molker*

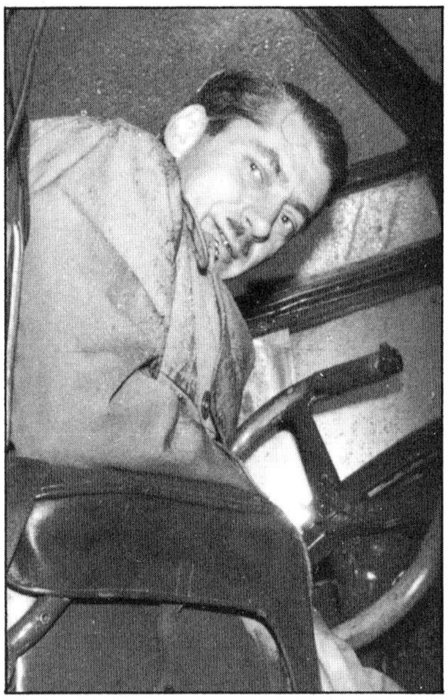
189

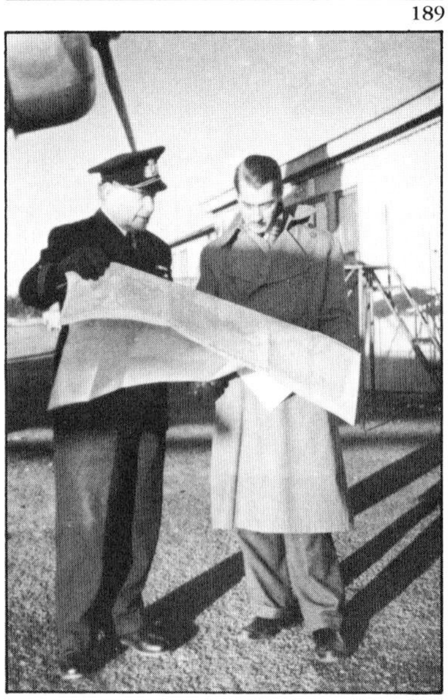
190

By late September, 1955, BUE was ready to depart. Gibbes engaged an experienced flight engineer/pilot for the ferry flight. The engineer, Gustav Adolf Jerdnell, was a Junkers expert. Gibbes' cousin, Alison Moxham, and her friend, Jean Cameron, came along for the ride. A formidable list of approvals and clearances had to be obtained before they could depart, but at last, on 8 October, they went.

The flight was one long adventure. The port engine began to misbehave in the early stages, and a fuel shortage forced Gibbes to land in the desert 160 kilometres from Karachi. An inspection at Karachi revealed that the port engine had a fractured crankshaft, and at enormous expense a complete crankshaft 'spider' was flown out from Sweden, and fitted by Jerdnell. The flight continued, through Delhi, Calcutta, Rangoon, Labuan and Zamboanga, with Gibbes battling officialdom all the way. They reached Goroka on 18 November, 1955.

The first JU52 was in PNG, but putting it to work was another matter. Both Qantas and MAL opposed the introduction of the Junkers, on the grounds that their DC3 fleets were more than capable of handling the work offering, particularly from Madang into the Highlands. However, following a meeting at DCA headquarters, Melbourne, attended by Gibbes and representatives of the other two companies, Director-General Donald Anderson gave his approval to their use. The first Junkers was registered as VH-BUU. Preliminary manuals were submitted to DCA early in 1956, and the department gave all possible assistance to the GSA engineers, who were planning extensive modifications to the JU52, to fit it for PNG operations.

The principal modification was the replacement of the original BMW engines with Pratt and Whitney R1340-S1H1-G geared engines turning three-bladed Hamilton Standard airscrews, as used in the re-engined Norsemans. It was a difficult modification. The particular model of the BMW fitted to BUU developed 800 hp, and the R1340 only 600 hp. But the BMW had been out of production for years, and spares were scarce and expensive. Clearly, different engines had to be installed if the JU52 was to be operated in PNG, and Gibbes had a number of SIHI-G engines on hand. There were obvious advantages, too, in using a common engine in both the Norseman and the JU52.

When Gibbes sold his company to MAL at the end of 1958, his engineering workshops were possibly the best in PNG, well equipped and staffed with eleven licenced engineers. In 1956 available facilities were insufficient to cope with the major task of changing over to the Pratt and Whitney engines, and on the advice of his then chief engineer, Eric Cross, Gibbes decided to send BUU to Australia for the job to be done.

The GSA charter licence allowed the company to operate only within PNG, but this did not stop Bob Gibbes from accepting paying passengers for the return flight, Goroka-Sydney-Goroka, at a fare (£60) far below that charged by Qantas, the only company with a legal right to carry paying passengers between PNG and Australia. BUU made the flight in June, 1956, with John McDermott at the controls.

Over the Papuan delta, the starboard BMW failed, and McDermott was forced to land at Horn Island. Gibbes' old wartime friend, Brian 'Blackjack' Walker, chief test pilot for de Havilland Australia, flew to the rescue in a Drover, with a Pratt and Whitney SIHI-G engine. The starboard BMW was removed, and replaced by the centre BMW, which was itself replaced by the Pratt and Whitney. McDermott completed the flight to Sydney with no further trouble. In September, Brian Walker left de Havilland, and joined GSA as a pilot. Walker had helped Bobby Gibbes avert disaster in the 1952 financial crisis, and before leaving de Havilland, he test-flew the re-engined BUU at Bankstown, to the satisfaction of DCA.

This flight brought Bob Gibbes into conflict with the new regional director, Vern Rule. Qantas complained about the illegal carriage of paying passengers in BUU, and Rule severely censured GSA. He was concerned at the casual attitude of some of the smaller operators towards DCA rules and regulations, and was determined to do something about it. Actually, it was fairly common practice to grant special permission to companies sending aircraft to Australia for annual C of A overhauls to carry paying passengers. Gibbes' offence was that he failed to ask for permission.

The engine conversion was reasonably successful in that the JU52 with R1340 engines *did* fly, although the loss of power was very apparent. The port and starboard wing mounts for the BMW engines were skewed, which resulted in the wing engines being offset by an angle of six degrees to the centre engine. Eric Cross planned to rubber-mount the R1340's on straightened mounts, but this proved impossible for technical and financial reasons. The result was that the R1340 wing engines vibrated badly on the rigid skewed mounts, and this engine when fitted to the JU52 proved much less reliable than when installed in the Norseman. Another limiting factor was the engine cowls. The original BMW cowls were close-fitted and streamlined, but they would not fit the Pratt and Whitneys, and bulky Wirraway cowls had to be substituted. They increased drag, and degraded the performance.

It proved very difficult for GSA to complete the overhaul and maintenance manuals to the standard required by DCA. The JU52 was long out of production, and the parent factory was located at Dassau, in East Germany. Gibbes' engineers had to start almost at scratch to produce the manuals, and in recognition DCA granted GSA a twelve-month extension, and allowed BUU to go into service.

On the strength of the successful conversion and flight testing of BUU, DCA granted approval for the two JU52's still waiting in Sweden to be engine-converted and flown out. Robin Gray went to Sweden to supervise the fitting of the replacement Pratt and Whitneys and the return ferry flight. Brian Walker followed. They were to be away for most of 1956, for the two remaining Junkers — later registered as VH-BUV and BUW — were in poor condition, and much work was required to fit them for the trans-world flight to PNG. None of the GSA engineers had any experience of the Junkers, and Gray and Walker looked round for qualified men. The obvious source of engineers trained on the JU52 were airlines that had operated these aircraft, and Gray was advised to contact the famous West German airline, Lufthansa.

Lufthansa had shortly before established an engineering workshop at Hamburg, in charge of chief engineer, Paul Raasch.

Raasch was a Berliner, born on the eve of the Great War. He was fortunate enough to be chosen as Lufthansa's first engineering apprentice when that airline began during the post-war depression that ravaged Germany and saw the rise to absolute power of Adolf Hitler and his National Socialist Party. When the Second World War began, Paul Raasch joined the Luftwaffe as an engineer in a unit flying the JU87 Stuka dive bomber. He served in the French campaign and at Crete before being sent to Africa as a member of the lst Luftwaffe unit of Rommel's Afrika Korps. Raasch was one of the last to leave when the Luftwaffe evacuated Africa. Sicily followed, then the Hungarian front, fighting against the Russians. Raasch finished the war as a paratrooper on the Austrian front, and finally as a prisoner-of-war. His engineering skills were put to work by the British after the collapse of Germany, and he finally rejoined Lufthansa as chief engineer.

Here was the man with the knowledge and experience to supervise the overhaul of BUV and BUW, and Robin Gray engaged him on the spot. Raasch had met Australians during the war — crews of captured aircraft — and he had long wanted to leave war-ravaged Germany, and migrate. He moved to Stockholm with

191. Peter Manser (standing) and Adrian Nisbet, 1957. *Photo: Sten-Erik Molker*

191

his wife and two children, and set to work. Raasch was soon joined by another German engineer, Eddie Halle, and later by a Swede of wide experience, Sten-Erik Molker. Between them, this highly-competent team got the two Junkers ready for the long flight. Tony Chadim arrived from Goroka to install the Lear radio equipment. A tall, slim Englishman, John Green, was engaged as co-pilot. Green was the ex-RAF jet pilot who subsequently flew the Cessna 190 for Dave Robertson of Madang Air Services.

The time for departure arrived. The aircraft were re-engined and thoroughly overhauled, maps, landing permits and paperwork were under control, spare parts assembled, and personnel cleared. It was decided to make the flight in formation, Robin Gray and Tony Chadim flying BUW, with Sten-Erik Molker and Chadim's wife and daughter as passengers and carrying all the spares, and Brian Walker and John Green flying BUV, with Paul Raasch and his family, and Green's and Halle's wives as passengers. Green's wife was eight months' pregnant! All personal effects also went into BUV. Theoretically, the loads of each aircraft were balanced, and each should be able to fly at the same speed at the same power settings, and with identical fuel consumption.

It was a momentous journey. They took off from Stockholm for Hamburg on 13 January, 1957, into bad weather that quickly deteriorated into a snowstorm, which forced them back after two hours, twenty minutes in the air. Leaking fuel tanks caused constant problems, and it was very difficult for the pilots to keep the two aeroplanes in close formation. BUV proved a faster aircraft than BUW: it had a slightly modified smooth-skinned wing leading edge, whereas BUW's wings were all-corrugated, in traditional Junkers fashion. And the flying techniques of the two pilots were different. 'Blackjack' Walker had spent years as a test pilot, and he was constantly experimenting with BUV in an effort to discover its characteristics. This sometimes proved disconcerting to his passengers, as Tony Chadim relates: 'BUV had a very close shave on the way from Beirut to Abadan. It was a long trip. We had to cross Syria through a narrow corridor, from there to Baghdad and on to Abadan. The Abadan control tower told us to hurry, because their airport lighting system was out of action. About half-way to Abadan, we caught up with, and overtook, BUV. It was way down, circling just above the ground over the most horrible rocky desert one could imagine, but then it started climbing again, so we went on and landed a little after the last light. Brian had to use his landing lights, but he made it, too. When his passengers came out of the aircraft, they were just about hysterical.

'What happened was that Brian was curious how much petrol one of his tanks held, and let it run dry. John Green was most impressed how calm and quick Brian was when all three engines stopped. To restart a Junkers 52 is quite a procedure: winding up small crank handles of oil coolers and carburettor heat controls, pumping and priming the engines and energizing the inertia starters, and all this time they were going down with not a square yard of flat ground within sight. They were about 100 feet [thirty metres] above the ground when Brian got the engines re-started. The German families refused to fly with Brian after that, and it took Robin a long time and a lot of diplomacy to change their minds. Brian had to promise to behave, and we were on our way again.

'The flight on VFR over such long distances was very interesting. Flying so low in a slow aircraft we certainly could see what thousands of airline passengers and even airline pilots inevitably have to miss. I was very lucky to be with Robin Gray, safe, steady and always polite and pleasant. Robin flew long patrols in Catalinas during the war, and the first thing I noticed when he took over the controls in the initial stages of our journey was that he made the Junkers fly some two or three knots faster with the same power settings. Until then, I never realised how wasteful it was to over-control an aircraft.

'Since Abadan, Brian's passengers had all available alarm clocks set up to remind him when it was time to change over to another tank. He retaliated by leaving it to the last possible moment to get up and turn the tap on the big tank in the fuselage, instead of leaving it to the engineers. And so, en route from Zamboanga to Sorong, with the vast area of blue tropical sea all around, he turned the tap with his big wrist — and broke the handle off!

'Paul Raasch and Eddie Halle sprang into action. They had less than ten minutes of fuel left, and in that time they had to dismantle the valve, switch it over and reassemble it without flooding the fuselage with petrol and blowing it up. In the

192. Long-range fuel tank inside VH-BUW. *Photo: Sten-Erik Molker*

192

meantime, Brian was calmly preparing to crash-land the Junkers in the shallow water near a tiny and solitary uninhabited island, with two children and three women, one of them eight months' pregnant, on board!'

They survived this emergency, and innumerable encounters with irate customs men and aviation officials, and arrived safely at Wewak on Friday, 8 February, 1957, after a flight of over 20,000 kilometres that had occupied ninety flying hours over twenty-six days through seventeen countries and landings on twenty different airports. All of Goroka turned out to cheer the two Junkers when they flew in formation over the town the following day.

BUV and BUW joined BUU in service, and were soon familiar sights at airports all over New Guinea. Despite the performance limitations imposed by the fitting of the Pratt and Whitney engines, the JU52's quickly proved themselves to be extremely versatile, and tough opponents for the DC3's. They were the best airdrop aircraft that ever operated in PNG. The big double flaps and low wing-loading permitted airdrops at 105 kph and allowed the Junkers to climb rapidly out of steep valleys. The aeroplanes experienced constant engine problems, and the radio installations, too, suffered from the excessive vibration. They were nonetheless a distinct success. Had the JU52 been available in numbers just after the war, it would have had as marked an effect on PNG aviation as did the DC3, but Gibbes introduced the Junkers at a time when the old days were about over. The GSA JU52's were not destined to have long lives.

GSA pilots in general liked the JU52, although they recognised its limitations. Most were glad that they had the chance to fly the fabled 'Iron Annie', survivor of an earlier, braver era in aviation. Says Adrian Nisbet: 'I was at Goroka the day the first JU52 arrived. I converted onto the JU, then was posted to Madang. Those JU's were terrible things, heavy as a brick, but magnificent aircraft for the job. Fancy getting into something like the old Maprik strip, about 1,600 feet [488 metres] long, with a three-engined aircraft carrying as much as a DC3 — 7,500 pounds [3402 kilograms] or so of payload. Qantas was stripping down their DC3's to try and get a bit more payload out of them to compete with the JU52.

'But they were terrible things, a great big wooden wheel, terribly spacious and heavy. Although we changed the BMW's to Pratt and Whitneys, they weren't very reliable, and if you dropped one engine with a full load there was no way you could hold height for long — you'd slowly descend. Extraordinary aircraft for their time. We carried all sorts of things in them...'

In May, 1957, a pilot with a remarkably varied aviation background joined GSA. Ivan Bennett was a stockman before the war, living on the back of a horse from the age of five. When war came, Bennett did not fancy the idea of descending from the saddle to foot-slog with the Army, so he joined the Air Force. He trained as a pilot at Sandgate and Narromine before being sent to Canada, under the Empire scheme. He graduated as a pilot in April, 1943, and was promptly made an instructor. He remained in Canada for two years, and was attending a Mosquito school in Nova Scotia when the war ended.

Bennett was demobilised in December, 1945. He subsequently flew for ANA and Butler Air Transport on DC3's before joining the Indonesian airline, Garuda, in 1952. For three years, he flew DC3's throughout the length and breadth of Indonesia, wherever a DC3 could go. It was pioneer flying of a type very much like that being done in PNG. Leaving Garuda in 1955, Bennett joined East-West Airlines, and in 1957, Gibbes Sepik Airways. He was an expert on the DC3, and approached his first flight in the JU52 with deep professional interest. He recalls: 'You didn't need a pilot's licence to fly the old JU — needed a plumber's licence. It was a bloody good aeroplane, though. The corrugated metal construction made them very strong, and gave them a much greater wing area than was apparent. Had great ability on short fields. The worst aspect of them was that Bobby took the BMW's out of them and put in Pratt and Whitney1340's. There was a loss of 200 hp per engine, and he was still flying them at an all- up weight of 24,000 pounds! [10,886 kilograms]. Should have been knocked down to 21,000 [9525 kilograms] all-up — would have given a much better performance, but there'd have been a loss of revenue, too. The JU52 was under- powered with the engine switch, took a hell of a long time to get to 10,000 feet [3048 metres]. The1340's were prone to blowing a pot in the JU. I had quite a number of engine failures.

'I did a lot of airdrops in the JU52. You could get down to sixty-five knots, made the job easy. One of the worst I did was at Simbai, went in to drop a roller to Jim

193. Eddy Halle and assistant work on an engine, Goroka, 1957. *Photo: Sten-Erik Molker*

193

McKinnon, who was building a strip there. Ken Davenport was my first officer. We landed at Aiome to get the roller — one of those metal things that you filled with water to give it weight. Empty, was about 500 pounds [227 kilograms]. Dave Robbie was going to drop it out of his Cessna 190, but DCA wouldn't let him.

'Jim gave us a detailed plan on how to fly up the Simbai strip and drop this roller. All very well in theory, but that Simbai valley is lousy. We didn't have room to complete a 360-degree turn in the old JU, and coming around I could see that we were going to be in deep trouble. So I gave the signal to drop the roller, quick. Had to get it out, had no option, and we just missed a ridge. Dropped the roller in the next valley, and it went in so deep that it took three weeks to dig it out. Jim McKinnon was most upset. Poor old Davenport was looking very pale, took four years off his life...'

J.K. Davenport was a young post-war pilot, commencing his flying training in late 1955 in Vancouver, Canada. He gained his commercial licence in October, 1956, and then went looking for a job. Says Davenport: 'I was ready for a commercial flying position, but to my surprise the airlines were not. A Canadian winter turn-down was in progress, and the Canadian and American Distant Early Warning Radar System was now completed, and numerous commercial pilots with heavy experience were on the market. There was little chance of employment in airlines, and during early 1957 I wrote to over 100 overseas light commercial airlines requesting consideration. The only favourable replies I had were from Australia, so I borrowed money from my mother, and arrived in Sydney in October, 1957. I had a hundred dollars in my pocket, and I did not know anyone in Australia.

'You will recall that Ansett and ANA merged during 1957, and quite a few surplus pilots joined Qantas. When I found they would not be taking on pilots with minimum hours, I asked and received advice and help from Brian Crapp, the Qantas Personnel Manager. He suggested I contact the air companies in New Guinea. Bobby Gibbes offered me a two-year contract, flying Norsemans, and eventually JU52/3m's.

'Peter Manser became my mentor on the Norseman. He had spent many years in New Guinea, and had a wealth of experience and I was grateful for his tuition. I learned to approach hills from the side, and if there was no down-draught, slide over the ridge into the next valley. He'd say, "Flying straight or over a hill, one day you will hit it." I learned to land uphill at Omkolai, and the grass aircraft carrier at Chimbu. Later I moved to Wewak, to be route qualified in the Sepik, and once again Peter Manser was assigned to the task.

'My first check out was a flight to Green River, some one hour fifteen minutes to the west of Wewak. Peter said, "If it gets too tough, turn back." We took off early in the morning, headed out over the small range of hills to the west of Wewak into the Sepik valley. There was 8/8's of low stratus cloud with occasional large cumulus. After numerous diversions around the offending clouds, I considered Manser's words, "If it gets too tough, turn back." Unfortunately, I didn't know which way back was. So we continued in formation, to the west. After approximately one hour fifteen minutes, VH-ASN spiralled through a small break in the clouds, and disappeared. I was so nervous I did the same thing. I found Peter on the ground, being unloaded, and when I taxied in behind him he said to me, "I told you to turn back when it got too tough, however, you're now checked out in the Sepik".

194. Sten-Erik Molker and friend. *Photo: Sten-Erik Molker*

194

'On one occasion I recall complaining to Dick Shaw, senior pilot at Wewak, that my engine was running rough, and blowing blue smoke. I was assured that this was normal for the Pratt and Whitney 1340. I continued to be suspicious, and continued to complain to Dick. Finally, he agreed to swap aircraft for the day. On departure next day, his Norseman had an engine failure, and Dick made an emergency landing at the small secondary airfield north of Wewak. I had the pleasure of flying over the mechanic, and picking up Dick. Little was said, except he would have his aircraft back, and I would be left to help the engineer change the cylinder in my aircraft, which I did. From there on, there was no more smoke.

'I later spent some time in Madang, flying the JU52. At first sight, I was impressed in a strange way, like does it really fly! I learned it did, and quite well, at least to 500 feet [152 metres]. It was common knowledge that the 1340 engine was not as reliable when installed in the JU52, because of excessive vibration. I recall numerous engine failures. The centre engine, while not causing asymetrics, sure did cause visibility problems. A few gallons of oil spilt on the windscreen always made an interesting landing.

'The brakes on the Junkers were air driven through a compressed air bottle system in the fuselage accessible on the ground only. A 3,000 psi bottle was connected without a regulator to a 1,000 psi bottle in the cockpit. The large bottle was usually removed at night and charged at the local machine shop. To apply brakes to both wheels the centre throttle was retarded to the idle position and applied in a rearward motion which would open up an air valve and supply air directly to both wheels. The left throttle at idle and applied in the reverse direction would apply brakes to the left wheel, the right throttle to the right wheel. If you were to have an engine failure on take-off, let's say the port engine failed, the aircraft would naturally swing to port. If the rudder could not compensate the swing, the starboard brake would need to be applied. To do this you would need to retard the starboard engine to apply braking. You then only had one engine at full power. This always made some interesting take-off scenarios.

'The flaps on the Junkers were also a trap for young and old. Beside the captain's seat was a large wooden laminated trim wheel and a lever which would select either trim or flaps. Bearing in mind there were no hydraulics on this aircraft, approaching an airfield you would pull the lever up and start winding down your flaps. As the flaps ran out the aircraft became tail-heavy. Therefore, down with the lever, now wind the trim wheel forward. If you happen to get the lever out of sequence you would find yourself on short final perfectly trimmed however with no flaps extended...'

Ken Davenport eventually left GSA and joined Qantas (where, at the time of writing, 1985, he was Flight Operations Line Director).

On 21 March, 1957, Peter Manser experienced the trauma that could follow an engine failure in a JU52 on take-off. On this day, Manser was at Wau at the controls of VH-BUW, about to depart for Lae with a heavy load of sawn timber. He opened the throttles of his three engines, and began his take-off run down the steep Wau strip. The port engine suddenly stopped, dead. Manser was at the point where he was almost airborne: on this precipitous strip he was committed to the take-off. But the Junkers swung sharply to port, and headed for the small wooden Qantas terminal building on the side of the strip, where the agent, Mrs Ivy Crawford, was at work. An eye-witness later described what happened: 'She saw the aircraft trundling down towards her, and ran for her life through the passenger room. The direction she took headed her into the path of the aircraft. Peter Manser sat up there in the cockpit, goggle-eyed, saying to himself, "If she keeps running that way, she's had it." There was nothing he could do. He was pounding out of control, urged on by thousands of pounds of aircraft and wood cargo.

'By this time, Manser was only a few yards away from the terminal and the running woman. She threw a look over her shoulder, jolted, wheeled, and doubled back like a League half-back. She dived over a high embankment, and disappeared just as the aircraft struck the terminal. The port wing caught the building about eight feet [two-and-a-half metres] from the corner. It sliced through wood, foundations, beams, nails, tin — the lot — and carried on. The roof of the building dropped like a limp hand.

'The aircraft still carried on. It reached the road, shattered its own structure, and snarled back on its own tracks. It ended the run at the embankment over which Mrs Crawford had disappeared.' Mrs Crawford collapsed from shock, and was taken to the Wau Hospital. Peter Manser escaped injury, but BUW was in a bad way indeed. The fuselage was broken in two, aft of the trailing edge of the wing, which was itself shattered, and there was other damage. At first sight, it looked as though BUW — which had only been in service for a short time — was a write-off.

Bob Gibbes flew to Wau with his Junkers specialist, Paul Raasch, for a post-mortem. BUW was a depressing sight, but Raasch was convinced that he could rebuild the broken bird if she could be transported to the facilities of the Goroka workshops, 483 road kilometres away.

This was a tall order. There was a good, if narrow, motor road from Wau to Lae, but the Highlands Highway to Goroka was still being constructed. The major rivers in the Markham valley were unbridged. The road up to Kassam Pass and the beginnings of the Highlands was little better than a track. Internal roads in the Eastern Highlands were being extended at an amazing rate by the tribesmen under the direction of Administration patrol officers and their police, backed to

195. Paul Raasch at Mendi where emergency engine repairs were carried out on GSA JU52. *Photo: Paul Raasch*

195

the hilt by District Commissioner Ian Downs, but they were in the main jeep roads, hardly suitable for heavy trucks.

Only one transport man was prepared to tackle the mammoth task. For a fee of £500 the road haulier and contractor, Jack Thick, undertook to bring BUW from Wau to Goroka.

Paul Raasch and a small team flew to Wau and began to dismantle BUW, while Jack Thick marshalled his resources. He modified two old ex-army diesel semi-trailers at Lae, and with a team of local drivers, drove to Wau late in August. BUW was stripped, ready to be loaded. Smaller parts, including the engines, had already been flown to Goroka. The broken fuselage was loaded onto one semi, and the two great metal wings, bound back-to-back, on another. When the two big trucks and accompanying jeeps rumbled off on the long journey, few of the crowd of townspeople watching gave Thick and Raasch much chance of succeeding.

Succeed they did, although they were two weeks on the road. Their problems began immediately after leaving Wau. The road through the Bulolo Gorge was narrow and twisting, with innumerable hairpin bends. Time and again the semi-trailers with their long, unyielding loads had to be winched around bends too sharp to negotiate normally. Rivers were forded, ravines crossed, the Markham valley defeated, mountains climbed, until at last the convoy arrived safely at Goroka, in mid-September, 1957.

In some respects, transporting BUW to Goroka was the simplest part of the task of rebuilding the aeroplane. Paul Raasch and his colleagues had little to help them. They had no original factory materials, specifications, plans, drawings or jigs. Spars, frames and the skin of BUW were made from unobtainable German wartime alloys, the precise properties of which were unknown. The work of reconstruction had to proceed step by cautious step, as each batch of engineering problems were solved, often by trial and error. When a suitable substitute — American Alclad — skinning material was obtained, it was in flat sheets. Special equipment had to be developed to roll in the corrugations that were an essential part of the design.

DCA watched every step of the reconstruction with close attention. The rebuild took five engineers a year to complete, and was a triumph. DCA tests proved that the tolerances achieved by Raasch and his team were actually superior to those of BUU and BUV. It cost Bobby Gibbes £8,000 to get BUW back into the air, but by the time BUW was on line again, Gibbes Sepik Airways had been sold to Mandated Airlines.

Bob Gibbes might have been able to survive the loss of Norseman hours and revenue had he been able to put more JU52's into the air. By 1957, DCA had accepted the JU52. Maintenance and overhaul manuals to the standard required by the department had finally been provided (after an exasperated Vern Rule had threatened to suspend all GSA operations if this was not done) and when DCA began to restrict the loads that could be carried on the DC3, the JU52 was a better proposition than ever for the customer. Robin Gray had located two JU52's in excellent condition in Laurenco Marques, Mozambique, in 1956. In March, 1958, Gibbes applied to DCA and the Customs authorities for licences to import and operate these aeroplanes. Customs licences were duly granted, and in April, Gray flew to Laurenco Marques to complete delivery arrangements.

A week after Gray's departure, Gibbes was stunned by the arrival of a telegram from Vern Rule: 'Am directed to advise that application for permit additional Junkers is to be deferred pending policy review in New Guinea.'

In September, 1957, Prime Minister Robert Menzies had announced a complete review of aviation policy in PNG. The vexed question of airline licences had still not been decided, and Dr Harold Poulton, Assistant Director-General Air Transport, was given the responsibility for the conduct of the review. Poulton announced that the review would begin in February, then in June, 1958, but by this time the fate of GSA was sealed.

Gibbes appealed to Senator Shane Paltridge, Minister for Aviation, against the decision to refuse him permission to import the two additional Junkers. He was told that the customs import licences had been issued in error. DCA did not want to see more JU52's in service. Gibbes was told that there was a possibility that the Australian Government would assist him to obtain cheap ex- RAAF Dakotas instead. Gibbes recalled Robin Gray, and abandoned the opportunity to purchase the additional Junkers that he so badly needed to improve his cash flow. He tried

196. VH-BNE drops supplies to Southern Highlands patrol, 1956. *Photo: James Sinclair*

196

desperately to obtain a decision on the RAAF aircraft, but by this time he had lost the initiative. It was too late. Company debts were mounting, revenues falling, staff unpaid. Gibbes tried everything — government loans, bank loans, the sale of his personal plantation, Tremearne, calling in overdue accounts. The crash of the Norseman, VH-GSA, and the death of Ron de Forest, was the last straw.

By July, 1958, Gibbes Sepik Airways was almost bankrupt. Norseman revenue had almost completely dried up, and Qantas had just introduced the DHC-3 Otter into PNG service. The Otter was a direct and potent competitor with much the same performance as the Norseman but of modern, all-metal construction. Qantas was still expanding, as was MAL. Patair was about to place a DC3 into service, and would soon introduce its first Piaggio 166. Other operators were bringing in Cessnas. Without development capital, GSA could not match these technical advances, and could not hope to survive.

The most likely buyer for GSA was Mandated Airlines. MAL was Qantas' principal competitor, and was after an airline licence, and a route to Australia. There was intense competition between the two companies. Qantas was especially (and understandably) jealous of its sole right to carry paying passengers between PNG and Australia, and in December, 1954, had complained bitterly to DCA when MAL charged fares to eighteen Lae residents who travelled to Sydney on VH-MAC, being flown down by Tom Deegan for annual C of A overhaul. MAL charged the passengers fifty per cent of the Qantas fare, but did not provide meals or airline-type seating. DCA sided with MAL, and altered the charter licence to allow the practice.

Later, MAL applied for and was granted permission to carry out its own C of A work at Lae, but did not immediately exercise the right. MAL major overhauls were being performed by Trans-Australia Airlines, and in 1958, by Qantas in Sydney, with MAL engineers responsible for maintenance between C's of A at Lae. Engines were sent to various firms in England, Hong Kong and Australia.

GSA workshop facilities and staff were thus valuable to MAL. Bobby Gibbes also let it be known that he would approach Reg Ansett and Arthur Butler if MAL was not interested in buying him out. Nobody in PNG airline operations wanted competition of that calibre, which was an additional spur to MAL.

The fixed assets of GSA, excluding all sundry creditors and current assets, were valued at £167,000 as at 26 August, 1958, by the firm of public accountants, A.E. James & Co. Gibbes offered his company to MAL for £150,000, or £120,000 if he paid all liabilities. B.B. Perriman, the W.R. Carpenter director, discussed the proposed acquisition of GSA with Director-General Anderson, and received an assurance that if the deal was concluded, all existing GSA route licences would remain in force and would be renewed. If desired, the licences would be transferred to MAL. Anderson continued: 'In addition, we would be prepared to undertake that no new licences would be issued to any other private airline operator in respect of any routes over which Gibbes Sepik Airways or Mandated Airlines currently operate on a regular basis, for so long as adequate services are being provided on these routes.

'As you know, the government has under review civil aviation policy within New Guinea, and firm decisions on this will be taken as soon as practicable. The question of rationalisation as between yourselves (in the event that you purchase Gibbes Sepik Airways) and Qantas Empire Airways is one of the matters which the minister proposes to submit for Cabinet consideration.

'I am unable to give you any assurances in respect of a service between Port Moresby and a mainland port because this is also a matter which is receiving the general consideration of the government. I think, however, it would be prudent to conduct your present negotiations on the assumption that such a request would be refused.'

It was under these understandings that GSA was finally taken over by MAL. The contract of sale specified that the assets being acquired included five Norseman and three JU52 aircraft in operating condition. It was typical of the drama that had accompanied every phase of the existence of GSA that at the moment the company ceased to exist as a separate entity, an accident occurred to one of the aircraft.

On 15 November, Ivan Bennett and Ken Davenport landed at Mendi in the JU52, VH-BUV. While taxying in to the landing bay, the starboard undercarriage axle snapped and the gear collapsed, causing substantial damage to the fuselage. Gibbes heard the news with dismay, but fortunately his solicitor, Norman White,

197. BUV in Sweden being prepared for the flight to New Guinea. *Photo: Paul Raasch*
198. Robin Gray. *Photo: Sten-Erik Molker*

197

198

had signed the MAL contract hours before. BUV was MAL property when the accident occurred. The damage was quickly rectified.

Most of the GSA staff joined MAL. For some time, GSA continued to operate under its own name as a part of the MAL organisation. Among those who went to MAL was Dennis Buchanan, owner of the tiny Territory Airlines.

199

200

201

202

204

199. From left: VH-BUU, VH-BUV and VH-BUW, the three JU52's of the GSA fleet at Goroka, 1957. *Photo: James Sinclair*
200. Junkers JU52/3m, VH-BUW after the accident at Wau in March, 1957. *Photo: Sten-Erik Molker*
201. Junkers JU52, VH-BUV at Tari, 1957. *Photo: James Sinclair*
202. Norseman VH-GSE. *Photo: Bill Sherwood*
203 & 204. Norseman VH-GSD which crashed at Ialibu in March, 1960, was later rebuilt and resumed flying again four months later. Peter Manser sits in what had been the cockpit. *Photo : John Bayles*

205

206

207

208

209

210

205 & 206. Fuselage of BUW en route to Goroka. *Photos: Paul Raasch*
207. GSA Norseman VH-BLM at Kainantu, 1956. *Photo: Sten-Erik Molker*
208 & 209. Qantas terminal building and VH-BUW after the accident at Wau in March, 1957. *Photos: Sten-Erik Molker*
210. VH-BUW under repair in the Goroka hangar. *Photo: Paul Raasch*

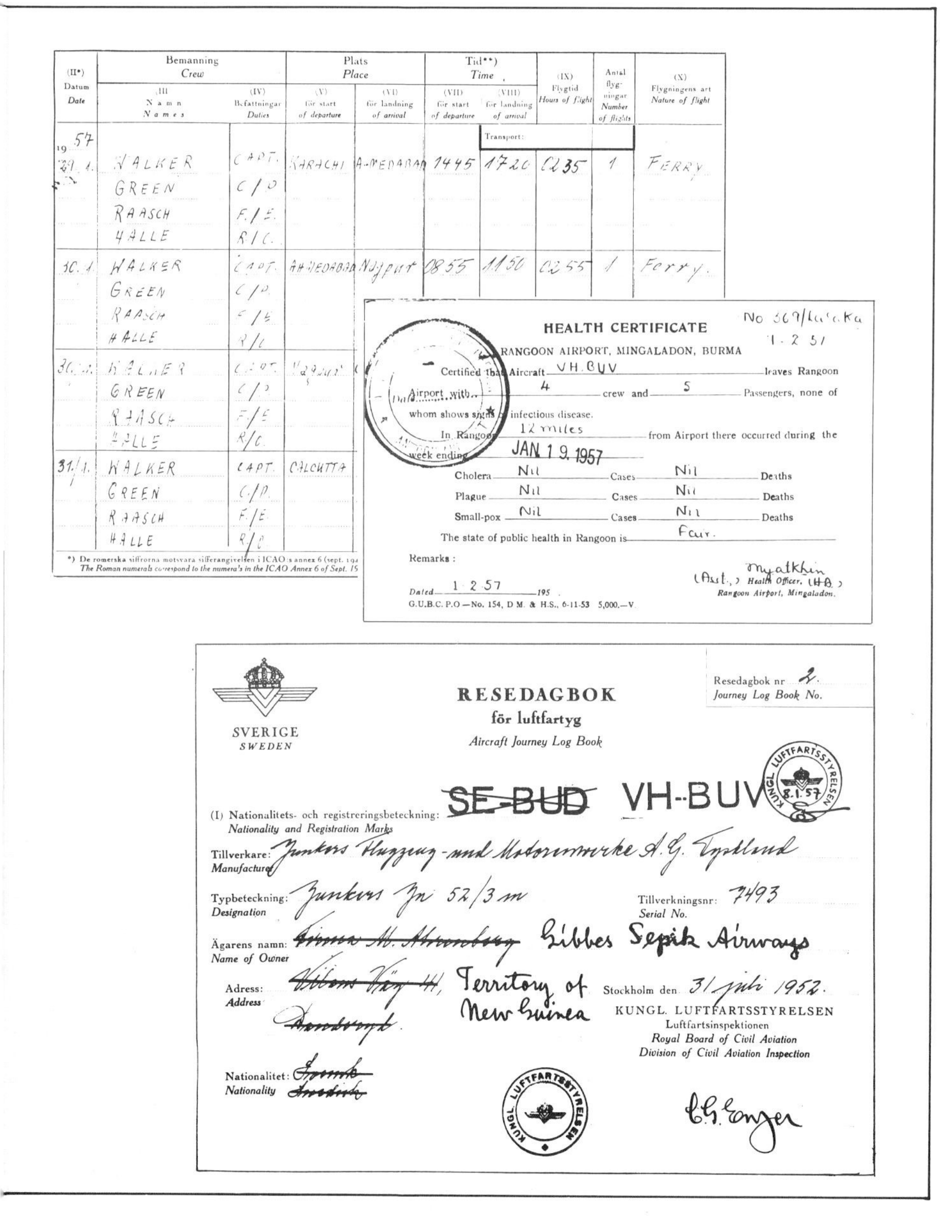

(II*) Datum Date	Bemanning Crew (III) Namn Names	(IV) Befattningar Duties	Plats Place (V) för start of departure	(VI) för landning of arrival	Tid**) Time (VII) för start of departure	(VIII) för landning of arrival	(IX) Flygtid Hours of flight	Antal flygningar Number of flights	(X) Flygningens art Nature of flight
19 57 29. 1.	WALKER	CAPT.	KARACHI	AHMEDABAD	1445	1720	0235	1	FERRY
	GREEN	C/O							
	RAASCH	F/E.							
	HALLE	R/C.							
30. 1.	WALKER	CAPT.	AHMEDABAD	NAGPUR	0855	1150	0255	1	Ferry.
	GREEN	C/O							
	RAASCH	F/E							
	HALLE	R/C							
30. 1.	WALKER	CAPT.	NAGPUR						
	GREEN	C/O							
	RAASCH	F/E							
	HALLE	R/C.							
31. 1.	WALKER	CAPT.	CALCUTTA						
	GREEN	C/O.							
	RAASCH	F/E.							
	HALLE	R/C							

*) De romerska siffrorna motsvara siffrangivelsen i ICAO:s annex 6 (sept. 194… The Roman numerals correspond to the numera's in the ICAO Annex 6 of Sept. 19…

HEALTH CERTIFICATE

No 369/ta/ca 1. 2. 57

RANGOON AIRPORT, MINGALADON, BURMA

Certified that Aircraft VH-BUV leaves Rangoon with 4 crew and 5 Passengers, none of whom shows signs of infectious disease.

In Rangoon 12 miles from Airport there occurred during the week ending JAN 19 1957

Cholera Nil Cases Nil Deaths
Plague Nil Cases Nil Deaths
Small-pox Nil Cases Nil Deaths

The state of public health in Rangoon is Fair.

Remarks :

Dated 1 2 57 195

(Asst.) Health Officer. (HB) Rangoon Airport, Mingaladon.

G.U.B.C. P.O.—No. 154, D.M. & H.S., 6-11-53 5,000.—V

SVERIGE SWEDEN

RESEDAGBOK för luftfartyg Aircraft Journey Log Book

Resedagbok nr 2. Journey Log Book No.

(I) Nationalitets- och registreringsbeteckning: Nationality and Registration Marks ~~SE-BUD~~ VH-BUV

Tillverkare: Manufacturer Junkers Flugzeug- und Motorenwerke A.G. Tyskland

Typbeteckning: Designation Junkers Ju 52/3 m

Tillverkningsnr: Serial No. 7493

Ägarens namn: Name of Owner ~~[illegible]~~ Gibbes Sepik Airways

Adress: Address ~~[illegible]~~ Territory of New Guinea

Stockholm den 31 juli 1952.

KUNGL. LUFTFARTSSTYRELSEN Luftfartsinspektionen Royal Board of Civil Aviation Division of Civil Aviation Inspection

Nationalitet: Nationality ~~Svensk~~ ~~Swedish~~

Page from Logbook for flight of JU52, VH-BUV on delivery flight from Sweden to Goroka, 1957. *Courtesy of Paul Raasch*
Health Certificate issued at Rangoon Airport during flight of JU52 from Sweden to Goroka, 1957. *Courtesy of Paul Raasch*
Original aircraft Log Book for VH-BUV with alterations after transfer of ownership to Gibbes Sepik Airways. *Courtesy of Paul Raasch*

The Helicopter Appears

The departure of Gibbes Sepik Airways from the aviation scene signified the approaching end of a colourful era in PNG aviation history. The last of the old wood-and-fabric biplanes that had ushered in New Guinea aviation before the war were about to vanish; the last of the Second World War military aircraft that had served the Territory so well would soon be gone — always excepting that exceptional aeroplane, the Douglas DC3. And yet it was an era that saw the introduction of that most marvellous of man's flying machines, the helicopter.

Just before GSA expired, another wonderful era in PNG aviation also came to an end. At precisely 12.12 p.m. on Thursday, 24 July, 1958, Captain Ken Montagu landed Qantas Catalina, VH-EBD *Island Patrol* on the smooth blue waters of Port Moresby harbour. It was the last regular Qantas airline service flight of the Catalina in PNG, and it simultaneously ended twenty years of Qantas flying-boat operations. It was a sad moment for Ken Montagu, with 4,000 Catalina flying hours behind him. Montagu was now the senior captain in Port Moresby, responsible for the operation of Catalinas VH-EBD and EBC, and Beaver floatplane, VH-EAU. Phil Oakley, the previous senior captain, had been transferred to Lae in 1957, as chief pilot of Qantas PNG operations. Ken Montagu remembers: 'This last service was the four-day flight to Rabaul, and at each port along the route we flew a paying-off pennant — great long thing a foot [thirty-one centimetres] long for every foot of the aircraft's length, and also a foot for every year of service. We presented the government stations a bottle of whisky and a thank-you note, and tried to make it a pleasant farewell. We had almost tearful departures from Kandrian and one or two other places. As you know, Qantas used the good offices of the Administration, their district commissioners, clerks, and other people at the different stations to do the Qantas paper-work, and they assisted us tremendously. Ralph Perrier, the station manager at Moresby at that time, and I, had decided that we should have a party after that last flight. Tried to interest Qantas management in Sydney, but simply couldn't get them to show the slightest interest — far too preoccupied with their various Constellation services, and no doubt thinking of the jets to come. So both Ralph and I subscribed both to the bottles of booze for the outstations, and the party...'

Montagu's sorrow at the passing of the Catalina was shared by the outstation people who had depended on the Catalina for so long. The Catalina was to be replaced by the DHC-3 Otter floatplane, and it did not take long to prove that the Otter was no Catalina.

It will be remembered that Qantas had tried to replace the Catalina on PNG internal services with the Sandringham, in 1952, without success. When the Sandringhams were withdrawn in 1954, the Catalinas again took over the internal routes. The Qantas management, however, looked upon this as a stop-gap solution. The Catalina was simply too expensive to maintain, and as early as 1956, Qantas had already decided to replace them with Otters.

In all, Qantas employed five Catalinas in PNG over the years: the two ex- Island Airways (W.R. Carpenter) machines, which in Qantas service were registered as VH-EBC *Island Chieftain*, and EBU *Island Warrior*, and VH- EBA, EBB and EBD. Of these Catalinas, EBB was scrapped in February, 1951, and EBA in July, 1953. EBU was withdrawn from service in November, 1952. It was fitting that *Island Chieftain*, the first Catalina to fly commercially in PNG, was one of the two last survivors.

The flying-boat unit in PNG was virtually autonomous, based in Port Moresby under the control of a station manager and a senior captain; part of the Qantas Lae-based organisation but by virtue of the very nature of the aeroplane and the areas it served, something apart. Operational and maintenance problems of flying-boats

211. Airdrop of supplies from a Catalina to Nomad patrol post. *Photo: R. Hoad*

211

are different to those of landplanes in important aspects.

One difficulty faced by all pilots in PNG in the early post-war years was obtaining night flying and instrument flying practice, for navigational aids were woefully lacking. Prior to 1954, the only radio navigation aids provided by DCA were non-directional beacons, at major airports only. The NDB was limited in its effectiveness because of the topography of PNG, but by 1955 distance-measuring equipment (DME) had been provided at most major airports, which made aerial navigation considerably easier, and safer.

When Fred Fox became senior captain at Port Moresby, he instituted regular three-monthly instrument flying training for the Catalina pilots. He remembers: 'The only aid we had in the "Cat" was automatic direction-finding, and to operate on the NDB. In those days, we had screens in the Catalinas and the DC3's to fit the cockpit. Very light-yellow tinted perspex screen, which went around inside the windshield and side windows, and the pilot wore light-blue goggles. The two colours together meant that you could see nothing outside. All you saw was black, but you could read the instruments, and see the controls, through the goggles.

'On one night flying practice, for some reason it was atrocious night flying, pilots either not touching down, holding the rate of descent too low, or hitting the water too hard. So after a time, I said, "This is no damned good. We'll set up a concentrated programme of instrument flying by day using the screens wherever we go, and we'll try it again at night in two weeks". So everywhere we went when it was possible, the pilot flying the "Cat" would put on his goggles, the co-pilot would taxi him out into position, he'd do a blind take-off, fly all the way over the next sector with the co-pilot guiding him, then be directed round on a normal circuit, told he was lined up on final and clear to land, then do the final stages of the landing still with his goggles on. The co-pilot would tell him all was clear, etc.

'Quite safe to land under those conditions. Night landing, or a flat-calm landing in a "Cat", involved setting up the airspeed at sixty-five knots indicated, with a 200-feet [sixty-one metre] rate of descent, and waiting until the aircraft touched the water. You are able to keep straight by using the directional gyro or turn needle, so no problem in doing a completely blind landing. Next time we had night flying, they were crash-hot...'

Phil Oakley continued the practice sessions when he became senior pilot on the departure of Fred Fox: 'It was a little bit exciting, really. We would lay down a flare path on the water, and DCA would put a boat out, and provide radio communications. It was pretty rough, and you couldn't see very much. It was a bit more civilised in Lae, with a runway and a double row of lights. The only hazard was that sometimes in the circuit, you'd be going in the opposite direction to the flying-fox colony, returning after dark to their nests. The odd DC3 hit the odd flying-fox, but fortunately without too much of a problem...'

During the second, post-Sandringham period of Catalina regular airline service operations in PNG, a number of additional Qantas pilots gained experience on the two Catalinas, EBC and EBD, and on the Beaver floatplane, VH-EAU, on permanent charter to the Australasian Petroleum Company. Two young first officers who went to Port Moresby in 1955 were John Simler, and Allan Terrell.

Terrell is today the Qantas Director of Flight Operations, his days of piloting Catalinas and the Beaver floatplane over the vast empty reaches of the Papuan Gulf delta a long way behind him. Simler is a senior executive of Heliservices, Hong Kong.

John Simler was born in London, and after service as a pilot in the RAF from 1946 to 1948 became a commercial pilot. He carried out many delivery ferry flights around Europe, and in October, 1953, the chance came to deliver a Percival Proctor to Australia. Simler stayed as he says 'to have a look at the place' and flew for a time for the pioneer outback aviator, Eddie Connellan, at Alice Springs, before joining Qantas in May, 1955. In June, he was sent to Port Moresby to join the Catalina unit, where he was to remain for two years. He remembers those days: 'I flew two route familiarisation flights to the Papuan delta with captains Phil Oakley and Marsh (Swampy) Burgess, before being handed to Captain Struan Robertson, as a very junior first officer for my first scheduled flight with Qantas. Naturally, I was very nervous, and Struan, who could be extremely stern when he wished, did nothing to dispel my nervousness.

'After the preliminaries of pre-flight, we taxied out into Fairfax Harbour, and

Captain Robertson did the take-off, and turned towards the north-east on our first leg towards Kerema. At this stage, I was doing my best to remember all the duties required of a first officer. After levelling off at 6,000 feet [1829 metres], the following conversation took place:

Captain R (looking very stern): 'Have you ever flown one of these before, son?'
First Officer S (almost standing to attention in the cockpit): 'Only on training, sir, Captain.'
Captain R: 'Well, son, normally I put the aircraft on auto-pilot now, but as a special favour, I shall let you fly it straight and level.'
First Officer S: 'Thank you very much, sir, Captain.'

'Whilst I was endeavouring to maintain heading, altitude etc., etc., Struan R. picked up his briefcase and started to search through it, at the same time growling to me, "And what height are we now, son?" I answered, now slightly more confident, '6,000 feet, sir.'

'At that moment, an extremely large .38 revolver was pulled out of the briefcase, the muzzle placed against my nostrils, followed by the advice to "Keep the bastard at that altitude, or else!" A great, if unusual, introduction to scheduled airline service.'

Qantas Catalinas flew with flight engineers, whose station was situated between the wings. The flight engineer — crews consisted of captain, first officer, flight engineer, purser and one of two accomplished Papuan cabin attendants, Anso and Mala — was responsible for engine operation — start, stop, fuel control, running maintenance, emergency procedures — and was invaluable in fuel management during the long-range search and rescue flights that the Catalinas were from time to time called on to perform. Flights of eight to ten hours in search of overdue small vessels were not uncommon. Says Simler: 'The Catalina, with its blisters on either side of the fuselage for observation, its slow cruising speed and its ability to land on water (albeit *not* under very rough sea conditions) and its extended range, was a most suitable vehicle for SAR work. The two Pratt and Whitney 1830 engines — same as in the DC3 — were very reliable, and would purr on forever, in spite of the rough pounding and immersions in salt water in heavy chop sea conditions, which took place on many take-offs.

'Our flight engineers did a great job, men like Tom Mitchell, Phil Heenan, Alan Roberts, Graham Bacon, Don Cranch, Keith Gordon. Keith is now chief flight engineer for Qantas in Sydney.

'At regular intervals on the Moresby-Rabaul service, we would carry Supreme Court judges on their circuits. Unless there were serious cases requiring some time to take down all the evidence, it was not unusual for the judge to carry out his function during the short Catalina stopover.

'I have a vivid recollection of carrying Judge Esme Bignold on one of these island services, and as we approached Esa'ala, he donned robe and wig. After taxying to the mooring buoy, the judge was transferred to a large double-hulled canoe which the flight engineer streamed by a short line, behind the Catalina. Whilst the crew transferred passengers in and out of the aircraft, Judge Bignold held court, and pronounced judgement, on the temporary floating courthouse...

'At that time, Qantas operated the two Catalinas — they were PB2B-2 types, straight flying-boats — on charter to APC, as well as on the scheduled services. Our operations were into the many rivers of the delta — the Omati, Wassi-Kussa, Bamu, Kikori etc. Additionally, a de Havilland Beaver floatplane, VH-EAU, was on full-time charter to APC and based at Middletown, the oil base camp just north of the Catalina alighting area at Kikori. The Beaver was manned by the Catalina first officers, myself and Allan Terrell, with occasional relief by one of the captains, Ken Montagu or Bill Nye...'

W.T.H. Nye was like so many of the Qantas Catalina pilots, RAAF-trained. He flew Sunderlands and Catalinas in PNG during the Pacific war and finished his RAAF career as chief flying instructor at Rathmines, in June, 1949. After a year as CFI for Kingsford Smith Aviation Services, Bill Nye was briefly associated with Overland Air Services, based in Condobolin, New South Wales. He then joined BCPA — British Commonwealth Pacific Airways — and flew DC6 airliners on the Pacific run, to Fiji, Honolulu, San Francisco and Vancouver.

BCPA merged with Qantas in June, 1954, and because of his earlier experience

212. Mechanics employed by World Wide Air Services. *Photo: Sten-Erik Molker*
213. Catalina of World Wide Air Services at the marine base, Port Moresby. *Photo: Sten-Erik Molker*

212

213

on flying-boats, Nye was offered a temporary position with the Qantas Flying- Boat Unit, at Port Moresby, while Ken Montagu was on four months' leave in Australia. The four months became two years. Bill Nye then commenced flying on the Qantas international routes, but was frequently recalled to Port Moresby for relieving duty, to the end of Catalina operations. Bill Nye is proud of the fact that in one period of a week, he flew the Sandringham, Catalina and Beaver floatplane, a most uncommon experience.

It will be recalled that Qantas had introduced the Beaver as a replacement for the unsuccessful Drover. Despite the initial misgivings of the pilots, who considered that going from three engines to one was a giant step backwards, the Beaver was quickly accepted for the superb little aeroplane it was.

The first Qantas pilot to be endorsed on the Beaver was Ron Roberts. Roberts was an ex-RAAF pilot, who spent most of his war flying Catalinas. He was discharged in 1946, and the following year joined Qantas as a first officer on Lancastrians on the Sydney-Karachi run. With his background, he was a natural for flying-boats, and in November, 1949, volunteered for a two-year posting as first officer on Catalinas at Port Moresby. He was endorsed on the Sea Otter in April, 1950, and was one of the few pilots who regularly flew that ponderous British biplane on APC charter work. Later, he gained his command on Catalinas.

Ron Roberts became a Sandringham captain in August, 1952, and alternated between this flying-boat and Catalinas until the Sandringhams were withdrawn. In December, 1954, he was sent to the de Havilland factory in Wellington, New Zealand, to be endorsed on the Beaver, and accept the first two of the four that Qantas had ordered, at Bankstown. He then carried out performance tests, and endorsed Jack Stammer and Dick Davis on the Beaver, who were to fly the aeroplanes back to PNG.

The first of the four Beavers, VH-EAU, was ready in December, 1954. The second two — EAT and EAS — were ready soon afterwards, and the fourth, EAV, in April, 1955. At the time of their delivery, more than 600 Beavers were flying throughout the world, and the type had earned a high reputation. Although single-engined, the 450 hp Pratt and Whitney Wasp Junior was one of the finest ever produced by that great company, and its reliability was legendary. With a capacity of seven passengers or 680 kilograms of freight and with notable short-field performance, the Beaver was an ideal light aeroplane for PNG conditions, apart from one serious drawback: the relatively short cabin, which restricted the type of loads that could be carried.

Dick Davis was on his second posting to Lae. He had first gone to PNG for Qantas in January, 1950, together with another young first officer, Dennis O'Malley. Both were after in-command hours — the great goal of the young professional aviator — so that they could aspire to higher things with Qantas. Only in New Guinea did the company operate light aircraft — Fox Moths and Dragons — that could be flown by a first officer, in command.

Both O'Malley and Davis started off in the usual Qantas way, as first officers on DC3's, but both were soon flying the de Havilland biplanes, and later the DHA-3 Drover. It will be remembered that Dick Davis was unlucky enough to crash at Menyamya in Dragon VH-URD, in December, 1951.

Shortly after this accident, Davis's two-year posting ended, and he accepted another term, on the promise of a DC3 command. This he achieved, in June, 1952. As recorded earlier, when Bill Forgan-Smith returned to Australia, his place as chief pilot at Lae was taken by Bruce Fawcett. After a year, Fawcett, too, resumed his career with Qantas in Australia, and at the end of 1953, Dick Davis moved into the position. He remained chief pilot until his own time in PNG came to an end, in June, 1955.

Jack Stammer and Dick Davis returned to Lae with the beautiful new Beavers, EAU and EAT, in January, 1955. Stammer, like most pilots who flew the Beaver, thought highly of it: 'Most of the pilots were against the Beaver when it was bought, myself included, it being a single-engined aircraft, and with the country we had to fly over. Of course, Bobby Gibbes had been doing it for a long time in his Norsemans. We weren't very happy, but the Beaver proved a magnificent little aeroplane for the country, and gave us great service.

'I'll never forget one trip, a special trip to Tari. They wanted some gelignite and detonators, and by this time things had tightened up a lot with DCA. You weren't supposed to carry detonators at all, but how they could ever get detonators out to

214. Pilot Jack Stammer in a Qantas Beaver. *Photo: Qantas*

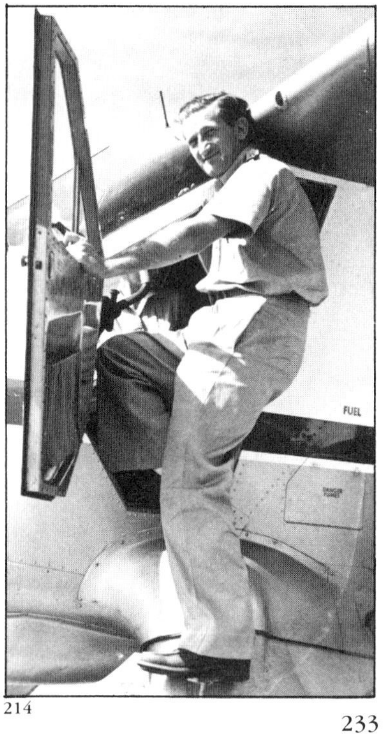

214

Tari without them going by air I don't know. DCA finally relented, and let us go, with the proviso that the detonators had to be carried by the traffic officer in the co-pilot's seat, on his lap. If the engine quit, his job was to throw them overboard, so that when we hit, the gelignite would just go up by itself. Seemed a bit funny to me, but that was their ruling and that's what we did.'

Perhaps the most unusual job that Jack Stammer did in the Beaver was in connection with the Australian film, *Walk Into Paradise* in July, 1955. This forgotten — and forgettable — epic was produced by Lee Robinson and starred the late Chips Rafferty as a rather venerable patrol officer, leading a patrol in new country. The story line called for a number of aerial sequences.

Now, Jack Stammer, a fine pilot, is a man of homely appearance, as he would be the first to admit. Dick Davis, on the other hand, far better fits the film star mould. Says Stammer: 'I did all the flying sequences for that film with the Beaver at Goroka strip. Interesting work to try and do what they wanted to have done, flying through the *pitpit* at the end of the strip to make it look as though it was a short strip, and so on. I was a bit cut that I did the flying, but Dick Davis stepped out of the cockpit as the pilot in the film. Lee Robinson explained to me that the public weren't quite ready yet for Chips Rafferty and me in the one film.

'I had met Chips during the war, when *Rats of Tobruk* was being made. I was involved in that, flying a Vultee Vengeance with the landing-gear down over the sandhills at Cronulla, making out it was a Stuka dive bomber. Three of us from Ferry Flight did that effort.'

At the time Qantas purchased the four Beavers, it was decided that one would be fitted with floats, and kept for full time APC charter operations. Beavers were designed to accept wheels, floats or skis, and the conversion to floatplane configuration was simple. The first of the Beavers to be flown as a floatplane was VH-EAT, although EAU was the one normally used. EAT was fitted with floats at the Qantas workshops in Lae, and was test-flown by Ron Roberts in March, 1955.

Roberts then flew EAT to Port Moresby, and in early April endorsed Ken Montagu on the aeroplane. Montagu was to take over the APC work with the floatplane, although he continued to fly Catalinas on occasion.

This conversion of Ken Montagu ended Ron Roberts' association both with seaplanes and with PNG. He went onto the superb L1049 Constellation, and then commanded Boeing 707 and 747 international airliners until his retirement from Qantas in June, 1979.

As stated above, the Wasp Junior engine fitted to the Beaver was remarkably reliable. However, even the finest of engines is but a piece of machinery, and the best of machines will sometimes break down. On 4 October, 1956, Ken Montagu was flying in EAU over the Papuan delta, with six passengers on board. Suddenly there was a loud bang from the engine. A piston had failed, and the engine was dead. Oil was flung onto the windscreen.

It was Montagu's prudent habit always to fly as high as conditions allowed, and when the engine blew he was at 1829 metres, which allowed him to send out a distress signal and glide for ten kilometres to a safe landing at the mouth of the Turama River.

Like most of the rivers of western Papua, the Turama was subject to a tidal bore, a great wall of water rushing upstream and fully capable of overturning a relatively unstable object like a Beaver floatplane. Montagu had no tide charts or bore predictions, but he knew that his aeroplane was exceedingly vulnerable where she was, and would have to be moved without delay.

Fortunately, there was an APC base camp on the Turama, but a long way upriver. Montagu contacted Port Moresby by radio and arranged for a message to be passed back through the excellent APC communications system to the base camp, requesting the despatch of two large powered canoes, together with a heavy baulk of timber. He passed another message to Qantas in Port Moresby, asking for a replacement engine, and two outboard motors.

Montagu's plan was to lash the baulk of timber across the floats of the Beaver, and attach the outboard motors. With the motors pushing and the powered canoes pulling, the Beaver could hopefully be towed upstream, away from the influence of the tidal bore. The replacement engine could then be fitted at the base camp.

The plan worked perfectly. The Catalina landed at the mouth of the Turama with the spare engine and a number of 25 hp Johnson outboards, and an engineering

215. Pilot Dick Davis in cockpit of a Qantas Beaver talks to Australian actor, the late Chips Rafferty. *Photo: Qantas*

215

team. The passengers were flown out in the Catalina, and the Beaver was prepared for the long tow up the Turama, against the powerful flow of the current. Says Montagu: 'We commenced driving the Beaver back up the river early in the morning. We ran aground a couple of times, but did no damage, and arrived at the base camp at midnight, very tired indeed. There not being any area to beach the Beaver, the engine was changed on the water. Quite a feat. Used big poles to make a tripod to hoist the engine, many double canoes and rafts of 44-gallon [200-litre] drums. Took an awful lot of effort. The engine had to be dismantled to fit it in the Catalina — fully assembled, you couldn't get enough men around the thing to lift it. I took a bit of a break in Moresby for a week, then went back when the job was finished. Test-flew the aircraft, and brought it back to Port, and went on with normal "ops"...'

Similar engine changes in the field had to be made to the Catalinas. Phil Oakley remembers a couple of difficult ones during his time as senior pilot, Port Moresby: 'We did an engine change on a "Cat "once, at Kikori. The ADO, old Clarrie Healy, had a big double canoe, with a small outboard motor on the back. With plenty of local labour, he'd service the "Cat" — bring out drums of fuel and a pump, bags of rice, whatever the load was. This canoe was brought under the wing of the Catalina for the engine change, and the engineers worked off that, and the top of the wing. Dropped a few bits in the water. Biggest problem was running-up the engine. Had to swing the tail of the "Cat" around, and tie it to a tree. But it was done. We had a similar operation at Kandrian, and one at Rabaul.

'That was a funny one. November, '56, when the Duke of Edinburgh came out to open the Melbourne Olympics. He did a tour of PNG afterwards. A Qantas DC3 was flying him around and we decided that since we had the regular Catalina Moresby-Rabaul service operating the same day, we could service search-and-rescue if required. But the only aircraft that needed S and R was us. I did a blower seal round about Jacquinot Bay, and we flew into Rabaul on one engine. Using drogues, we managed to moor up, and it was clearly an engine change. We didn't fancy doing it on the harbour, there would have been great difficulties. Somebody told me there was a wartime Japanese slipway in somebody's back yard — can't remember who it was. I met this bloke in the New Guinea Club, and put it to him. Told him it would be very noisy, what with engineers working, engines running etc, but he told me to go ahead.

'So that's where the "Cat" was beached. An engine was flown in in a DC3, and the change was done in his back yard. Left the yard and his drive in a fair mess, but the company fixed it up. Would you believe I was back there two weeks later on one engine, and had to ask him again? We went six-and-a-half years in "Cats" without an engine failure, then had six in six months. All quite unrelated. Not the same trouble twice.

'The Catalina could hold altitude on one engine very well, at civil weight. 28,500 pounds [12,928 kilograms] was maximum take-off weight, and she'd hold...'

When the Duke of Edinburgh visited PNG, DCA authorities considered that DC3 aircraft were 'too great a risk' to be used to transport the august visitor into the Highlands. A spokesman announced that 'conditions under which the airline companies flew the public into the Highlands in DC3's were not good enough for the duke'. Territory DC3's carried some 3175 kilograms more weight than that permissable in an aircraft used by royalty. DC3's were permitted to operate at 12,202 kilograms all-up weight into Goroka, whereas English authorities demanded that the all-up weight into such a strip should not exceed 8618 kilograms.

The Catalinas were nearing the end of their useful lives, and hull corrosion was becoming a serious problem. But this was a robust aeroplane. Phil Oakley remembers: 'The only time I ever remember damage to a Catalina from an open-sea landing was once when we popped a couple of rivets, in a very heavy slop at Daru. Every time you did a heavy landing, you'd pull the floorboards and have a look. Funny sight — little blue hole with a spurt of water coming in. Tom Mitchell, one of the early flight engineers, always carried extra pencils with him. He'd screw a pencil down the hole, and break it off. That'd get you home, put a new rivet in then.

'I ran aground once. We'd had a terrible day, doing APC charters. We picked up a badly-burned native. A bomb scow had exploded, and he was in bad shape. He died, and we covered him up, there were other people on board. Then I got a call to land at Romilly Sound, to pick up a bloke injured with shotgun wounds, a marital triangle problem. I landed, and taxied in over water and got in OK, and put the injured bloke on.

'When we started to taxi out, we started to hit mud. I knew that the take-off area, where I was headed was OK, so I just kept going. It was only soft mud. Kept on bumping over it until we got to good water, and took off. No damage.

'Probably the roughest landing ever made in a Catalina in New Guinea was in 1956. A British yachtsman, Danny Weil, arrived in Moresby, and he went off, single-handed, in his yacht down through the Papuan Gulf and Torres Strait. Would not listen to the locals, went out at night, and inevitably struck a reef, hard and fast, at low tide. I went out with a crew in a Catalina, and we located him. An RAAF Hercules was coming up to drop him some supplies and a raft, so we decided to stick around until the "Herc" arrived.

'We stayed as long as we could, but had to return for fuel before the Hercules arrived, and the second "Cat" came out, with Captain Marsh Burgess. He found the tide was rising fast, and the yacht was swamped. Danny Weil was in a lot of trouble. He had left the yacht and swum to the raft which the RAAF had dropped, and was drifting off fast in the wind.

'Burgess took a calculated and courageous decision to land in that rough sea, along the swell. Picked Weil up, and managed to take off again. Marsh was awarded a Royal Humane Society certificate for that effort — and the Catalina wasn't damaged.

'Initially, majors on "Cats" were done at Rose Bay, in Sydney. We would fly them down there. Then, in 1956, Orm Denny conceived the idea of doing them in Lae. Don McMaster, who was then the chief operating pilot in Qantas, came up to have a look at this and it was cleared, with some reservations, because you would get a pretty strong south-east swell rolling in to Lae sometimes.

'The "Cats" were flown across and landed in the Huon Gulf, just off the end of the strip. A little warping buoy had been dropped off the end of the strip. We'd moor to this, and they would fit wheels to the Catalina, put the towline on, and the beaching gear, and pull her up and tow her up the street to the Qantas hangar. It worked well, and avoided having to set up major maintenance facilities in Moresby.

'The main problem was the major corrosion repairs which had to be done. The old "Cat" was gradually powdering away. It used to sit in salt water like a great big battery. Corrosion developed under the seams particularly, and this was what ultimately led to their demise. Could not be repaired, and the thing finally became uneconomic...'

The Beaver had proved to be an excellent aeroplane for PNG conditions, limited only by its carrying capacity, and it is somewhat surprising that when the decision was finally taken to get rid of the Catalina, it was decided, too, to replace the Beaver. On 20 August, 1957, the vice-chairman of Qantas, W.C. Taylor, wrote to Senator Shane Paltridge: 'Since 1946, we have been operating Catalinas on behalf of the Administration...the stage has now been reached where because of excessive corrosion, it is now totally uneconomic to carry out the major repairs needed to keep them in operation. Our board recently considered and approved a proposal to purchase Otters as a replacement, either as a float or landplane. Four are to be bought as replacements for two Catalinas and four Beavers currently used on Highlands operations. Four Otters and spares costing Canadian $476,000 (A£224,000) as against A£126,000 from the sale of the Catalinas and Beavers...we must order within the next few weeks to meet Administration requirements next year...'

The proposal was approved by DCA. The Otters were ordered. The Beavers were sold between April, 1958, and April, 1959, and went on flying, but the two Catalinas had come to the end of the road. No airline wanted them, and as mentioned at the beginning of this chapter, Ken Montagu piloted VH-EBD on her last regular airline flight in July, 1958. A month later, on 27 August, 1958 — his birthday — Montagu flew EBD to Lae, where with the other Catalina, the famous *Island Chieftain*, she was stripped of her engines and other useful components, broken up, and sold to Arthur Scott, for £80. In March, 1959, the two Catalinas were melted down into over twelve tonnes of aluminium ingots, an ingnominious end to the careers of two notable flying-boats.

Before leaving PNG in May, 1959, Ken Montagu presented the mounted instrument panel and control yoke of EBD complete with the two aileron wheels to the museum in Port Moresby, as a memorial to the wonderful Catalina boats that had served the Territory so well. However, PNG had still not seen the last of the Catalina. After Qantas left PNG in 1960, Trans-Australia Airlines operated Catalinas, a story that will be told later in this book. And another company flew Catalinas in Papua during the final years of the Qantas presence. This was World Wide Air Services, whose parent company was Bahamas Helicopters Ltd.

The helicopter was a strange novelty indeed to most commercial aviators in the early 1950s. Operating standards for these radical new aircraft were still being developed in most countries of the world, including Australia. The first commercially successful helicopter, the Bell 47, first flew in 1945 and received American Civil Aviation Administration approval in March, 1946, and the considerably larger Sikorsky S-51, was certified in August, 1946. The standards that had been so carefully developed over many years governing the operation of conventional aircraft were of little use for the helicopter.

Bahamas Helicopters Ltd was formed in the Bahamas by ex-military pilots, and in 1952 began to operate Bell 47D-1 helicopters for the Royal Dutch Shell Company in Dutch New Guinea. In Papua, the Australasian Petroleum Company studied the use being made of the helicopter by Shell with keen interest. Although limited in size — it was a three-seater — the Bell could perform a variety of tasks impossible for a conventional light aircraft, particularly in servicing oil exploration teams in difficult country, in aerial survey work, and in emergencies. APC began talks with Bahamas Helicopters, and with DCA.

In common with aviation departments in most countries, DCA had little but theoretical knowledge of helicopters. Helicopter routes for Queensland, New South Wales and Tasmania were under consideration, and the department had information on the operation of various helicopter services in the United States. Commercial helicopter operations had yet to begin in Australia, and helicopters were unknown in PNG before 1953.

As a result of the APC approach, and following DCA approval, Bahamas Helicopters decided to extend its operations into PNG. APC was extending the scope of its oil exploration work. Early in 1953, APC and the associated Island Exploration Company, applied for additional prospecting permits over huge areas near the Dutch New Guinea border, bringing the total permit areas held to some 83,550 square kilometres. A contract was concluded with a Canadian company for an aerial magnetometer survey over the border areas, using a specially converted Catalina. APC now arranged for Bahamas Helicopters to bring in four Bell 47D-1's to aid the overall exploration task.

In April, the Bahamas managing director, Knute Flint, visited Port Moresby for discussions with APC and the Regional Director, John Arthur, and his officers. The veteran V.H. Augenson, from DCA headquarters, Melbourne, was also present. On 5 May, Flint, Arthur and Augenson visited the site proposed for the APC helicopter base, at Badili. Augenson reported to the Director-General: 'The earliest service for APC could commence at the end of September, 1953, and the base at Badili would be used for repairs and overhaul of helicopters which would normally be operating between APC field depots and camps. The Regional Director stated that operation of a public helicopter service would not be permitted from Badili, only from Jacksons...'

John Arthur immediately requested from DCA headquarters all available 'helicopter administrative information'. In October, four pilots, four engineers and three crated Bell 47D-1 helicopters, American-registered, arrived in Port Moresby from Dutch New Guinea. The helicopters were swiftly assembled at the Badili base, where a hangar, eighteen by nine metres, had been erected. The first Bell was test-flown on 26 November, by Bahamas chief pilot 'Sandy' Wheeler, a Californian. The other two followed, and the three helicopters flew in formation over Port Moresby on 29 November, watched by thousands of residents. One spectator reported seeing one of the Bells hovering over a high ridge, and landing a passenger: 'The helicopter flew away and left the passenger standing on the ridge like a shag on a rock. He was there only a few minutes before heavy rain fell. As far as I know, he is there still, because the helicopter did not return for him...'

The helicopters quite fascinated the blasé residents of Port Moresby, who had considered that they had seen all that aviation had to offer. The pilots put on an impressive display of helicopter manoeuvres typical for the breed, but new and

216. World Wide Sikorsky S-58 helicopter, VH-WWE, 1957. *Photo: British Petroleum Co. Ltd.*

216

incredible to onlookers. The Bahamas managing director, Knute Flint, announced that the Bells were short range helicopters, capable of lifting up to 181 kilograms at a cruising speed of just under 160 kph. He admitted that the company had had fourteen accidents since commencing operations, but there had been no serious injuries.

Two of the Bells left Port Moresby on 3 December for the APC base at Everill Junction, on the Fly River, and the third followed soon afterwards. John Arthur had still not received the requested 'helicopter administrative information' from Melbourne. He wrote again to the Director-General: 'The helicopters will operate from Everill Junction on aerial seismic operations for APC...they will fly to Port Moresby approximately every three months for overhaul. In the absence of advice, this office has tentatively specified that a helicopter alighting area shall consist of an area of ground or water 300 feet square [about twenty-eight metres square]...'

A Bahamas Helicopters director, Arni L. Sumarlidson, arrived in Port Moresby as resident manager, and the APC work was soon smoothly under way. All of the operating personnel were initially American, but on 29 January, 1954, an Australian, Robert Wilson, was presented with his helicopter 'wings' at a ceremony in Port Moresby. Bob Wilson was trained by 'Sandy' Wheeler, and as far as can be ascertained was the first Australian to fly helicopters in PNG.

Bahamas Helicopters suffered two Bell 47 crashes in the first year of operations for APC, the second in September, 1954, when a tin of emergency rations being dropped to a field party on the Aramia River, rolled in the cockpit and fouled the controls. The helicopter was soon repaired.

In April, 1955, the Chief Examiner of Airmen for DCA, Frank Shanahan, crashed in a Bahamas Bell 47 while gaining flying experience with helicopters. The tail rotor broke when the Bell was at a height of three metres, and the helicopter immediately went into a violent spin. Somehow, Shanahan managed to control the wildly spinning machine, and put it down. He was slightly scratched, and the Bell was badly damaged, the main rotor being shattered. Arni Sumarlidson said that Shanahan had done a 'remarkable job' in landing the helicopter while it was spinning. In May, another Bell crashed on the Morehead River. Both were swiftly repaired by the remarkably efficient Bahamas engineering team.

By mid-1955, the four Bahamas Bell 47D-1 helicopters in Papua (the fourth was brought in to fly for the New Guinea Resources Prospecting Company) had operated for more than 6,000 flying hours since their introduction, and had amply proved their value in the specialised field of oil exploration. Despite their limited performance, the 47D-1's, operating at altitudes up to 610 metres and with disposable loads of between 113 and 181 kilograms, had achieved remarkable results. One Bell had in thirty days flown 320 hours, and moved 88,452 kilograms including 198 passengers over ninety-six kilometres. In one period of four days, another Bell had accurately located thirteen kilometres of new road through dense rainforest in limestone country, a task which would have taken two survey parties each of six Europeans and eighty carriers, six weeks to accomplish. APC experience in general indicated that one Bell 47D-1 could do the work of between seventy and 120 carriers in a given time. Expensive though the helicopter was to purchase and operate, the figures spoke for themselves.

The work of the helicopters had been closely watched by senior officers of the PNG Administration, as well as DCA. There could be no doubt that the helicopter had an important part to play in general PNG aviation. In February, 1955, a conference was held between senior officers of the Administration and DCA. Following this, John Arthur announced that the meeting had 'worked out what services helicopters could be applied to in the Territory, and what work the Administration would be able to give a commercial operator'. The information gathered at the talks would be passed on 'to an airline company which was considering helicopter operations,' Arthur continued. He declined to name the company.

The company was Qantas. The PNG Administration had promised 'complete support' to Qantas, and the company was studying the economics of introducing the Sikorsky S-55 as being the most suitable helicopter for Territory conditions.

The S-55 was a much larger helicopter than the Bell 47. It was one of the best and most reliable helicopters ever built, and was constructed under licence from the American parent company in Britain, Japan and France. The S-55 first flew in 1949, had a normal payload of 680 kilograms and could carry up to seven passen-

217. Mechanic Jim Dean services a World Wide Catalina. *Photo: Sten-Erik Molker*

217

gers, plus pilot. It was powered by yet another version of the Pratt and Whitney R1340 engine.

The S-55 was considered by Qantas and DCA as comparable to the Beaver on a disposable load basis. As against the Beaver, the S-55 was forty-eight to fifty-six kph slower, and had a much higher fuel consumption. Overhaul and maintenance requirements were more frequent and expensive, but on the other hand the S-55 used an alighting area that cost only two per cent of a conventional airfield. The S-55 at US$146,000 basic cost at least twice as much as a Beaver at $60,000, and the direct operating costs were approximately four times greater. Establishment, training and ground engineering costs for the S-55 would also be high — possibly five times that of the Beaver.

After long consideration, Qantas finally decided not to enter the helicopter field, and DCA announced that further information about the operation of helicopters in PNG would be sought. In the meantime, Bahamas Helicopters Ltd announced that they had assigned their APC contract to World Wide Air Services Ltd, a company newly-registered in Papua. The head of World Wide was Arni Sumarlidson, who was also president of World Wide Helicopters Inc. of California (with headquarters in Paris). The chief pilot of World Wide was Robert Wilson. Eighty per cent of World Wide staff were Australians.

In October, 1955, Sumarlidson announced a £125,000 expansion programme, involving the purchase of two Catalinas*. They would be used principally for APC work, and were presently being specially modified in California. The flying- boats would also be available for limited private charter work. World Wide would also bring in a Republic Seebee RC-3 amphibian, mainly for internal company use. Doubts about the future of the Qantas Catalinas had forced APC to turn to World Wide.

The two World Wide Catalinas were PBY-5 models, amphibians that had been converted to straight flying-boats by removing the landing wheels and covering the wheel-wells with duralium plate. They had 'clipper' bows — the original nose gun turret had been removed and replaced by a long, streamlined nose. The first, which was eventually given the Australian registration VH-WWB, arrived in February, 1956. Permission to import the second was refused.

The Catalina was something of a nuisance to the World Wide engineering staff, who were accustomed to working on helicopters. A small workshop was established at the side of the Port Moresby Marine Base with a little nose-in hangar, equipped with block and tackle for engine removals and sufficient tools and facilities for major overhauls. An engineer, Bruce Evans, was given responsibility for Catalina maintenance.

The World Wide Catalina was worked very hard indeed. Although Qantas continued to charter its Catalinas to APC as they were available, the scheduled flying for the Administration took precedence, and APC relied principally upon World Wide. In October, 1956, Bruce Evans was sent to Chittagong to look after a World Wide Inc. Catalina operating in the Bay of Bengal, piloted by Frank Minjoy. This aeroplane had been working in PNG with VH-WWB. Another engineer was required at Port Moresby, for WWB — which was suffering badly from corrosion — was due for a C of A overhaul in February, 1957. Bob Wilson was now manager of World Wide Air Services, and he was fortunate to obtain the services of another of those many-talented aircraft engineers of wide experience who seem always to gravitate to PNG aviation.

Phil Bennett was a Queenslander, RAAF-trained on a variety of Service aircraft. He served in England and the Middle East during the Second World War, and afterwards worked for the US Army Air Corps at Guam, for Trans-Australia Airlines, and for the small Southern Airlines. He accumulated many licences and special qualifications, including a metal airframe ticket. Phil Bennett remembers: 'I'd been working in Melbourne for eight years, and I wanted to get out. It was too cold. Then Jimmy Darbyshire went up to fly the Catalina for World Wide in Port Moresby, and I asked him to have a look up there for me. He wrote and said they wanted a bloke for the Catalina, an A-licenced bloke for the C of A overhaul coming up. Bob Wilson came down to see me, and I got the job. Got to Moresby the week before the Melbourne Olympic Games.

'I had never had anything to do with flying-boats before, everything was strange to me. That "Cat" was a terrible machine to handle, you had to pull it out of the water. Those converted PBY-5 boats were the worst of the lot for corrosion. We

*According to DCA records, the Catalinas were actually purchased by APC.

nicknamed this one "Corrosion Kate", and it was in a pretty bad way, particularly between the planing surfaces, where the skin overlapped. The plates had a sort of fabric between them, covered with a yellow chromate mixed with a putty-like stuff, for waterproofing. This would peel off, the plates would corrode, and they would leak. You were forever patching them. Forever drilling rivets out and patching them. I trained two Mekeo boys to do this, and they were invaluable.

'Well, the thing was due for C of A in January, '57. We needed lots of spares and bits, particularly skins, because a lot of the skin was compounded curves. You had to have the right equipment to handle those — like a panel-beater on cars. So Bob Wilson sent me down to New Zealand. We bought all the Air Force spares — they had a lot of "Cats" at one time. I arranged for the stuff we needed for the C of A to be airfreighted back.

'But we didn't have enough metal workers for the C of A. "Corrosion Kate" was going to take a lot of fixing. We used to get metal workers from the helicopter shop, but they weren't keen on the "Cat". So Knute Flint went south, and contracted with the Bristol Aircraft Company at Bankstown for six blokes for six weeks working sixty hours each week. They were all good, real sheet-metal workers. Just as well — the spares from New Zealand didn't turn up. So these blokes made up panels, etc, did a beautiful job, and stripped it all down ready for re-painting. The heliport electrician did up all the instruments, we did the controls...she was out for two months, and APC was getting pretty anxious.

'So we got the old "Cat" going, and in the meantime Frank Minjoy brought this other "Cat" down from the Bay of Bengal. Bruce Evans went south to New Zealand to get those Air Force spares. They eventually arrived, all sorts of things, lot of skins, even spare floats. Frank test-flew "Corrosion Kate", and they put her back on the run and were very pleased with her...'

For all of 1956, World Wide Air Services had kept asking for a permit to import the second PBY-5, which was still waiting in California. APC was drilling new wells at Kuru, Barikewa and Morehead, and desperately wanted greater aerial uplift capacity. However, permission to bring in the second Catalina was still refused. In late July, it was announced that World Wide would shortly be importing three new helicopters, Sikorsky S-58's, capable of lifting up to 1814 kilograms — giants, compared with the little Bell 47D-1's. These new machines would be used during the APC and Island Exploration 1957 drilling programme. The three helicopters were worth almost £1 million. Five Australian pilots were sent overseas for training on the Sikorsky's. World Wide manager, Bob Wilson, announced: 'They are the largest helicopters in the world. They can lift a large sedan car with four people inside over hundreds of miles. The helicopters are to be used to fly APC rigs and crews into the drilling sites. This means that APC can do without costly roads. All the rig will be flown to the site, and the only way in and out will be by helicopter. We have flown rigs into other sites throughout the world...'

The first two S-58's arrived in late February, 1957, just after the Catalina, VH-WWB, was put back into service following the C of A overhaul. The third S-58 arrived soon after. The S-58 was a development of the S-51, but much larger, and was also produced under licence in Britain and France. It was fortunate for APC that the big Sikorsky's arrived when they did, for on Sunday, 16 June, 1957, WWB was lost.

Bob Wilson and Jim Darbyshire had flown from Port Moresby to the Kikori River in WWB, to pick up an APC party. The weather over the Papuan delta was dreary. A light, drizzling rain fell, and the sea was smooth and glassy, a dangerous combination. It is difficult for a seaplane pilot to judge his position in relation to the sea surface with accuracy when attempting to land under such conditions. Bob Wilson later described what happened: 'We came in, and had gone about 300 yards [274 metres] when things went wrong. The aircraft bucked, shuddered and slewed around. Both Darbyshire and myself were thrown about the cockpit. At one stage, the forward thrust threw me against the throttle and she opened out.

'However, we corrected that quickly. We got to the rear blister and were adjusting the life-saving jackets when she gave a final lurch, and sank. The whole thing took about three minutes...'

Phil Bennett was having a quiet Sunday drink at the Port Moresby RSL Club when he heard the news. He recalls: 'Somebody raced in and yelled, "They've sunk the Catalina!" She had skidded in on the glassy sea, and broke open in the front. Nobody's fault, just the conditions. Dick Turner, who had a trade store at

218. Ok Tedi Geological party supply drop area. *Photo: APC*
219. Storepedoes for Ok Tedi Geological party supply drop stacked on the government wharf at Kikori. *Photo: APC*

218

219

Kikori, towed the "Cat" to an island where the Kikori divides, with his big launch. She finally settled down on a mud-bank there. At high tide, the wing was just level with the water.

'I went out to salvage what I could. I had a bomb scow to work from. The tail of "Corrosion Kate" was sticking up out of the water, and I got the tailplane off, pulled the reduction gear out of the engines at low tide and recovered the props, but that's about all we could get. She was all silted up, and I suppose she's still there, on that mud-flat in the Kikori.'

Following the loss of WWB, DCA immediately granted permission to World Wide to bring in the second Catalina, from California, and Phil Bennett left for Israel to look at three Catalinas for sale there, fine aeroplanes with very low hours. World Wide Inc. had a contract for survey work in the North Sea. Bennett recommended the purchase of one of the Catalinas, and it was used on the North Sea contract, horribly damaging work for a flying-boat originally designed for smooth-water operations. Bennett remembers that this Catalina was fit only for scrap after the North Sea adventure.

Phil Bennett returned to Port Moresby to find that the second PBY-5 had arrived, and was at work. This Catalina was registered as VH-WWC, and was in every respect identical to the lost WWB. Bennett had by this time had enough of Catalinas, and he returned to Australia. But he was soon back, as chief engineer for Papuan Air Transport.

World Wide did not last for much longer in PNG. The great APC oil exploration programme was beginning to wind down. Throughout 1957, the three S-58's performed admirably, and were credited with saving countless thousands of pounds for the company. The helicopter enabled APC to erect a complete new rig at Komewu, western Papua, in only five weeks, saving months — perhaps years — of road construction. An APC spokesman said that helicopters 'had overcome one of the worst obstacles of oil exploration in Papua: access to drilling sites'.

The APC exploration programme failed to discover a payable oil field in Papua, although Pari No.1 well did begin producing oil and condensate in November, 1958. But great economic benefits flowed to Papua from the massive APC presence, and there were other, less tangible benefits. APC oil exploration teams penetrated into many unknown corners of Papua, and brought back information of great value to the Administration.

One of the most notable of the APC expeditions was that led by geologist Keith Llewellyn accompanied by geologist John Zehnder and field assistant Frank Duke. This party was sent out to examine the almost completely unexplored territory lying between the new outpost of Tari, in the Southern Highlands District, and the Strickland River. All of this country was uncontrolled, but known to be inhabited by warlike tribes in considerable numbers. The Administration therefore provided an escort of thirteen members of the Royal Papua and New Guinea Constabulary, under D.J. Clancy, Assistant District Officer.

The APC expedition left Tari on 20 April, 1954. It was one of the biggest ever sent out in PNG. One hundred and fifty carriers had been recruited from local tribes, and amongst the mass of supplies and equipment they bore on their backs was a transceiver unit. So great a body of men could only be supplied by airdrops, and radio communications were thus essential.

Early in May, John Zehnder left the main expedition to visit an extraordinary high valley called Lavani, cradled by great limestone peaks at a height of more than 2134 metres. Oil traces had been reported from Lavani, and Zehnder collected samples. He remained overnight in Lavani, and was visited by a few tribesmen — the valley was sparsely populated — wearing the curious human- hair wigs characteristic of this region. Next day, Zehnder caught up with the expedition.

They carried on, through the mighty Strickland Gorge, and down the river to its junction with the Fly, and to a waiting APC vessel, which they reached on 25 June. During the long and dangerous journey, one police constable and ten carriers died by drowning in the flooded Strickland River.

While Keith Llewellyn and his companions were fighting their way through the perilous depths of the Strickland Gorge, a combined DCA-Administration team was conducting an ambitious series of reconnaissance flights over the major remaining areas of uncontrolled country in the interior of PNG. Australia's Minister for Territories, Paul Hasluck, had set the end of 1955 as the target date for bringing these areas under Administration control. United Nations pressure on

220. Loading a portable driller on side racks of a helicopter at Everill Junction. *Photo: APC*
221. World Wide mechanic, Jim Dean. *Photo: Sten-Erik Molker*

220

221

Australia to accomplish this had been steadily increasing. To those who knew the realities of travel and exploration in the PNG interior, the target date was an impossible goal (and in fact it would be many more years before the last of the restricted territory was brought under control) but the effort had to be made. The matter was urgent, and the decision was taken to conduct a series of aerial reconnaissance flights over the remaining territory, to pin-point population concentrations, eliminate unpopulated areas, and discover airstrip sites. Many months, even years, of arduous foot-patrolling would thus be saved.

To get the programme started, DCA made a 'penetration survey' in a DC3 from Kikori to the north-west limits of the Southern Highlands, and the mountain country to the west of Mount Hagen. In September, 1953, following this survey, another series were made in the Western Highlands and Madang districts, in conjunction with a number of ground patrols. The flights were made in a DCA Anson, piloted by John Arthur. Accompanying him was the Director of District Services and Native Affairs, Alan Roberts, and the Western Highlands District Commissioner, R.R. Cole. Four possible strip sites were located during these flights.

In May and June, 1954, a second series of flights were made, over the almost unknown country between Tari and the Strickland Gorge, that had just been traversed by the APC expedition. During one of the flights, the Anson flew into the high valley of Lavani, and upon the return of the party to Port Moresby, it was announced in the *South Pacific Post* that the survey had revealed the presence of 'densely populated valleys' that had increased the population of PNG by 100,000 tribesmen. Alan Roberts' description of the spectacular Lavani valley as 'a better Shangri-la than the one found in Dutch New Guinea during the war' (actually the Grand Valley of the Baliem, discovered by the Third Archbold Expedition of 1938 with the aid of the Catalina amphibian, *Guba*) was seized upon by the press, and in an amazing outburst of lurid, distorted reporting, the world was told of the discovery of a fabulous 'lost valley' inhabited by an enormous population of previously unknown people. When the APC expedition returned to Port Moresby soon afterwards, and reporters learned that John Zehnder had actually entered the valley, he became an instant celebrity.

It was following these flights that John Arthur conceived the idea of supplying the Southern Highlands from a new airstrip to be built at Vaimuru, on the coast near Kikori, rather than from Lae and Madang as hitherto. The plan had to be abandoned when it was found to be impossible to develop Vaimuru to the required DC3 all-weather standard.

By mid-1957 the APC oil exploration programme was entering its final phase. The World Wide manager, Bob Wilson, was transferred to the Paris headquarters in July, and flew there in the Catalina that had been working in the Bay of Bengal. Frank Minjoy became PNG manager, and one of his first actions was to announce the departure of the last three Bell 47D-1 helicopters for Columbia and India, in September. The first of the Bells had been sent to Pakistan in February. 'There was no further work for the Bells in PNG,' said Wilson, 'but the three Sikorsky's would be retained'.

In April, 1958, World Wide purchased the DHC-2 Beaver, VH-EAT from Qantas. It was re-registered as VH-FBV and operated on floats. However, in July Minjoy announced that APC was not renewing the contract for the three S-58's and they would be withdrawn in October. The APC manager in PNG was transferred to Persia. Vacuum Oil Company and British Petroleum Exploration Company, partners in the Australasian Petroleum Company, said that it would be 'unlikely' that any further funds would be committed to the search for oil in Papua. In August, Island Exploration Company began issuing notices of dismissal to employees, some of whom had been working for the company for twenty years.

It certainly looked as though an APC withdrawal was now inevitable, but there was a last-minute reprieve. Fresh capital was raised and it was decided that APC would make one last effort. The Sikorsky contract was renewed and the World Wide Catalina and Beaver continued flying.

In March, 1959, an S-58 was en route from Era drill-site to Bwata site when the engine failed. The pilot engaged the emergency auto-rotate mechanism, and just failed to reach one of the rough emergency landing clearings located every kilometre or so along the route. The big helicopter settled into the Era River, about ten kilometres from the Puri oilfield. It fell over onto its port side, damaging

222. Pilot George Strohlenberg in World Wide Sikorsky S-58 helicopter. *Photo: Sten-Erik Molker*

222

the tail cone and tail pylon. The pilot and his three passengers escaped, and swam to the river bank. Another S-58 carried in a canoe, in which they paddled to Puri.

The salvage of the Sikorsky was described by a World Wide spokesman as heart-breaking: 'When we arrived, only a small part of the rotor blade was visible above the surface of the river. We used derricks, a diesel winch weighing thousands of pounds and another S-58 to help in the operation. When we raised it a little, we found that there was about two tons [about two tonnes] of silt in the fuselage. We had to shovel this out before we could do anything.

'We'd get the helicopter above the level of the water, and night would fall. During the night there were great claps of thunder, inches of rain, and bolts of lightning. We had to leave the site every night, and when we returned in the morning we would find that the helicopter was again submerged. The river level often rose fifteen feet [four-and-a-half metres] overnight, so we would have to start all over again.

'At one stage, we placed the legs of the derrick we used to lift the helicopter in forty-four gallon [200-litre] drums, to stop them from sinking in the mud. When we lifted the helicopter, the weight of the aircraft drove those forty-four gallon drums six feet [nearly two metres] into the mud like a couple of pistons. And when we finally got it out of the river, we found there was little of it left that was any good...'

It took seven days to pull the S-58 out of the Era. It was shipped to Port Moresby, but was beyond repair. Another was immediately ordered — at a cost of £250,000. The new APC well at Bwata was now under development, and two of the Bell 47D-1's were brought back. The other two arrived in June, 1959.

There was another incident involving an S-58 in May. The helicopter was flying in heavy drilling equipment to Bwata, suspended beneath the belly. The equipment — the base of a rig — slipped from its securing hook, and fell for hundreds of metres into dense bush. It was the twenty-first load to be lost in such a way since World Wide operations began.

With the failure of the Bwata well, APC finally began preparations to withdraw. The first great phase of post-war oil exploration in Papua was almost at an end, and helicopter work ceased in January, 1961. World Wide was informed that the helicopter contract would not be renewed, and in March the Bell and Sikorsky machines were dismantled. Frank Minjoy made the last flight in a World Wide helicopter in PNG on 23 March, in a Bell, over Port Moresby.

The World Wide helicopter fleet was shipped out on 2 May, 1961, on the Norwegian motor vessel, *Thorsisle*, bound for California. The helicopters left PNG in thirty-two crates, each of the three largest weighing 6350 kilograms and measuring 11.6 metres long by 3.6 metres wide. Each carried the main fuselage of an S-58. It was the end of an amazingly successful helicopter operation over a long period of years in some of the most inhospitable country in the world, an operation which saw many crashes and loss of equipment, but which was accomplished without loss of life. At the very beginning of the helicopter era in the southern hemisphere, World Wide had showed what could be done by professionals using the right equipment. When World Wide left PNG, the helicopter was no longer an oddity, but a thoroughly proven piece of equipment, and over the following years it was to become a familiar sight throughout the skies of the Territory.

The World Wide Beaver, VH-FBV, was sold to a company in Dutch New Guinea. The surviving PBY-5 Catalina, VH-WWC, passed into the hands of Trans-Australia Airlines after that company entered PNG. Like its companion, WWB, VH-WWC was to come to a sticky end, a story which will be told in Volume Two.

223. Steward, Mara, about to pick up mooring for Beaver floatplane at APC camp. Pilot, Ken Montagu. *Photo: Ken Montagu*
224. World Wide S-58 helicopter. *Photo: Sten-Erik Molker*

223

224

225

226

227

228

225. Sikorski S-58 helicopter, VH-WWE. *Photo: APC*
226. Qantas Beaver floatplane, VH-EAT, at Middletown APC camp site piloted by Ron Roberts. *Photo: Ron Roberts*
227. World Wide Sikorsky S-58 helicopter, VH-WWF, being assembled, 1957. *Photo: APC*
228. Bell 478 helicopter, VH-WWD, of World Wide Air Services. *Photo: APC*
229. Loading Bell 47 helicopter parts into a C47 transport bound for Port Moresby. *Photo: APC*
230. Qantas Catalina being beached at Lae, 1957. *Photo: Qantas*

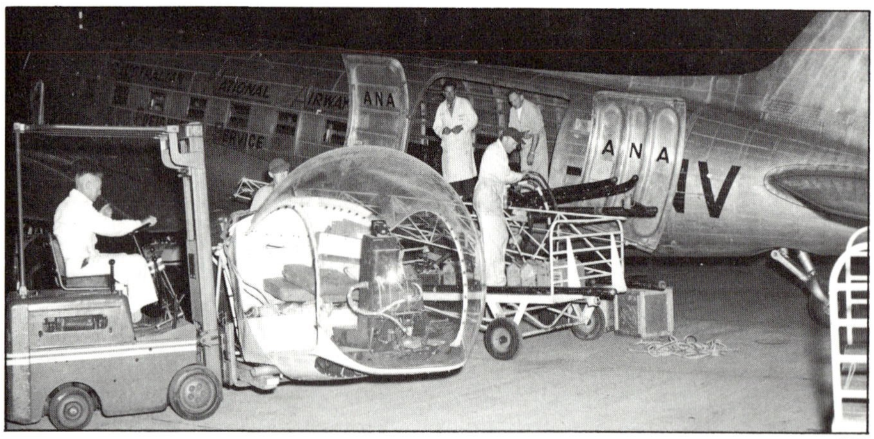

229

230

Carsair

One of the last jobs that Phil Bennett completed before leaving World Wide Air Services was a major repair to an Avro Anson C19,of Carsair Air Service, Port Moresby.

The C19 was the biggest and best of the famed Anson series. It had a re-designed, roomier fuselage, hydraulic operation of the undercarriage and flaps, more powerful Cheetah XV engines, and all-metal wings. The Anson, registered as VH-BIX, blew a tyre on take-off from Aroa Plantation. The aeroplane slewed violently, and the undercarriage leg was pushed right through the metal wing. Bennett recalls: 'World Wide took on the repair job for the insurance company, and sent me up to fix it. I went out, pulled all the crumpled metal off, went back to town and made up ribs and stuff, and rebuilt the thing. Then Bill Vink flew it out to Moresby, and it went to Brisbane. Carsair was just about finished in PNG...'

Robert Glen Carswell, owner of Carsair Air Service, was an ex-Qantas engineer, highly qualified. He held an Aircraft Maintenance Engineer's Licence in categories B, C and D. Carswell had been a member of the Qantas rescue team led by Orm Denny who plucked the refugees from Mount Hagen in two DH86 biplanes in May, 1942, following the Japanese invasion. After the war, Carswell left Qantas, and set up an aircraft engineering business in partnership with another engineer, W.R. Dalgleish, at Archerfield, Brisbane. Bob Carswell was also a licenced commercial pilot.

In 1955, Carswell decided to move to PNG and establish an RPT service out of Port Moresby. It appeared to him an opportune time. Papua was as yet sparsely served by the air companies. It was common knowledge that Qantas intended to retire the Catalinas. Goilala Air Service was still flying, but on a very small scale. Papuan Air Transport was operating efficiently with its small Anson fleet, but all available routes had not yet been covered.

In June, 1955, Bob Carswell visited Port Moresby and discussed his plans with the Regional Director, John Arthur. In a letter to the Director-General, Arthur wrote: 'We have very definitely advised Mr Carswell that we will not recommend any routes for him now served by regular and reliable operators, and he has come to New Guinea on the clear understanding that he must locate his own work outside the air transport system of these operators. We suggested to him Port Moresby-Milne Bay-Vivigani-Kiriwina-Baibara, after Qantas Catalina operations stop...a direct Moresby-Goroka service will be very profitable for Carsair. It is not being operated by other air services...'

On 27 July, 1955, Carswell applied for an airline licence to use four Ansons — two flying, two in reserve — each radio-equipped, on two routes: Port Moresby-Goroka, and Port Moresby-Kiriwina, with stops. As recorded earlier in this book, DCA was not yet issuing airline licences to small operators and the application was refused. However, interim permission was given to operate between Moresby and Goroka, Goroka and Bereina, Goroka and Yule Island, and Moresby-Baibara, pending the issue of a charter and aerial work licence. The approval was valid for twenty-eight days only, from 17 October, 1955.

DCA Headquarters in Melbourne were not convinced of the need for another operator in Papua. As previously noted, the rationalisation of all PNG air services was at this time under study. The issue of a charter licence to Carsair was delayed while the application was considered, and John Arthur granted another twenty-eight day extension. This went on for months before Charter Licence No.181 was finally granted, on 30 April, 1956. The extensions had been granted by Arthur 'without prejudice to the decision of the Director- General on the application', but Vern Rule, the new Regional Director, was informed that the issue of the extensions had 'seriously prejudiced' the position of the department. Rule was instructed to refer any future cases of this nature to Melbourne in the first instance.

Bob Carswell commenced his PNG operations with two Avro Ansons, VH-BIF and AKB. A third, BID, was sold to Goilala Air Service in November, 1955. Carswell's first two pilots were Jan Kingma and Brian Everingham.

Carsair had its first accident in April, 1956. Bob Carswell was at the controls of BIF when an engine faltered on take-off from the dangerous Tapini strip. The Anson swung, and left the strip. One wheel was torn off, and the other collapsed under the additional load. The Anson skidded through *kunai* grass over two-and-a-half metres high, and crashed into a fence. None of the six men on board, including co-pilot C.L. Searle, were injured, although Carswell had a close shave when a propeller blade from the faulty motor tore through the side of the fuselage, only just missing his legs.

In 1956, Ernie Reid, proprietor of Goilala Air Service, died. It will be remembered that the company had actually been owned by Dr Alexander Casselberg, but in February, 1954, he withdrew in favour of Reid. On Reid's death, his widow attempted to carry on. Bob Carswell had been maintaining the GAS aeroplanes with another engineer, T. Whillians, but he severed his connection with the company, and Whillians left PNG.

Goilala Air Service was never a large concern. Between 1952 and 1957, the company put six Ansons into service, seldom more than one at the same time. As previously recorded, VH-BDD was totally destroyed in a crash at Embessa in February, 1953. The second GAS Anson, AVY, was withdrawn from service in 1956. Three others, BID, BMA and BMM, were retired in 1956 and 1957 following deterioration of the wingspar. The sixth, EFR, was sold to Patair in 1958.

Ernie Reid also owned a Percival Gull, VH-UTP. In August, 1954, Laurie Crowley piloted the Gull in the Redex Navigation and Reliability Trial for light aircraft, accompanied by Bill Englebach, of MAL.

By mid-1957, Goilala Air Service had folded. Carsair was now making considerable inroads into Patair's business. Bob Carswell had increased his fleet. He brought in a Lockheed 10B Electra, VH-CMA, in December, 1955. He sold the Anson, AKB, to Joh Bjelke-Petersen, the future Premier of Queensland, in May, 1956. Another Anson, BUY, joined the fleet in June. A Norseman, VH-ASS, was purchased from Gibbes Sepik Airways early in 1957. In February of that year, a Beechcraft F17D, VH-ACU, was added, and the Anson C19, BIX. Pilots working for Carsair included Keith Kaufman, Andy McCulloch, Jerry Foude, and Bill Vink.

231. Pilot Bill Vink. *Photo: Bill Vink*

231

Vink was born in Holland. He entered the timber industry as a trainee executive as a young man, but his true love was flying. In February, 1939, he became a student pilot, as a member of the Netherlands East Indies Army Air Service. He trained on Koolhoven biplanes and Lockheed12's and in October, 1940, won his wings. The Pacific war was clearly coming, and Bill Vink was selected for dive bomber training. The Netherlands Government had on order from America a number of Douglas TBD-1 torpedo-bombers, and pending their arrival, Vink and his fellow trainees used Curtiss Falcon fighters as a makeshift. The Falcon was a low-powered export version of the Curtiss Hawk 75, and when the Japanese attacked the East Indies, the Dutch warplanes, which included poorly armed Glenn Martin bombers and ancient Fokker C-X biplanes, were totally outclassed, and were shot out of the sky. The Douglas dive bombers never did arrive: the ships in which they were being transported were sunk by the Japanese.

Vink and the surviving members of his squadron became prisoners-of-war. They were shuttled from camp to camp in the East Indies, Singapore and Siam as the fortunes of war changed. After their release at the war's end, they were sent to Biak for re-training — they had remained members of the Air Force throughout their years in POW camps. Bill Vink was granted home leave in 1949 — ten years after coming to the East Indies.

Bill Vink had run out of a future. The Indonesians gained their independence from the Netherlands, and Vink resigned from the Air Force. He decided to migrate to Australia, and arrived there in October, 1950. He was thirty-one years old.

It was a bad time for a foreign-born pilot to find work in aviation in Australia. Vink tried, and failed. He became an interpreter with the Department of Immigration, then worked for two years for the Snowy Mountains Authority before becoming a farmer, near Bankstown, Sydney. He still had the urge to fly, and since

Bankstown was not far away, he began the long struggle to obtain his Australian commercial pilot's licence. More than three years had passed since his last licence renewal, and he had to start again, from the beginning. Bill Vink quickly found that great changes in aviation training had occurred in the fifteen years since his student days. The relatively simple subjects he had studied then, in Dutch, had become vastly more complicated, and now had to be mastered in English. He persevered, and was eventually granted his licence.

Vink was already too old for regular airline service, and he had to take what work he could find. He ferried aircraft for the Royal Aero Club, Bankstown, delivered newspapers throughout outback New South Wales in Austers, crop-dusted in Tiger Moths, and flew a Percival Gull on charter work out of Bourke, in far western New South Wales. Then one day his luck changed. Bill Vink recalls: 'I was in Brisbane, and I got a phone call from Bob Carswell. He wanted an Anson flown to New Guinea. He had a pilot's licence, but he was an engineer, hadn't done much flying. So I went with him to Port Moresby, and he said, "You can stay and work for me if you like." So I did. This was the beginning of a stay of just over ten years in New Guinea, from December, 1956, to February, 1967.

'I did two years with Bob. I was broken in by his training captain, Jan Kingma, a Dutchman like me. We were doing a run to Goroka three times a week, and twice a week to Wau. We went to Milne Bay, Cape Rodney, Abau, Baibara, Gurney, Milne Bay, and also into Tapini, in opposition to Patair. Tapini and Moresby were the only strips we were allowed into that Patair serviced...'

After Goilala Air Service folded, Vern Rule granted Patair the sole right to service Tapini, as a measure of protection against the growing Carsair presence. Bob Carswell had the backing of Burns, Philp (New Guinea) Ltd, one of the most influential firms in the Territory. BP's were the Carsair booking agents at Port Moresby, Goroka, Wau, Samarai and Daru. Carswell decided to object to the granting of a Tapini monopoly to Patair. Patair was charging seven pence per pound on freight from Port Moresby into Tapini, and Carswell considered he could carry freight for five pence. He went to Melbourne, and put his case to the Director-General.

As a result, Vern Rule was instructed to authorise Carsair operations into Tapini. This he did, in March, 1957. Predictably, Patair began to bleed.

Aviation is a business. Specialised, certainly, but a business nonetheless. Competition between airlines, large and small, has always been intense — even ruthless — and if a minor operator is to survive, he must use all avenues open to him to protect his interests, hence Carswell's visit to Melbourne, and the Patair response. Patair appealed to the Regional Director. The largest single customer at Tapini was the Administration, and on Vern Rule's recommendation, it was decided to split government business equally between Carsair and Patair. Carsair was also instructed to lift its freight rate to seven pence per pound.

Patair had demonstrated from the beginning that it was prepared to fight for what it saw as its legitimate market. Vern Rule wrote to the Director-General, commenting on the squabble: 'Papuan Air Transport has been established for six years, and is wanting to expand, but is demanding protection. The newcomer, Carsair Air Services, with a more versatile fleet (but just as obsolete as PAT's) and with apparently strong capital backing is anxious and willing to compete on all routes and has obtained the business of one of the most powerful firms in this Territory...the crying need is for an operator with sufficient resources to equip with aeroplanes of later vintage than pre-war Ansons...'

Patair was soon to embark upon a replacement programme that would answer Rule's complaint. The company wanted the protection of an airline licence, however. In a letter to the Minister for Civil Aviation in May, 1957, Cliff Jackson, the Patair manager, wrote: 'We are prepared to invest £100,000 in new aircraft and equipment, bringing the total sum of our investment to £150,000...if this step has Government support...'

The Carsair fleet was certainly more versatile than the Patair Ansons, but as Rule had noted, the aeroplanes were just as ancient. The Lockheed 10B Electra, VH-CMA, was in its day one of the best transports on the market, but that day was in the past. The Electra first flew in 1934, and production was terminated in 1941. In December, 1957, Carswell imported another old Lockheed, a model 12A, VH-ABH. The 12A was a scaled-down version of the 10B, and was often referred to as the Electra Junior. It, too, was manufactured before the war.

Bill Vink flew the Carsair Lockheed 10B, and wasn't impressed.

'I did one trip in the Lockheed 10, to Daru, and couldn't get the thing above 1,000 feet [305 metres]. The oil pressure went up too high if you went above 1,000 feet. We had the same trouble with the metal Anson, BIX. You couldn't get above 8,000 feet [2438 metres]. Going to Goroka, you had to be sure that the weather was fine, otherwise she just couldn't do it. The old Anson, BIF, was different — I've had that to 18,000 feet [5486 metres]. She still had the bomb doors in! If you had a bit of rough weather, if you had to go high to get over the top, you'd hit the bumps and the bomb doors would open, and that cold air would be coming in...'

Bob Carswell continued to fight for extra routes for his aeroplanes. In June, 1957, he lodged an application for an extension of his operations from Vaimuru to Horn Island. It was vigorously opposed by Vern Rule, on the grounds that the sea crossing would be too hazardous in the types of aircraft that Carsair flew. Carswell applied for a Madang-Goroka route, and this, too, was refused. He declared his intention to purchase a DC3, once second-hand prices fell.

Nothing seemed to go right for Bob Carswell. Like many other small airline operators, he was a rugged individualist, and resented DCA regulation of his affairs, constantly battling officialdom. Vern Rule wrote to the Director- General: 'There has never been a time in the past two years when we could not have closed him down. We have bent over backwards to help...he can't see this, he believes DCA requirements are red tape, designed to frustrate him...'

In 1958, Qantas brought in its four Otters to replace the Catalinas and Beavers. Bob Carswell was particularly concerned about the introduction of the Otters, and in August, 1958, informed Vern Rule that he would withdraw from Papuan operations 'before Christmas'.

He was as good as his word. Bill Vink recalls: 'Because of the competition, Bob dropped out. He didn't sell out, or anything. Just stopped operating, said Qantas was going to push us out of business. Carsair stopped in October, 1958. Bob took the metal Anson and the Lockheeds down to Brisbane, but just left the old Ansons there. The laminated spars were gone, anyway. There were only three Carsair pilots left. Brian Everingham got a job with Madang Air Services, Jerry Foude went to Ansett, and I started with Patair. Their DC3 had just come when I joined them...'

Bob Carswell continued in business in Australia under the name, Carsair Air Service. He acquired more aeroplanes, including a pair of DH89 Rapide biplanes, and in 1960 commenced operations in the Northern Territory, hauling fresh buffalo meat. He never returned to PNG.

Papuan Air Transport now had a clear road, and was ready to expand. Before this could happen, the Ansons had to be replaced. Main-spar deterioration in a tropical climate doomed them, and they had to go. The company asked for approval to operate a DC3. The Director-General was somewhat reluctant to grant this request. A major government decision on future air operations in PNG was pending, and permitting Patair to operate a DC3, an aircraft of altogether higher capacity than the Anson, would upset the status quo. On the other hand, Papua was now poorly served by air companies. Goilala Air Services and Carsair had gone, and Qantas was in the throes of discarding the Catalinas and Beavers, and introducing the Otters. Qantas services within Papua had declined. A need existed, and Patair was ready and willing to satisfy it. The company had provided a safe and efficient service with the Ansons. Withholding approval to replace them with better aeroplanes would finish Patair, and the Director- General approved the acquisition of a DC3.

Until this time, Patair had operated almost entirely within Papua, but in December, 1957, a direct weekly service between Port Moresby and Madang was announced. In the following February, regular weekend charters between the capital and Wau began. Services to Mendi, Mount Hagen and Garaina were commenced. Qantas and MAL began to feel the competition.

To help locate the right aircraft for its future operations, Patair appointed Bill Cavanagh, a Sydney-based aircraft consulting engineer, as technical advisor. A suitable DC3 was located. It was an ex-RAAF aeroplane, and was purchased from the Commonwealth Disposals Commission. Cliff Jackson remembers: 'This DC3, VH-PAT, was rebuilt to our specifications and standards for passenger travel and comfort which were unprecedented in Papua New Guinea. We introduced the first forward-facing seats in DC3's, with cabin service attended by flight attendants. We were the first to have flight attendants...'

VH-PAT, the new Patair flagship, arrived in Port Moresby on 10 October, 1958. Cliff Jackson was reported in the *South Pacific Post* as saying: 'Two more will arrive next year. The company's present fleet of Ansons will be replaced within a year, and we are now negotiating to buy two Beaver aircraft which will be used on 'dromes which the DC3 cannot service. These 'dromes include Tapini, Woitape, Yule Island and Aroa...'

DCA was somewhat alarmed by Patair's ambitious plans, and both Qantas and MAL complained that their operations would be effected. The Director-General therefore had the Patair charter licence amended, prohibiting the company from operating their DC3 to or from Mount Hagen, Goroka, Kainantu, Madang, Bulolo and Wau, without prior approval.

There was an extraordinary sequel. The Patair directors sought legal advice, which held that the Director-General had exceeded his powers. Patair applied to the Full Bench of the High Court in Sydney for an order nisi, claiming that Director-General Anderson had acted 'arbitrarily, capriciously and unreasonably' in varying the charter licence. It was a measure of the lengths to which the company was prepared to go in defence of its interests. The order nisi was granted, requiring the Director-General to show cause why a writ of mandamus should not be issued commanding him to determine a Patair application for a licence to operate the DC3 'within PNG', and to issue air traffic control clearances for flights to and from the banned airstrips.

These proceedings came at an awkward time for DCA. On 12 November, 1958, the regional director, Vern Rule, died suddenly in Port Moresby. He was only forty- four years old. R.M. Green took over as acting regional director, until the appointment of the veteran Arthur Affleck, in December, 1959.

The Patair case did not go to a hearing, but was settled by negotiation. DCA agreed to allow Patair 'reasonable utilisation' of the DC3 within Papua, and on special occasions on the New Guinea side. It was a sensible compromise. Nor did DCA attempt to retaliate for this bold clipping of its wings. On the recommendation of Bill Cavanagh, Patair had decided to purchase a new light aircraft of unusual design for its PNG operations: a Piaggio P166. No aircraft of Italian manufacture had yet been admitted to the Australian register. Patair was allowed to proceed with the purchase, on the understanding that the Piaggio would have to be evaluated by DCA in Melbourne before a C of A would be issued.

The Piaggio had been chosen following an evaluation of a Beaver chartered from Australia, and a careful consideration of the standard American designs. It was a bold decision. The P166 was an all-metal, shoulder gull-wing monoplane with twin Lycoming pusher engines each of 340 hp. It was developed from the beautiful P136 amphibian, and retained retractable landing-gear stowing in the undersurface of the wing root. It was distinctive in appearance, truly Italianate and dramatically different from current American designs. In 1958, it was still unproven in service. The first flight of the prototype had taken place only in November, 1957. Patair was one of the first companies to place a firm order for this lovely aeroplane, which was to remain in production for many years, but on a tiny scale, less than 150 of all models having been manufactured by 1981.

It will be recalled from Chapter 5, The Helicopter Appears, that Phil Bennett joined Patair as chief engineer early in 1959, after leaving World Wide Air Services. It was shortly after this that the company made the decision to purchase the Piaggio P166. Phil Bennett remembers: 'They said one day, "We're going to get a couple of Piaggios. Italian things." They sent me over to the Piaggio factory for ten days to learn something about them. You needed six months' experience with new aircraft before DCA would give you a licence, but they'd give an authorisation in the meantime...

'Away I go...ended up being there seven weeks. The aircraft was ready, but some of the radio stuff was wrong, had to send it back and get some more. It was a very good aeroplane. Excellent engineering.

'The Patair chief pilot, Henk Van Santen, came over to fly it back. I flew back with him, with a Piaggio engineer, Luigi Lureli. No problems, except for a bit of HF radio trouble. We did some demonstration flying at Singapore, and changed the oil to summer grade. We did more demonstration flights at Alice Springs. Cliff Jackson met us in Melbourne, and I went back to Moresby then, and they did a survey in Queensland for a week, and DCA looked at it for certification...'

Van Santen left Italy on 28 August, 1959, and arrived in Melbourne on 9

September. The DCA evaluation trials were entirely successful — the Department later acquired a P166 for service in the PNG region — and after final demonstration flights in Melbourne and Sydney, Ansett Transport Industries announced they would be purchasing a pair of P166's.

In Port Moresby, Patair prepared the public for the arrival of the new aeroplane, registered as VH-PAP, with large advertisements in the *South Pacific Post*: 'Papuan Air Transport introduces a New Concept of Flying to the Territory...When you fly by PAT in its new Piaggio, you get individual reclining seating, temperature control, and the complete absence of noise and vibration...The Piaggio P166 was selected after exhaustive analysis of all available aircraft as the most appropriate aircraft for Territory conditions...'

Patair had come a long way in a short time. The company had faced bankruptcy following the loss of the two Ansons in one week in December, 1954, but had fought back to achieve a commanding position in Papuan operations, servicing a wide network of plantation airstrips and outstations. Great things were expected of the Piaggio, which represented a large capital investment for what was still a small company.

Van Santen flew the Piaggio to Port Moresby on the last leg of the delivery flight on 8 October, accompanied by Cliff Jackson. The arrival was reported in the *'Post'*: 'With a high whine and tail, Papuan Air Transport's new twin- engined Italian aircraft, the Piaggio, streaked into Jacksons Airport at about 200 miles per hour [about 322 kph], and landed before 300 people. She is roomy, fast and comfortable. She came into the airport with six plush seats for passengers and luggage piled high to the roof.

'As he climbed from his seat, Captain Van Santen said, "She's the best aircraft this country has seen for a long time. We got here from Cairns in two hours forty-five minutes — that's at a speed of about 190 miles an hour [about 306 kph]".

'Manager of PAT, Mr Cliff Jackson, said that the company had already ordered a second Piaggio. "She is due to arrive here under the command of an Italian crew next February," he said. "There is already too much work for this one. We have 1,400 hours work a year for her, and you are flat out fitting in 1,000 hours a year in the Terrritory because of weather..." '

With the coming of the Piaggio, the death-knell sounded for the last of the Patair Ansons, that had founded the company's fortunes. Only two were left. One, VH-AOA, was a derelict when the company presented it to a Boroko kindergarten as a plaything; the other, VH-EFR, was in operation up to the week before the Piaggio arrived. Said a report in the *'Post'*: 'Does anyone want an Anson? Manager of Papuan Air Transport, Mr Cliff Jackson, who is near hysterics, said yesterday that he will give it to anyone or any organisation that wants it. And it goes!

' "She's about the only one left in the Territory," said Mr Jackson, "and one of the few still flying in the Commonwealth. But Ansons, mobile or otherwise, are not easy to give away. We offered it to the Technical Training College...it will have its engines in working order — in fact, everything will work...but the Technical College tells us, frankly, they can't see any use for it. Maybe the Ela Beach kindergarten can use it...if that misses, well, we'll just take it away and burn it".'

Late in October, EFR was towed to a Six-Mile junkyard, and laid to rest. It was destined for the Boroko RSL Club, but while awaiting the final move the old Anson was heavily raided by bands of thieves, who reduced it to a pitiable state. It was a sad ending to the career of a notable Anson. It was in this aeroplane that a Papuan mother had given birth to unexpected triplets, in May, 1958. Henk Van Santen had flown to Yule Island with Dr R. Roderique, to bring the woman to hospital in Port Moresby. As Van Santen told it: 'We were coming in to land at Moresby when Dr Roderique told me to fly on — the woman was giving birth to her baby. So I went back over the sea to avoid the turbulence, and flew around waiting for the birth to take place.

'Then I got the OK sign, and down I went. I was about to land when the doctor said, "Go up again. There's another one coming." So I went up again...I was just about to land for the third time, and the Doctor said, "Sorry, you'd better go up again — there's another one!" So we flew up and around again until that one was born...it was fantastic. We have had births in aircraft before, but never three. I have never heard of that happening in any other part of the world...'

The coming of the Piaggio also spelled 'finish' to the faithful old Waco biplane, VH-UYD, the first of the Patair fleet. She was sold south, and left Port Moresby on 16 November, 1959.

232

233

234

235

236

232. Patair Piaggio 166, VH-PAP, 1959. *Photo:* Pacific Islands Monthly
233. Patair Piaggios. *Photo: Bill Vink*
234. Anson VH-BDD operated by Goilala Air Services. *Photo: Geoff Goodall*
235. Patair Piaggio P166. *Photo: Cliff Jackson*
236. Patair Piaggio P166. *Photo:* Pacific Islands Monthly

Territory Airlines Grows

Patair in 1959 was clearly going places. Few observers of the PNG aviation scene then would have placed the fledgling Territory Airlines in the same league, but it was in this year that Dennis Buchanan finally became a full-time airline operator, and began the steady expansion that was to bring his company to the forefront within a surprisingly brief span of years.

It will be recalled that Buchanan joined MAL together with most of the Gibbes Sepik Airways personnel after Bob Gibbes sold his airline. Gibbes himself began another PNG career as a hotelier and businessman. He played no further active part in airline development, although he retained his close personal links with Buchanan, and in later years often flew for him. Gibbes never lost his interest in flying — he kept his commercial licence, and owned a variety of personal aircraft at different times.

The arrangement that Buchanan had with Gibbes was based on their personal relationship, and could scarcely have survived the change of ownership of GSA. To MAL, Territory Airlines, no matter how tiny, was a competitor, and it was inconceivable that the company would permit an employee to run an opposition enterprise. As Buchanan remembers: 'Carpenters asked me to stay on when they took over GSA. I stayed for about six months, then they said, "You've got your own airline, you'd better go." So I did. Then they found they couldn't do the paperwork, and they asked me to come back, part time! So I did, until August, '59, which gave me money and wasn't a draw on TAL, which helped TAL a lot, too.

'By this time, I had my first new aeroplane, a Cessna 180. I went to the bank to borrow the money; £7,000. It was second-hand really, but it only had 400 hours on it and was virtually new. I bought it in America, and it was shipped out in two big crates to Madang. Bill Sherwood and I went to Madang, and we stayed with Father Joe Walachy in his old shack on the side of the airstrip, and the three of us put it together in his old fallen-down hangar...'

William Sherwood had joined TAL as a pilot in March, 1959. TAL employed only one other pilot then, Gerd Franck, and had one operational aircraft, the Cessna 170, VH-GKA. Adrian Nisbet flew for TAL on occasion, when Gerd Franck was out of hours. Sherwood was employed sight unseen, and Dennis Buchanan got a shock when he first saw his new pilot. Says Bill Sherwood: 'I arrived in Goroka in a Qantas DC3, and Dennis was there to meet me. His first comment was, "Christ, not another big one!" Gerd Franck was six feet two, [about 188 centimetres], and well, I'm six three-and-a-half [192 centimetres], and weigh fourteen stone [about 89 kilograms]. You can imagine the little 170 with Gerd in it and myself — probably only about 350 pounds [159 kilograms] of cargo space remaining. When Dennis got another pilot a bit later, he made sure he was a little one!

'I served my time as an electrician, in Fiji. Got very restive, and came back to Australia in 1956 and took up flying. Did all my training at the Royal Aero Club, Bankstown. Did six months' instructing there, then got a job at the South Coast Aero Club, just south of Wollongong, flying Saturdays and Sundays. Got a job with International Harvesters, Sydney, as a salesman, and there I met Noel Evans [Noel Peacock had changed his name to Evans]. A bit later, Noel got a letter from Dennis Buchanan, offering him his old job back with TAL. Noel didn't accept, but wrote back recommending me. So I joined TAL.

'I arrived just before the 180 reached Madang from America. Dennis and I went down to assemble it. Father Joe had bent the back of the Divine Word Cessna 180, VH-SVD, when he deliberately ground-looped the aeroplane at Koge, to stop from going over the end, and I helped him with the repair. We took all the rivets out just behind the back window, and put a new tailpiece on, and then, having helped him

237. Bill Sherwood (left) and Neale Williams. *Photo: Bill Sherwood*

237

to do that, he helped Dennis and me to put our 180 together.

'First time I flew the aircraft was on 22 April. I was endorsed onto it by Adrian Nisbet. Dennis then sent the chief pilot, Gerd Franck, up to live in Mount Hagen and gave him the 180, which was registered as VH-GKB. It was a nice aeroplane, painted green and black. I operated out of Goroka in the 170.'

GKB was a model 180A, and was based at Mount Hagen primarily to combat Madang Air Services, in the Western Highlands. Bill Sherwood and the 170 flew mainly into the Southern Highlands airstrips — long flights for such a small aeroplane. After the Dragon, VH-AON, was sold, Buchanan dropped the weekly Goroka-Lae service, which had been flown by that aeroplane, a move he was later to regret when DCA refused to allow him to fly into Lae. As Buchanan recalls: 'They cut me out of Lae, and made me a Highlands-based show. I was given approval by DCA to base the 180 at Mount Hagen in competition with Madang Air Services. They were our competitors in the Highlands to quite an extent. They were buying Cessna 182's, and there was a race to see who would base an aeroplane at Hagen. Bob Green, the Regional Director, had helped us and given us approval, but MAS put an aeroplane in a month or so before us. But our allegiance was more with the Southern Highlands. I was friendly with Bob Gibbes, Bob Cole was DC, and Des Clancy was out there, and I'd known him since '49. Des Clifton-Bassett became DC later on, and he was another old friend. So we actually did mostly Southern Highlands work with the 180 at Hagen, and left the Western Highlands to Madang Air Services.'

Madang Air Services was continuing to prosper with its modern, all-Cessna fleet, although the 190, VH-BVD, crashed at Kompiam in June, 1958. Damage was extensive, but following an on-site inspection by Bill Cavanagh of Aeronautical Laboratories — the Patair consultant — and John Pople of Bristol Aviation Company, the 190 was dismantled, towed to Wabag by Land Rover, flown to Lae by DC3, and shipped to Sydney where it was repaired, and flown back to service by Dave Robertson.

1958 was a bad year for aircraft accidents in PNG. They were few in number, but four men died. As recorded previously, Ron de Forest was killed in a Norseman crash in July. Just weeks earlier, a Lockheed Hudson of Adastra Aerial Surveys had plunged into the sea off Lae, killing all three crewmen — the third fatal Hudson crash at Lae since 1948.

Adastra was a long-established company and had close links with PNG aviation. Adastra Airways was originally formed in 1930 by F.W. Follett and H.T. (Bunny) Hammond. Hammond subsequently flew for Holden's Air Transport Service in Lae, and later for Guinea Airways. In 1944, Frank Follett formed Adastra Aerial Surveys as a subsidiary company. Adastra was the leader in aerial survey techniques in Australia, and over the following years operated a considerable fleet of survey aircraft, initially war-surplus Ansons, and then Lockheed Hudsons.

When Guinea Air Traders folded, Adastra purchased the Lockheed Hudson, VH-BLA, which GAT had leased from Lionel Van Praag. Van Praag joined Adastra, and later became chief pilot. The Hudson was re-registered as VH-AGG after being extensively modified for the aerial survey role. The second Adastra Hudson was VH-AGO, an ex-RAAF aeroplane. With other Adastra aircraft, the Hudsons flew extensively throughout Australia and PNG on survey contracts. Adastra aircraft were based at Lae for considerable periods of time.

Early in 1957, Adastra began an extensive photographic survey of PNG and adjacent areas. On 1 July, AGO crashed while carrying out a baulked approach at Horn Island, after a flight over Cape York Peninsula. All six on board perished.

The Lae tragedy occurred on 8 June, 1958. VH-AGG had been based at Lae over the preceeding eighteen months, and the aeroplane and its different crews were a familiar part of the local aviation scene. For six months prior to the crash, the pilot had been Alan Motteram, ex-RAAF, the navigator Patrick Murphy, and the photographer Gordon Murrell. Both Murphy and Murrell were in their early twenties.

Just after dawn that morning, the Hudson took off on a routine photographic flight. Early starts were part of the game for Adastra crews. Photography was carried out from a height of 7620 metres, and it was only during the post-dawn hours that the atmosphere was clear enough to work. On this day, the weather commenced to deteriorate, and Motteram soon turned back. Gusty winds buffetted the Hudson over the Huon Gulf on the approach to Lae. The last message from Captain Motteram, advising that he was going in to land, was heard by the crew of the scheduled Skymaster of the Qantas Bird of Paradise service shortly before 8.20 a.m.

Observers on the beach at Voco Point, who were in the act of launching a speedboat for a session of water-skiing, noticed the Hudson when it was over the wreck of the *Tenyo Maru*, on final. They were surprised to see that the starboard motor was feathered. Brian Jaques later told a reporter from the Lae newspaper, *New Guinea Courier*, of what followed: 'It was flying jerkily, and was losing altitude rapidly. As it approached the airstrip it came down lower and lower, with one wing dipped. Suddenly, the one motor operating gave out a terrific roar, and the plane screwed around and went down into the sea, one wing hitting the water first. The roaring continued until the plane went under. On contact with the water, the plane seemed to break in two. The tail part remained afloat briefly before sinking...'

The speedboat crew raced to the scene, and found the bodies of Murphy and Murrell floating on the surface, one still strapped into a seat. Two cannisters of film and a flying boot were recovered, but that was all. Motteram's body was never found, nor was the Lockheed raised, despite several attempts.

Opinion among experienced pilots held that Captain Motteram had probably decided to use the aborted flight as an opportunity to practice asymmetric flying — flying and landing the Hudson on one engine, after a simulated engine failure. This was normal practice on twin-engined aircraft, and the Hudson was notoriously difficult to handle under asymmetric conditions.

One of the early Adastra pilots at Lae was G.A. Glassey, a New Zealander who after two years service with the Army of Occupation in Japan, joined the RAAF, in 1949. He became a Mosquito photo-reconnaissance pilot, and later an instructor. Dick Glassey loved the RAAF, but he finally decided to resign his commission and enter civilian life. He remembers: 'I joined Adastra Aerial Surveys in Sydney. A good job, but not a married man's job. My wife found it rather difficult, for we had constant moves around Australia. On 29 June, 1955, we came up to New Guinea. The Hudson was a good photo-survey aeroplane really, it was fairly speedy, faster than the DC3, and was good to fly, more manoeuverable. It had a reputation as a killer, but that was more because the teaching of asymmetrics wasn't particularly well developed in civilian aviation at the time.

'One of the first things I had to do when I went out on the Hudson was practise asymmetrics. You had to know about that. One day I was doing an approach on Charleville with one pulled full back, and the other full forward, at about 600 feet [about 183 metres]. I had to practise a go-around, so I opened up with full power on the light one and started lowering the undercarriage when the light one cut out. The first thing I did was knock the other one back in, but that only compounded it, and I had to wind trim in to get control. Fortunately, I recovered the aircraft with the loss of about 400 feet [about 122 metres]. The two blokes with me, well...

'I was fairly sure I was going to get control of it, but it was horrifying getting control. I had to get power off, then get the trim in, and accept the height loss. Luckily, I was used to going round in a Mosquito at 600 feet...we'd have been in big trouble if we'd been lower down...

'I was getting pretty desperate with flying for Adastra, all that shifting around. In two years, my wife and I probably had no more than three or four months together the whole time. The airline industry wasn't recruiting to any great extent then, but MAL was starting to expand. I approached Ron Horlock one day, and he said, "How many hours?" I said, '2,000.' Then Tom Deegan, the chief pilot, said, "All right"...

'In those days, you used to have some concern for the job you were doing. I told Tom I couldn't give him a date until Adastra had got a replacement for me. Adastra people in Sydney said, "Goodness, we'll give you extra money," but they understood when I said, 'No.' I told Tom it would be two or three months, and he said, "That would be all right — couldn't leave their aeroplane just sitting there — do the right thing by them." That was typical of Tom. "In the meantime," he said, "we'll do your conversion here on the DC3, so that when you come in you can start flying."

'He did the conversion on regular flights! The first flight I had in a DC3 in New Guinea — I'd flown them briefly in the Air Force — was on the Madang- Wewak courier one Monday morning. Tom picked me up at Madang, put me in the right-

238. 'Blue' Russell at Nomane with TAL Cessna 185, VH-GKE. *Photo: Jasper Maskelyne*

238

hand seat to Wewak and back — take-off, landing, the lot. There'd be hell to pay if you did it today. DCA couldn't have caught us, there was a full crew on the aeroplane, they weren't to know who was sitting in the seat. I was pretty horrified, as a matter of fact, yet I admired Tom for doing it. It wasn't unsafe. He had final control of the aeroplane, he was in the left-hand seat. I was an experienced pilot, I had done the ground course, and I did know how to handle twin-engined aircraft. Tom was an excellent pilot, he flew an aeroplane with perfection...

'So that started my endorsement on the DC3, and it went on during the three months I was waiting for a replacement pilot until I had enough hours to go to DCA. That was February, '56...'

Dick Glassey was to become the head of the Ansett-MAL Light Aircraft Division based at Mount Hagen in later years, competing head-on with the growing Territory Airlines fleet for the lion's share of the huge Highlands market. By that time, Madang Air Services had been absorbed by MAL, in the great rearrangement of PNG aviation that followed the withdrawal of Qantas.

TAL's expansion was initially slow, governed by the availability of finance for the purchase of additional aeroplanes. There were still only two aeroplanes in TAL livery until late in 1959. Both Gerd Franck in the 180 at Mount Hagen and Bill Sherwood in the little 170 at Goroka flew long and hard, principally into the Southern Highlands, piling up the hours. Both were single men, and that was how Dennis Buchanan liked it:

'We had a lot to do, and we demanded a fair bit more from our pilots than some felt they should do. We wanted single pilots who'd get out and about, do flights on demand, not worry about schedules of work. Cash flow paid off the first 180, GKB, and it was hard flying that did that...'

So successfully did GKB operate that Buchanan was able to look around for a second 180 towards the end of 1959. Bill Sherwood earned a lot of revenue in the 170, too, despite his size, and its limited capacity.

It sometimes happened that Franck and Sherwood swapped aeroplanes when one or the other was in for maintenance. Franck will not forget one early flight in the 170 out of Mount Hagen. The little aeroplane was loaded to capacity with the very big Gerd Franck; the Western Highlands DC, Ian Skinner, no lightweight; two other passengers, plus baggage. Says Franck: 'The take-off was rather spectacular in that the aircraft hogged the ground forever, and when the end of the strip came all I could do was lower an extra notch of flap to get her airborne. Luckily, she left the ground, but it took two miles [over three kilometres] at between five and ten feet [one-and-a-half to three metres] off the ground, with the stall warning sounding, to get proper flying speed. Mr Skinner was not impressed. Because of my inexperience, I had most definitely overloaded the aircraft for Highlands flying. One learned to be very cautious of loads, and good in judging what was on board...' Most pilots, understandably, have a keen interest in the mechanical aspects of the aeroplanes they fly. Perhaps because of his background before becoming a pilot, Bill Sherwood was more than usually interested in aircraft engineering, and while flying for TAL, and encouraged by a DCA engineering inspector, Roy English, he studied for, and passed, the exams for an engine licence. Sherwood then began to study for his airframe ticket (which he gained after leaving TAL) and as a result was able to take over the routine maintenance of the two company aircraft from MAL, who had continued the agreement Buchanan had made with Bobby Gibbes. It suited Sherwood: 'I was happy. I was involved in flying, and doing quite a lot of the maintenance — running maintenance, not complete overhauls and that. When I joined TAL, Sam 'One-Eye' and another boy used to change the plugs and what-have-you, under MAL supervision. 'One-Eye' was very good.

'I had a few side-lines, too. I opened the first trade store at Ialibu, and later on opened another one at Koroba. In 1960, I started the first bus service in Madang — six buses, a workshop, a taxi too...sold this business to Jack Thick.

'I remember an incident one beautiful sunny Saturday afternoon, in the old 170. Now, it was a very simple aeroplane. Had a fixed-pitch propeller, and only had a throttle and a mixture control, which also shut the engine down by cutting the fuel off. Very reliable four-cylinder engine, and very easy to fly. The 180 was the next step up. Constant-speed propeller, three controls — throttle, engine speed and mixture control, which again shut off the engine. The 170 had 145 hp, and the 180 used a Continental six-cylinder 470, and had 225 hp and that's a lot of difference.

'On this fine Saturday, Jack Thick's daughter married John Edgar. They were to have their honeymoon in the Kainantu Hotel. The wedding was at about 1 p.m. and a party in Goroka was everybody's party then — there must have been 150 people at the strip to see them off with me in the 170.

'Off I went, got down just the other side of Finintegu, and the weather started to go bad. I hopped over a ridge, got into a tiny little valley, and I flew round and round, trying to get out of the bloody thing. Went round and round for about half an hour. John and his new wife were beginning to understand the situation. Eventually, the little mini-gap that I'd used to get into this valley opened enough for me to pop back out. I radioed through, and when we got back to Goroka the whole 150 were back at the strip, and the party carried on as though nothing had happened. But it had been a nasty half-hour, and I had a few that night to take the edge off my nerves.

'You can't fly in New Guinea and not recognise the fact that it is dangerous, particularly in single-engined aircraft. You take off from one aerodrome and land at the next, and generally nowhere to land in between. I remember adding up something in the order of forty-five or fifty light aircraft pilots in New Guinea at one time later on, and in one period of twelve months, seven of them were killed. Not a good batting average. I was in my late twenties when I started flying for Dennis, a bit older than most, and I was a deep thinker. I observed all the incidents in areas where I was flying. Most incidents weren't in areas where you'd expect them, like in hard gaps, like Tari. There was one accident half-way down the Wahgi valley, and it's ten miles [over sixteen kilometres] across. Why was this fellow flying so close to the hills on the left-hand side? I got the message. I found that by flying down the centre of the valleys I only wasted a minute or so over flying down the sides to save time. Going from Mendi to Hagen, instead of flying close against Mount Ialibu over the trees as some did, I'd fly over the *kunai* line, which gave me a fifty-fifty chance of surviving if the motor stopped, and it only took a minute or so longer.

'I remember another incident in the old 170. In June, '59, a fellow called Ben Kidd came up to New Guinea with a Cessna 180 float amphibian, VH-RER. Very well equipped aeroplane, had a lot of the direction-finding gear usually fitted to larger aircraft. He worked for Reynolds Aluminium, and was doing a geographical survey for potential hydro-electric sources in connection with the Cape York Peninsula bauxite scheme.

'There's a little strip in Lake Murray called Pangoa, almost water-to-water. Ben turned the 180 amphibian on its back landing there in the wet season. Got some of the local lads to turn it right way up. One of the wing struts was bent in the overturn. To get him out, they got a piece of Norseman wing strut, which was steel and slightly smaller in diameter to the Cessna strut, which was aluminium. They bored about ten holes in the Norseman strut, and brazed on ten washers. The idea was we were to fly down, hacksaw the bent strut off, slip the steel strut in, hand-drill ten holes through, and nut-and-bolt them up tight, and this would give the strength needed in the wing strut to fly it back to Goroka for proper repair in the old GSA workshops, now being run by MAL.

239. John Bayles (left) and Bill Sherwood. *Photo: Bill Sherwood*

239

'We took off in the ever-faithful 170, GKA — me, Ben Kidd, and Jack Eifler, a German airframe man working for MAL, with his box of tools. It was 17 June. We called in to Mendi and filled the tanks right up, and off we went. Long flight, and the weather was not too good, had to go to 12,000 feet [3657 metres] to get over the mountains, and it was eight-eight cloud all the way...

'There were holes in the cloud over Lake Murray, and we let down. We found Pangoa after twenty minutes searching and landed, and boy, was it soft. Runway only about 1,500 to 2,000 feet [457 to 610 metres]. When I looked at the amphibian, I could see that one of the main left-hand undercarriage struts was sitting dead flat — the oil pressure in it which gives it a bit of bounce had collapsed. That was the reason for him tipping over. The nosewheel on the front of the float hinged downwards, and folded back into the nose, and because of the collapsed strut on the left-hand side, this nosewheel when extended had dug into the ground because it was so soft, and turned the aeroplane over on its back.

'OK, we set to, sawed the strut in half, put the bracing piece in, and put the ten bolts in. I was a bit horrified to see that on the 180 the strut is attached to the main spar, but there is a rear spar which has merely a fore- and-aft function, doesn't carry much up-and-down load. It's only made up of three or four pieces of onesixteenths aluminimum to give it strength. This had buckled upwards about a quarter of an inch [about six millimetres]...

'Ben taxied the aircraft with the left-hand strut very low and attempted to take off, and the damned thing dug in again, before he had much speed. I'll never forget the sight — it stood right up on its nose, seemed to stand there for half a minute, and came crashing down, on its wheels, fortunately. He tried it again, and the same thing happened.

'We finally had to give it away until the wet season ended, to give the strip a chance to dry out. Ben went back a month or so later, and the surface was even worse, so he harnessed the locals and cut through the trees at the end of the strip and pulled the 180 into Lake Murray, and flew it out in bad weather to Erave. He landed with about ten minutes' fuel left — he was very lucky.'

In September, 1959, Gerd Franck left TAL, to fly for Patair. Bill Sherwood was now the only — 'chief' — pilot, and another was urgently required to fly the 180. Sherwood had a pilot friend in Australia, John Bayles, and he wrote, asking Bayles to find a suitably qualified man. Says Bayles: 'This I did. Me. I went up in October, 1959. The 180 was being operated from Minj by Adrian Nisbet. I went to live with Adrian on his coffee plantation there while he endorsed me on the aircraft, and route-endorsed me through the Highlands. In mid-November, 1959, I moved with the aircraft to Mount Hagen, and operated from there, mainly into the Southern Highlands. I continued flying for TAL until the end of February, 1960, when I was employed by Mandated Airlines to fly the Norsemans they had acquired from Gibbes Sepik Airways.

'Just after I joined MAL, the Norseman VH-GSD* crashed after take-off from Ialibu in the Southern Highlands. Engine failure. The pilot, John Sutcliffe, wasn't hurt. At the time Peter Manser was training me on the Norseman, and we were sent into Ialibu with a team of the old Gibbes engineers to see what could be salvaged. We pulled the aircraft back onto its wheels, removed the engine and the wings, cut the fuselage into sections, and carried the whole lot to Mount Hagen, where it was loaded onto a JU52 and flown to Goroka for a rebuild. The aeroplane was flying again in about four months. It weighed over 100 pounds [over forty-five kilograms] heavier, due I believe to the welding again together of the fuselage sections...'

MAL had bad luck with other ex-Gibbes Norsemans. The only fatal aeroplane accident in PNG during 1960 was of a Norseman, VH-GSC. The pilot, Desmond Gleeson, had been with MAL just four months. He was en route from Minj to Mendi with a load of freight and one native passenger. He crashed in bad weather into dense bush at a height of 2285 metres, about 1.5 kilometres from the Kuli mission airstrip. Although the crash occurred so close to Kuli, so rugged was the terrain that it was two days before GSC was found, with the bodies of Gleeson and his passenger still in the crumpled cockpit. The search was conducted by only three aircraft, a DC3 and two Cessnas (one piloted by Bob Gibbes), so restricted and dangerous was the mountain terrain. Forty people in several search parties covered 1036 square kilometres of mountains, looking for the Norseman. The cause of the accident was presumed to be pilot error; Gleeson had not transmitted any distress calls prior to the crash.

The JU52's did not long survive MAL ownership. On 17 October, 1959, Captain Dave Wills and co-pilot Mick McGee crashed at Baiyer River while making a downwind landing in VH-GSH (BUU). The Junkers was extensively damaged, and was left at Baiyer River, where its weathered bones remain to this day. GSW, the former BUW, which had crashed at Wau in March, 1957, and the last JU52, GSS (BUV) were withdrawn from service in 1960. It was no longer economical to keep such obsolete aeroplanes flying.

In December, 1959, Dennis Buchanan registered another C180A, VH-GKC. This aeroplane, too, was purchased second-hand in America, but it was shipped to Bankstown, Sydney, for assembly. It was ready for test-flying when Buchanan and Bill Sherwood arrived to take delivery. After the test flight, on 10 December, they flew GKC back to Goroka in two days, via Thursday Island, Daru and Port Moresby. GKC was unusual in being completely painted in a striking red and grey — most Cessnas were left in the natural shining metal, with details in colour.

Another pilot was required to fly the second 180, and this time Dennis Buchanan made sure he was small enough to leave a comfortable margin for cargo. Neville Hargreaves weighed about 64 kilograms and was in Bill Sherwood's

*BNE in GSA service. The GSA Norsemans acquired by MAL had their registrations changed to GSB (BHG), GSC (BLM), GSE (ASN) and GSF (BNL)

240. TAL Cessna 170, VH-GKB. *Photo: Bill Sherwood*
241. TAL Cessna 170, VH-GKC with, from left, Bill Sherwood, Neville Hargreaves and John Bayles. *Photo: Bill Sherwood*

240

241

words 'a real beaut bloke, one of Nature's gentlemen. He arrived with his wife, and Dennis sent him to Mendi. He got the ADO to build him a really good *pitpit* house, but they put it right at the top end of the Mendi strip. There was always a south-easterly breeze blowing up the strip, and every time an aircraft took off, the house was covered in dust. How his wife ever managed to get the washing clean and dry, I'll never know...'

Early operators of Cessna 180's had initial problems with the engines, including TAL. Dennis Buchanan recalls: 'The 180's were our backbone. They were the best type. The missions operated them, too. Father Joe, a good old friend, had his 180 and he helped us a lot, supported us, helped us with spare parts, anything technically wrong with our aircraft, he'd fly up...

'We had an engine blow up in the first 180, a few months after we got it. We didn't know those Continental engines, and we hadn't been running this one up properly. Heavy oil in Highlands conditions used to overload the oil pump, and it used to start growling the metal off the casing, and metal fragments would go through the filter. We didn't know where this metal was coming from, but it didn't look too good. Looked like we'd have to buy a new engine, and I didn't know where I was going to find the money, about three grand.

'I was a bit worried, so I called up Father Joe. He came up, had a look and said, "That's the oil pump. Nothing wrong. Take the pump off, and we'll have a look." Sure enough, the metal had been reamed out. We put in a new pump, he told us how to run the engine up properly, and that was the end of that.'

Bill Sherwood remembers another 180 engine problem, that nearly brought disaster to John Bayles: 'John came in to Goroka from his base at Hagen with GKB. He asked me to have a quick look at the engine, there was a little bit of oil coming out of the cowls. Continental engines are very oil-tight, so this was unusual.

'The 180 had an auto-drain oil system, a sump plug with a pipe coming out from beneath the aircraft. To make it easy to drain the oil out of the sump, you just undid a cock on the side of it, and the oil would run out. But unfortunately, because it weighed a bit more than just a straight sump-plug, it used to vibrate and crack the alloy sump pan. So we used to remove it, and have a straight sump-plug.

'When I took the bottom engine cowl off, the oil appeared to be coming from the sump plug. So I got a spanner and undid the lock wire. I barely put any pressure on the spanner, and the whole sump plug fell on the ground, and about one and a half gallons [about seven litres] of sump oil followed it. When I looked at the plug you could see a crack in the weld — one eighth of an inch [over three millimetres] of metal was all that was hanging on. I am in no doubt that in the next five minutes of flying, that plug would have fallen out. So John Bayles was lucky...'

Territory Airlines continued to prosper. Colin Hay was engaged as a pilot, and Dennis Buchanan decided to buy another aeroplane, this time a new one, a Cessna 180C, the latest model. It was purchased from Rex Aviation, Sydney, and was ferried to Goroka by Bobby Gibbes. Bill Sherwood remembers that delivery flight: 'You know how impatient Bobby is to get on with the job. He was coming from Thursday Island to Port Moresby, and suffered radio failure. DCA rules say that if you have radio failure, you go to the nearest aerodrome that can fix it up. The radio failure happened near Vaimuru, but Bob reckoned Goroka was the nearest place, so he came straight on. The authorities weren't too impressed, but by the time they sorted things out, Bob was up at his farm, Tremearne. GKD arrived on 21 May, 1960...

'Our pilots were all good, experienced, but our operations manual at the time didn't have too much laid down about power settings or anything like that, so when I became chief pilot by straight succession after Gerd Franck left, I attempted to get some uniformity. The "proof was in the pudding". In the time I was there we had no incidents, engine failures or suspected engine failures or anything.

'I left TAL on 16 June, 1960, and joined TAA. Like all pilots, I wanted to fly the big ones. Just after I left, Neale Williams joined TAL, another experienced pilot. I'd flown hard for Dennis, got on very well with him, and respected him immensely. Hard businessman, but a good one. Fair to his pilots. You really felt you were achieving something...'

Bill Sherwood joined Trans-Australia Airlines in PNG, not Australia. A momentous era in post-Second World War aviation in PNG had come to an end. Qantas was about to pull out. The aviation scene was about to change, drastically, as powerful rivals fought to fill the vacuum created by the withdrawal of the giant.

242

243

244

245

246

242. Lockheed Hudsons belonging to Adastra Aerial Surveys at Mascot. *Photo:* Pacific Islands Monthly
243. TAL Cessna 185 on Wonenara airstrip, 1960. *Photo: Jasper Maskelyne*
244. TAL Cessna 185, VH-GKE at Nomane — first landing, 1961. *Photo: Jasper Maskelyne*
245. Cessna 180. *Photo: Jasper Maskelyne*
246. Territory Airlines logo and agency sign.

24

6

The End Of The Beginning

Qantas Bows Out

Qantas left PNG by direction, not by choice. This is not the place for an examination of Australia's Two Airline Policy, which resulted in the emergence of the government-owned Trans-Australia Airlines and the privately-owned Ansett-ANA as the major domestic carriers within Australia. Suffice to say that the Two Airline Policy, born in November, 1951, continues to the present day. It has had a profound effect upon the aviation industry, in PNG as well as Australia, and upon Qantas. Qantas was the sole Australian international flag-carrier, and when the Commonwealth Government decided that the company must concentrate entirely on its overseas operations, withdrawal from PNG was an inescapable result.

The Two Airline Policy gave government protection to both TAA and Ansett-ANA, and ensured that each had comparable equipment, and an equitable share of the domestic market. It was probably inevitable, however, that the tough-minded, ambitious and able R.M. Ansett would generally make the running. As a private businessman, he had more freedom of action than did the TAA management, and he was relentless and effective in his playing of the deadly serious game of airline politics.

Ansett began to look at the possibility of competing with Qantas on the Australia-PNG route as early as 1956. The time appeared opportune. Qantas had been in PNG for a long time and had done an excellent job, but in late years had often been the target of criticism and complaint. Qantas was the leader, highly visible, and with the demise of Trans Oceanic Airways in 1952 had enjoyed a monopoly of the Australia-PNG route. Every time Qantas altered timetables or changed local arrangements, some person or organisation was sure to be adversely affected, and would publicly protest. People were appreciative when Qantas introduced Skymasters on the Bird of Paradise service, but by late 1956 these aeroplanes were aging, and the service was under great pressure. Extra Skymasters had on occasion to be provided to overcome passenger back- lags. From the end of 1955, Qantas promised to introduce the Super Constellation on the Bird of Paradise run, when Port Moresby and Lae aerodromes were capable of accommodating them, promises that were starting to wear thin. Internally, the withdawal of the Dragons, the Drover fiasco, and the on-again, off-again Catalina saga had won few friends for the company.

On 11 April, 1957, Qantas pilots on overseas services went on strike for improved pay and conditions, to give them parity with other international airline pilots. The strike was not called off until 20 April, when the pilots agreed to submit their claims to arbitration. Although Qantas internal flights in PNG were not affected, the Bird of Paradise service was.

Pacific air services in general were seriously disrupted by the pilots' strike. The public perception of the airline pilot as a superior being, above the concerns of ordinary mortals, was shattered, and the Qantas reputation suffered heavily. In PNG, the reaction was fierce. Hundreds of passengers and thousands of kilograms of freight and mail were delayed, and the influential Port Moresby Town Advisory Council angrily passed a motion calling on the Administration to invite another airline to extend its operations to PNG, and end the Qantas monopoly.

At this time, Australian National Airways was sending the occasional Bristol Freighter to PNG on special charter with cattle, and it was at first hoped that ANA might fit the bill, but in August, the company was taken over by Ansett Transport Industries.

Ansett had already submitted an application for a licence to run a service between Cairns and Port Moresby, relying on Section 92 of the Commonwealth Constitution, which laid down that there should be no interference with free trade

247. Qantas DC3, VH-EBU, lands at Kavieng, August, 1960. *Photo: Qantas*

between the States. The application was strongly supported by the Port Moresby TAC, and an Ansett director, Captain S.C. Middlemiss, promised that if it was successful, the company would place 'the most modern type of Convair' on the run. The Convair Metropolitan was capable of operating into the existing Jacksons strip, and would provide 'a better and speedier service' than that offered by the old Qantas Skymasters.

The Ansett application was generally supported in PNG. It was discussed in the Legislative Council, where the nominated member and influential businessman, R. Bunting, described the Qantas monopoly as 'a calamity of the first magnitude'. The member for New Guinea Mainland, George Whittaker, told Council that Qantas had lost its efficiency since coming under Commonwealth Government ownership.

The Ansett case appeared to be strengthened when the Minister for Territories, Paul Hasluck, was delayed at Cairns when the Qantas DC4 on which he was travelling developed a problem with the undercarriage. He subsequently told the Port Moresby Chamber of Commerce that the government was 'not opposed' to the introduction of a private airline on the PNG-Australia route. He was reported as saying that although DCA might be against it, no policy decision had been made by Cabinet. He went on: 'As a matter of fact, while I was waiting at Cairns I seriously considered a connection between the Territory and the privately operated services in Australia. This is something which no doubt would benefit the Territory...'

Despite this evidence of sympathy in high places, the Ansett application was refused. The Air Navigation Act was amended to defeat the Ansett reliance on Section 92. The takeover of ANA, however, had catapulted Ansett Airways from a distant third in the Australian domestic aviation stakes to the position of chief competitor to TAA, with an even larger aircraft fleet, and R.M. Ansett now had an even greater incentive to obtain new routes. Another application was lodged, and an Ansett spokesman made an appeal for public support in PNG. In short order, more than 2,000 signatures to a petition calling for a second air service were collected.

Qantas was under increasing pressure, but before a decision on the latest Ansett application was made, Prime Minister Menzies announced that PNG air services were under the 'complete review' referred to in an earlier chapter. In another surprise move, the aviation veteran, C.A. Butler, revealed that he had made a personal application for a licence to operate an Australia-PNG service using Vickers Viscount 810 turbo-prop airliners. Arthur Butler's company, Butler Air Transport, had recently been taken over by Ansett Transport Industries after a long and bloody fight.

Thus far, the question of Ansett competition on PNG internal routes had not been raised, but it was pretty obvious that the writing was on the wall. Qantas really was in an unfortunate position. A major upgrading of Jacksons Airport to allow the introduction of the Super Constellation — which would answer public criticisms — had been delayed because of government financing problems. A Qantas spokesman claimed that the delay had forced the company to continue operating the now uneconomic and unpopular DC4 Skymasters, and had cost £1 million in lost DC4 sales. Although the PNG Administration never wavered in its support of Qantas, it had been obliged to pass on recommendations for a second air service from three local powerful organisations. It was at this unhappy stage that Qantas was forced, for reasons already discussed, to announce the withdrawal of the Catalina services and the introduction of the replacement single-engined float Otter. There were howls of rage and dismay from all sections of the community in the districts affected, and a lot of public criticism of Qantas, most of it unfair. The editor of the *Rabaul Times* used the controversy as an opportunity to side-swipe Qantas, and support a local operator who had just started a tiny charter service in Rabaul, called Kenair.

Kenair was begun by Harold Kenny, former senior instructor with the Royal Aero Club of Queensland, in March, 1958, using a single Republic RC-3 Seebee four- place, high-wing, pusher-engined amphibian, VH-KNA, imported from America. The Seebee was an attractive little aeroplane with a respectable range — 805 kilometres — and a cruising speed of 153 kph, and Kenny held high hopes for the success of his venture. However, within two months the new service had been so poorly patronised that Kenny was considering a return to Australia. The *Rabaul*

248. Handover of Otter aircraft to Qantas at de Havilland factory, Bankstown, May, 1958. *Photo: Qantas*

248

Times editor called for a government subsidy to save Kenair. 'The presence of the Seebee at Rabaul for emergencies would far outweigh in value the amount of public money required,' said the editor. He went on: 'The Commonwealth has seen fit to subsidise the operation of the Qantas Catalina services because they bring communications to isolated stations which would be otherwise at a big disadvantage. To subsidise the present charter service at Rabaul would not, therefore, be a precedent...'

A Qantas spokesman was quick to point out that the company no longer received a subsidy for any of its PNG services, and in fact operated some 'at a heavy loss. The Catalina service subsidy was cut out five years ago...there is no subsidy even on mail, which is carried at normal freight rates...'

Kenny did not get a subsidy, and his service was short-lived. Qantas did open an office and branch at Rabaul as a concession to local feeling. Hitherto, Burns, Philp & Co. had been the Qantas agents in Rabaul.

Although the tide of public opinion was running against Qantas, the airline did not lack defenders. In the thrust of public debate, the services, usually uneconomic, that Qantas continued to provide to isolated outstations ignored by commercial companies were in danger of being overlooked. DCA stepped up efforts to complete the Jacksons Airport improvements, and in May, 1958, Vern Rule announced that the work would be completed by Christmas, in time for the annual invasion of schoolchildren returning home from Australia for the holidays. Super Constellations would be able to use the strip in January, 1959. He added that both TAA and Cathay Pacific had also lodged applications for licences to operate between Australia and PNG. A week later, Rule advised that the Australian Government had approved the purchase by Qantas of four Lockheed Electra turbo-props for the Bird of Paradise service, which would later replace the Super Constellations. Qantas was doing everything it could to fight off the challenges of its rivals.

There was another fuss, however, when Qantas announced that after the Super Constellation services began, its PNG headquarters would be moved from Lae to Port Moresby where the service would in future terminate. The proposed move was naturally welcomed in Port Moresby, but was badly received in Lae. The decision, though, appeared inevitable. The extraordinary delay in coming to a final decision on the future of Malahang had resulted in the comparative neglect of Lae, which DCA considered was incapable of accepting aeroplanes of the weight of the Super Constellation. The position was well summarised in *Pacific Islands Monthly*: 'When Super Constellations and later Electras begin using Port Moresby airstrip at the beginning of next year, it is inevitably going to mean a new centre of gravity for air matters in PNG...To date, the air centre has been Lae, but from next year it is going to be Moresby...'

Qantas had enjoyed a long period of accident-free operations on its internal services, but now, when the future of the airline in PNG seemed in doubt, came three in a row, which did not help matters.

Lae was closed to DC4 operations for six weeks from late March, 1958, to enable repairs to be made to the main runway. The Skymasters were diverted to Nadzab while the work was being done. Nadzab had been a colossal air base during the Pacific war, with 129 kilometres of surfaced strips, taxiways and parking areas. There were nine airstrips, the main one 2134 metres long and capable of taking any civil transport then in use. The strip had been kept in repair by DCA, and was the one used by the Skymasters. Passengers were flown to and from Lae by DC3's, which could still use the airstrip there.

On 28 March, Captain J.A. Bird came in to land at Nadzab in DC4, VH-EBN. Just before touchdown, the hydraulic system failed, and the main brakes and nosewheel steering ceased to function. Bird immediately applied the emergency brakes, and the Skymaster slewed off the runway into the *kunai*, and all tyres burst. The passengers were shaken but unhurt, and no other damage was sustained by the DC4. It was an emergency coolly handled by Captain Bird, but the incident was widely discussed in PNG.

The second accident was more embarrassing for Qantas, for it involved one of the new DHC-3 Otters. The first two Otters were handed over by de Havillands at Bankstown on 27 May, 1958. One, VH-EAW, was a float amphibian version, the other, EAX, a conventional landplane. Ken Montagu and Bill Nye went to Bankstown to be endorsed onto the new Otters, and fly them back to PNG.

The handover of the two gleaming new Otters was a gala occasion, before a crowd of some 500 people including representatives from newspapers, radio and the new-fangled television. As the cameras clicked, the de Havilland managing director, Lester Brain, handed over the logbooks to C.O. Turner, managing director of Qantas in a formal ceremony. Much was expected of the Otter, particularly the amphibian, about to attempt the impossible job of handling the work that had been the province of the Catalina.

Montagu and Nye flew the Otters to PNG. Montagu converted a number of pilots onto the new aeroplanes, and they went into regular service, the amphibian on the Port Moresby-Abau-Samarai run, and to Kerema, Kikori and Daru; EAX on Beaver routes, including Port Moresby-Tapini.

The third Otter, VH-EAY, arrived in July, and the fourth, EAZ, in August, 1958.

On 3 August, Ken Montagu took off from Jacksons in VH-EAY, bound for Tapini with two passengers. Tapini was one of the most difficult strips in PNG, with a steep ridge at one end, a fifty-five-metre sheer drop at the other, and mountains all around. Pilots joked that when coming in to land at Tapini, the best technique was to aim about a metre below the lip of the gorge, depending on the prevailing up-draught to lift the aeroplane onto the beginning of the strip. Tapini claimed a number of victims over the years, but few as experienced as Ken Montagu.

Just before touchdown, a down-draught caught the Otter. A Qantas spokesman later described what followed: 'She bumped hard, and the bump damaged her undercarriage. On the second bump, the undercarriage collapsed and the aircraft belly-landed. She slewed and skidded for about 150 yards [137 metres]. Damage was extensive, but we feel that the aircraft can be repaired...'

EAY was repaired, and was soon back into service. The incident was bad luck for Ken Montagu, whose long and successful career in PNG aviation was coming to an end. In May, 1959, he left the Territory to be converted onto the Lockheed Electra, and fly on the Qantas international routes.

A few weeks after the departure of Ken Montagu, the Qantas area manager in PNG, Captain Orm Denny, also left, on retirement. It was somehow symbolic that Denny should be leaving now, when Qantas itself was on the verge of retiring from the Territory, for he had been a pioneer New Guinea aviator, and an early Qantas pilot. He was probably the most respected figure in PNG aviation. Denny was replaced as area manager by Captain P.J. Miller, another RAAF-trained pilot. Miller had been one of the RAAF officers who joined Qantas during the war, to fly Lockheed Lodestars for the Americans on the Brisbane-Port Moresby courier service. His entire subsequent aviation career was spent with Qantas. Phil Miller came to PNG in February, 1956, as chief pilot, New Guinea Internal, and in June, 1958, was appointed assistant area manager.

In October, 1958, came the third accident to a Qantas aircraft that year, and again it involved an experienced pilot, John Simler. As previously recorded, Simler spent his first two years in PNG as a first officer on the Catalina and the float Beaver, and in 1957 was sent to Sydney to be re-familiarised on the DC3, which included a three-week attachment to TAA for refresher night flying. Returning to Lae, he was command-checked on the DC3 in January, 1958. Simler remembers: 'Having been used to sea-level operations on water, I now found myself on land operations, initially at low level coastal strips but soon followed by an introduction to strips at altitude, of limited length (unlike most water areas) and at various angles of slope. The only thing was that at least the land strips did not move up and down or sideways! — as did a lot of the sea areas.

'The DC3 was, like the Catalina, a very nice aircraft to fly, and I found the Highlands flying extremely interesting, but was necessarily cautious in this new environment. Qantas training was, and no doubt still is, of the highest standard, and no pilot was allowed to operate over any sector in PNG until he had proven to the training captains his thorough knowledge of the route to be flown, and the various escape routes in case he was caught out in bad weather...local knowledge, and a cautious attitude, were the only ways to go. No pressure was ever exerted by the company if one couldn't maintain schedule due to weather or airstrip unsuitability. The captain's decision was never queried...

'I remember the highlights of my DC3 days — my first sight of Mount Wilhelm covered in snow...a beautifully clear morning, and a snow-capped peak in the middle of the New Guinea Highlands...a clear day for my first run down to Honiara from Rabaul, over the areas of the historic Coral Sea battles...and the lowlights...

'On 18 October, I landed at Wau, with first officer Bill Taylor in the right- hand

249. Qantas Otter, VH-EAY, which crashed at Tapini. *Photo:* Pacific Islands Monthly

249

seat of VH-EAQ. There had been a particularly heavy night rainfall, which unfortunately had not been reported, and we slid off the top end of the airstrip whilst attempting a ground-loop in an effort to control the path of the aircraft on the wet surface. EAQ was badly damaged...'

Following an inspection by Qantas accident investigator, Bill Burns, Simler and Taylor were found to be at fault, which naturally upset them, given the state of Wau strip at the time. As Bill Taylor says: 'The main damage was to our pride. We came to grief, and John got demoted to first officer for six months, and I got a severe reprimand. Well, it's all water under the bridge now...'

Taylor was an old Qantas hand, an ex-RAAF navigator who trained in Fairey Battles and flew in Douglas C39 and C53 freighters, and Lockheed Lodestars, during the Pacific war. He spent a lot of time in the New Guinea theatre, and joined Qantas in December, 1945, as a radio operator. In June, 1954, he was re-mustered as a pilot, and a year later was posted to PNG, as a DC3 first officer. Later, he flew the Beaver and the Otter, and in mid-1960, got his DC3 command.

There was a catastrophic DC3 crash at Wau in October, 1960. An RAAF aeroplane with eleven men on board made too steep an approach to the famous old airstrip — dangerous if not treated warily, with its slope of one in twelve — and bounced heavily on touchdown. The pilot had never before landed at Wau, and he opened his throttles and attempted to fly around for another approach. Any commercial pilot could have told him that this was impossible: the DC3 simply did not have the power to outclimb the slope of the strip. The DC3 stalled, and crashed, narrowly missing houses at the top of the strip. It broke in half and in seconds was a sea of flame, molten duralium running down the strip in long, hissing silver streams, greasy black smoke staining the pale blue sky. Incredibly, the escape drill pounded into all RAAF aircrew from the time of their enlistment worked; all on board got out of the inferno, although two crewmen were badly burned.

There was a lot of criticism of RAAF flying methods in PNG following this crash. The fiasco at Finintegu in March, 1952 — previously described — was resurrected. One pilot said: 'It's all right for the Air Force to be so darned sure of itself as long as the people flying their aircraft are the only ones who get hurt. But this latest crash shows just how big a danger is created for civilians when Air Force pilots, who, while they are highly competent, are unfortunately unfamiliar with Territory flying conditions...what is deplored is Air Force policy which permits men to fly into airstrips with which they are not familiar...'

The RAAF DC3 was a total loss; the remains were bulldozed off the strip into the Little Wau Creek. John Simler's aeroplane, VH-EAQ, was however salvaged. The right undercarriage had collapsed and the wing damaged in the incident. In a remarkable feat of improvisation, reminiscent of the salvage of the GSA JU52, VH-BUW, Qantas engineers dismantled the DC3, sent off the massive wing sections by road, built a special free-wheel bogey, loaded the fuselage on board, manoeuvred the bogey around the hairpin bends in the Bulolo Gorge, and got everything to the Lae workshops where EAQ was rebuilt, and put back into the line.

Despite the uncertainty over its future in PNG, Qantas pressed on with the introduction of the Super Constellation. A new, temporary terminal was hastily constructed at Jacksons. The airstrip improvements were completed, just in time. The first Super Constellation — a superb aeroplane, the ultimate development of the world-girdling piston airliner — landed at Jacksons on 21 January, 1959. The Bird of Paradise service was now re-arranged. The Super Constellation could lift many more passengers than the old Skymaster, and the service was reduced from six and sometimes seven flights per week, to four. The move was not popular in PNG — it resulted in the provision of only a few more seats per week, and drastically reduced frequencies.

On board the inaugural flight of the Super Constellation was the Director-General, D.G. Anderson, and a party of DCA officials. In 1958, the Minister for Civil Aviation, Senator Shane Paltridge, had visited PNG to investigate the need for a second airline, following the Ansett and Butler applications. He made a statement that appeared to leave the question wide open: 'I cannot say whether Mr Ansett or Mr Butler, or both, will get a licence. It is possible that no new licences will be issued at all. It may be necessary to let Qantas carry on as the sole operator to New Guinea.'

In January, 1959, Paltridge authorised Qantas to operate a Sydney-Port Moresby-Manila-Far East service, thus making Jacksons a true international airport. The international service began on 2 March, and increased the number of weekly Super Constellation flights to Moresby from four to five per week.

Senator Paltridge had soon to make a recommendation to Cabinet on the PNG aviation issue. His recommendation would of course depend to a considerable extent on advice from his Department, and D.G. Anderson announced that the purpose of his visit was to 'find the needs of the public...as they are pioneering up here, the people, we believe, are entitled to the best service and deserve the best standards at the most reasonable cost. Who runs the service is of no concern to us. After my visit, my recommendations will go to the Minister, who must be the one to decide what services you will have.'

Anderson emphasised that the department would give 'no consideration to vested interests' in its decision on whether PNG would receive a second airline. The decision 'one way or another' would be made without consideration for companies already operating.

Anderson made a very significant statement on future aviation development in PNG during his visit. The basis of this development would be the provision of DC3 or replacement aircraft standard aerodromes. Experience had shown that it was becoming increasingly difficult to earn the revenue that the high capital investment in modern aircraft demanded. All development therefore had to be related to the traffic offering. In Australia, the policy was to provide aerodromes for the operation of aeroplanes to the size of the Super Constellation and in the near future, the Boeing 707 jet, into capital cities only. For intermediate and rural services, aerodromes of DC3 or replacement — Fokker Friendship, Handley Page Herald, Super Broussard — standard would be provided. He went on: 'Here, in PNG, it is even more emphatically clear that this policy must apply. Port Moresby alone of all the centres is capable of providing the traffic essential to the operation of heavy aircraft, and no other aerodrome in the Territory can be so developed.'

Anderson went on to warn that it would be a long time before the DC3 was replaced in PNG, and any proposed increase in DC3 numbers would be examined 'most carefully'. He paid tribute to the foresight of the PNG Administration in developing DC3-standard aerodromes, of which there were now forty-two. He predicted that a decision on the future pattern of air services would be made 'in about two months'.

Despite the earlier announcement by Qantas, and Anderson's comments, the headquarters of the company remained at Lae after the commencement of Super Constellation services. Passengers were transferred to and from Port Moresby by Skymaster. There was no escaping the fact that Lae could no longer be considered the aviation capital of PNG, and there was bitter comment from local civic bodies. Qantas got the blame for a situation beyond its control.

The decision predicted by Anderson was further delayed, and rumour and speculation on the position of Qantas continued to sweep PNG. During a debate on the Australian National Airlines Bill in Federal Parliament in April, 1959, Gough Whitlam defended Qantas in his usual forthright fashion, claiming that action was being taken to allow Ansett-ANA, and no other airline, to operate between Australia and PNG in competition with Qantas. He said: 'Will Ansett-ANA merely fly to Port Moresby, Lae and Rabaul, or will it also provide feeder services?...if it is fair for the company which has pioneered the services and developed the Territory to carry the burden of feeder routes, it is fair enough for its competitor also to play some part in developing the Territory...

'If Ansett-ANA were allowed into the Territory,' Whitlam continued, 'it should not be allowed to pick the eyes out of internal services, but should "take the bad with the good".'

Qantas carried on regardless. A new Super Constellation weekly service from Port Moresby to Tokyo via Manila and Hong Kong was introduced in August, 1959, and an additional weekly flight on the Bird of Paradise service in September. The DRUM column in the *South Pacific Post* predicted that Ansett-ANA would be in PNG 'in January. But don't bet on it. The Government don't know about it yet. Or rather, to be quite, quite perfectly factual, the Government hasn't told anyone about it yet.' In November, R.M. Ansett, in Sydney, said that his company was 'pressing' the Commonwealth Government for licences to operate services to PNG, and throughout the Territory's internal network. He continued: 'We would also agree to TAA being allowed to operate a similar service to ours in New Guinea, but we are against one company operating to the Territory without opposition.'

250. Ken Montagu. *Photo: Ken Montagu*

250

Following this statement, Reg Ansett sent his personal assistant and Ansett director, John West Lau, to PNG, to 'find out at first hand what the feeling is about a new airline to compete with Qantas Empire Airways'. John West Lau told the *South Pacific Post* that his company's chances of obtaining government approval were improving with each month. If successful, Ansett-ANA would employ Douglas DC6B airliners between Sydney and Port Moresby, and would also provide a service from Cairns.

Qantas doggedly went on developing its services, even at this late hour. On 8 December, 1959, Captain Ian Ralfe landed the Lockheed Electra, *Pacific Explorer*, at Jacksons on a proving flight. A new and highly-effective electric flare path had been hastily installed by DCA before the Electra's arrival, and approach lighting was provided soon afterwards. Ian Ralfe praised the new aids at Jacksons, which for the first time allowed full night flying in PNG.

1959 was almost gone, yet no decision on the future of Qantas had been announced. It created great difficulties for the local management of the company, as Phil Miller remembers: 'During the whole of my time in New Guinea, the threat of Qantas's withdrawal from the area hung over our heads. This created a situation where it was extremely difficult to obtain management approval to spend money, and to do many other things which would have been straightforward if we had had an unclouded future ahead of us. This attitude by top management was fully understandable, in the circumstances...

'There was a continuing debate, too, as to whether Qantas headquarters should be based in Lae or Port Moresby. There were sound arguments in favour of both, but the operational aspects were considered paramount, Lae was the centre of operations, so we stayed there. Bob Weekes was Area Traffic Superintendent in these final days, Bert Gould was Area Accountant, and Jack Prince, Area Engineer. Ralph Perrier was our manager in Port Moresby, Allan Wallace was at Madang, and Pat Henry was our man at Rabaul...'

The year ended with a scathing attack on Senator Paltridge by the Deputy Leader of the Federal Opposition, Arthur Calwell, no friend of R.M. Ansett. In a radio broadcast, Calwell said that Ansett had demanded 'with truculent demeanour' to be allowed to fly to PNG. He added: 'The Minister for Civil Aviation, who can be very tough and domineering when challenged about his failure to protect the nation's interest against Mr Ansett, will ultimately surrender to yet another Ansett demand. Senator Paltridge is certainly more the "Minister for Ansett" than he is Minister for Civil Aviation.'

That Ansett-ANA would win out seemed to be a foregone conclusion. For the first time, Territorians read in their newspapers of what eventually did happen. The issue of *South Pacific Post* for 2 February, 1960, had the front page headline: 'Qantas May Quit the Territory'. Qantas would go, and Ansett-ANA and TAA would take over air services to, and within, PNG: 'The Federal Government is anxious to take Qantas out of the Territory as soon as possible, as this would give Qantas an opportunity to standardise its fleet on purely international lines.'

Now, when Qantas looked set to go, there was a belated recognition of what the company had achieved in PNG, as Territorians began to contemplate a future without Qantas. An editorial in the *New Guinea Times Courier* on 3 February exactly expressed the fears of many: 'Is it true that Qantas will be completely removed from New Guinea where it has performed a sterling service for fifteen years?

'What will happen to the essential non-profitable low density traffic routes in New Guinea if TAA or Ansett-ANA or both take over from Qantas?

'Qantas in recent years has come under fire from the Press — but only in the same sense as the Press has criticised stores which over-charge, or public bodies that commit some flagrant breaches. In no wise would this criticism infer that Qantas should quit.

'In fact, the *Times Courier* now places it on record that it is alarmed that Qantas may be completely removed from New Guinea at this critical stage of the Territory's development.

'For more than twelve months the whispering speculation about the future of the New Guinea air trade has gone on beneath the inscrutable facades of Canberra and aviation circles...the Government takes the view that whoever operates the paying routes of New Guinea must make it an obligation to take over internal operations as well.

'But would a private enterprise airline be capable of operating the non- profitable routes to the same extent as a Government-backed concern which sees internal air development as a project of major importance? Why should it? — how could it? — take the responsibility of being an all-in-one road, rail and air benefactor for a country which it owes precisely nothing? The best it can do is 'rationalise' internal air services to the point where they are economic for the whole business?'

Only time would demonstrate whether the fears expressed in the editorial were well grounded. For Qantas was indeed to withdraw. On 12 April, 1960, it was announced in Canberra that permission had been given to both TAA and Ansett-ANA to operate competitive air services between Australia and PNG — effectively, an extension of Australia's Two Airline Policy to the Territory. Cabinet also gave approval for TAA to take over all internal PNG services operated by Qantas. TAA would buy Qantas PNG assets necessary to operate these services, as well as the regional services to Hollandia and the Solomon Islands.

Said *South Pacific Post*: 'After five years of procrastination, the Government has decided who will operate Australian-Territory air services. Qantas will be replaced on the Moresby run by Ansett-ANA and TAA, and internally by TAA.

'It is impossible to assess the extent to which aviation, and the development that follows improved aviation services, have been retarded in this country by the Government's dilly-dallying...

'Qantas must be congratulated for its tremendous contributions to post-war Territory development, and most people will farewell the airline with sincere regret. But now we are on the threshold of a new aviation era...'

In his book, *Wings to the World*, the late Sir Hudson Fysh made it clear that Qantas regretted having to leave PNG. He wrote: 'We maintained that under our original charter from the Commonwealth Government these services were our right as overseas services.

'Papua was an Australian Territory, but New Guinea was under international mandate, and of course we pointed to the international aspect of our services to Hollandia and the Solomons.

'During the latter1950's, when Senator Paltridge was Minister for Civil Aviation and D.G. Anderson was Director-General, pressure was exerted by Ansett-ANA to enter the New Guinea area. That company had great hopes of extending overseas from Port Moresby to Hong Kong, as they had inherited the ANA shareholding in Cathay Pacific Airways, centred in Hong Kong.

'New Guinea had become a magnificent training ground for our young pilots before they entered on our main international routes, and we fought hard against losing the New Guinea services...

'It seemed, as in other instances, that the Government was determined to do something for Ansett, a company full of vigor, and bent on expansion.

'However, there was the Government's Two Airline Policy to be reckoned with. Ansett-ANA could not go to New Guinea unless TAA did the same. This was evident. It ended by our losing the battle, and the Government deciding to relieve us of our services in New Guinea and the Australian Territories.

'We were required to assess the value of our undertaking and to sell to TAA, which would then operate to New Guinea in conjunction with Ansett-ANA, and operate internally. We had pleasant negotiations with Sir Giles Chippendall, Chairman of TAA, and John Ryland, his general manager. We agreed on a price of £700,000, and assisted in the transfer of our New Guinea staff to TAA...'

There has been constant speculation over the years in aviation circles on the true reasons for the support given by the Liberal-Country Party Australian Government to R.M. Ansett and his company. It is unlikely that any public statement will ever be made. A favourite theory is that Ansett actually did the government a favour by taking over the private enterprise ANA — with TAA one of the two original partners in the Two Airline Policy — when it was on the point of collapse following the death of its founder, Sir Ivan Holyman, in January, 1957. One good turn deserves another. Politics played a major part. Liberal-Country Party governments in Australia have always supported private enterprise; Labor governments traditionally favour public ownership of such undertakings as airlines, as an article of faith.

The Qantas investment in PNG had grown to almost £1 million at the time of the withdrawal, including eight DC3's and four DHC-3 Otters, workshops, hangars, staff and passenger accommodation and other assets. Staff numbered 620, and the route network included forty-three ports of call over some 10,461 route kilometres. In the last full year of operation, Qantas flew 2,056,813 kilometres, and carried 47,400 passengers within PNG. The Qantas act would be a hard one to follow.

A great deal of preparatory work had to be done before the transfer of Qantas operations within PNG could be effected. The TAA Director of Operations, Captain R.E. Bailey, visited PNG with three other company officers in May to study the situation, and make recommendations to his management. He announced that seven TAA pilots would soon be coming to PNG to serve with Qantas crews in the interim period before the takeover, for route familiarisation and training. TAA would set dates for the commencement of internal services 'soon', said Captain Bailey, but they would be unlikely to begin until after the Australia-PNG service started operating.

There was another visitor to PNG in May, whose arrival attracted little publicity but whose presence was an omen for the future.

During the Qantas era in PNG, relations between pilots and airline managements had been mostly cordial, based on mutually-agreed contracts and conditions of employment, and little affected by formal negotiation. Disagreements were usually settled amicably, and without recourse to processes of law.

There was, though, a basic difference between Qantas pilots and those of the resident companies like Mandated Airlines, Territory Airlines and Papuan Air Transport. Most Qantas pilots served for one or perhaps two terms in PNG, and then returned to Australia to continue their careers with the company on the international routes. There were exceptions, as this account has shown, but in general Qantas pilots spent only a few years in the Territory. Qantas management regarded PNG as a grand training ground for the young pilot, but PNG was, after all, but a small part of the Qantas empire. Pilots serving in PNG worked under plainly defined, and accepted, conditions.

The resident companies operated only within PNG. There was no career structure available outside PNG, and some Territory pilots spent most of their working lives in the country. Such men developed an intimate knowledge of PNG, and some looked upon the junior Qantas pilot, in the Territory primarily for experience, with tolerance at best. A good many belonged to pilots' associations, but industrial action by pilots within PNG to improve pay and conditions was unheard-of in the early days.

The position was about to change. Pilot unionism was alive and well in Australia, and now the Australian domestic airlines were about to enter PNG. In May, 1960, the manager of the Australian Federation of Air Pilots, B.I. Croft, visited PNG to investigate the conditions under which pilots worked. He was accompanied by Captain W.R. Forster, a senior TAA pilot, who was chairman of the committee which investigated general standards. Croft announced to the Press: 'We are primarily interested in finding out what conditions TAA pilots can expect to find when they start their Territory training in about three weeks. We want to decide what sort of contract they should have on internal routes, as conditions here are quite different to those in Australia. We also want to check the operational standards of aircraft being used...the single- engined aircraft, which is widely used, does not have the same safety factor as twin-engined aircraft...'

It was fair warning that the carefree days of old were no more. The pilots' association was to be vigilant in protecting the rights of members, and another factor was introduced into the aviation industry in PNG.

The arrangements announced by the Australian Government appeared to leave Ansett-ANA out in the cold insofar as PNG internal services were concerned. TAA and Ansett-ANA were to have equal rights on the Australia-PNG run, but Ansett would have no internal route structure, whereas TAA would inherit the very comprehensive network of services built up over the years by Qantas.

This would leave Ansett-ANA at a very real disadvantage. It had been decided that both airlines would use Douglas DC6B piston-engined airliners on the Australia-PNG route, pending the introduction of more modern aircraft. This was primarily due to the Two Airline Policy, which required both TAA and Ansett-ANA aircraft fleets be comparable, and resulted in the Australian Government imposing a cross-charter agreement by which two of the four Ansett-ANA DC6B's were exchanged for three TAA Vickers Viscount 700's, short-range airliners not suitable for the Australia-PNG run. But Lae airstrip was considered by DCA to be unsuitable for DC6B operations, and the service would therefore terminate at Port Moresby. This would mean that the large number of Ansett passengers who would want to travel on to Lae, Rabaul, Madang and other centres would have to transfer from Ansett to TAA when they reached Port Moresby — a situation which a man like Reg Ansett was hardly likely to tolerate.

The obvious way out was for Ansett to purchase an existing Territory company with an established route network. Only MAL filled the bill.

R.M. Ansett visited PNG in May, 1960, with a party of Ansett Transport Industry executives. Just before his arrival, the *Australian Financial Review* had published an article, claiming that Ansett was in the process of negotiating a merger with Mandated Airlines for a figure of £500,000, in ATI shares. The *Review* claimed that this would make MAL the largest holder of ATI shares, which was unacceptable to R.M. Ansett.

In Port Moresby, Ansett denied the *Financial Review* report. He admitted that his company had had 'tentative discussions' with MAL twelve months previously, but that was all. His assistant, John West Lau, said that the Ansett party was in PNG purely on a survey trip, to acquaint themselves with operating conditions — R.M. Ansett had last been in the Territory in 1941. When asked whether Ansett was considering buying any private operator other than MAL, West Lau said that this had not been considered. 'We are very good friends with MAL, and we hope that we will work very closely with them,' he added.

Already battle-lines were being drawn. R.M. Ansett said in Lae that it had never been intended to terminate the Ansett-ANA service in Port Moresby. 'We have an immediate licence to fly to Lae, Rabaul, Madang and Goroka...if we get permission to use Lae, our aircraft will operate out of that port.' He went on to say that his company would use DC3's on PNG internal routes for the first eighteen months, but these would then be replaced by the new French Super Broussard twenty-passenger turbo-prop aircraft, six of which were on order.

TAA was swift to answer Ansett's claims. The company said that only TAA had the franchise to operate internal routes in PNG, and the Minister for Civil Aviation, Senator Paltridge, confirmed this, adding that Ansett had made no application to him for permission to buy Super Broussards. And to show Territorians that TAA, too, cared, the company announced that a fast turbo-prop Fokker Friendship would be sent to PNG on a proving flight to major airports. If successful, the Friendship would possibly be used on some internal services. Unfortunately, the tragic crash in Australia of the TAA Friendship, *Abel Tasman*, early in June with the loss of twenty-nine lives caused the proving flight to be cancelled, but chairman, Sir Giles Chippendall, said that 'in no way' had the confidence of TAA in the Friendship been shaken.

R.M. Ansett decided to settle the matter of the suitability of Lae for the DC6B once and for all. With the approval of DCA, he sent a DC6B to PNG on a proving flight. On board was Director of Flying Operations, James Stone, and other DCA officials. The big Douglas landed without incident at Lae, and Stone subsequently agreed that the tests had been 'most favourable'. Subsequently, both Ansett-ANA and TAA were given approval to take their DC6B's to Lae.

Qantas was to cease operating its Australia-PNG Bird of Paradise service at the beginning of July, 1960. At the last possible moment, the right to carry passengers from Port Moresby to Sydney and return on the Far East Electra service was withdrawn from Qantas, probably because the authorities felt the magnificent turbo-prop Electra would provide unfair competition for the aging DC6B's about to be introduced by TAA and Ansett-ANA.

The last Qantas Bird of Paradise flight out of Port Moresby was made on 8 July, 1960. The captain of the Super Constellation was J.G. Morton, who had been a member of the crew of the DC3, VH-AFA, which had inaugurated the service on 2 April, 1945. TAA made the first flight in a DC6B from Sydney to Port Moresby on 9 July, and the Ansett-ANA service began two days later.

The last Qantas Skymaster flight from Sydney to Lae via Port Moresby was made on 8 July. The pilot was Captain J.R. Christensen, who had flown the inaugural DC4 flight in 1950.

251. Interior of a DC3 carrying passengers from Madang and Goroka to Port Moresby. *Photo:* Pacific Islands Monthly
252. Welcoming party greets Dutch officials on the inaugural flight of the Kroonduif Dakota into Lae, July, 1959. District Officer, Horrie Niall, is fifth from left (hat in hand) with the author, James Sinclair, to the right of him in white socks. *Photo:* Pacific Islands Monthly
253. Qantas Otter floatplane, 1958. *Photo: Qantas*
254. Qantas Lockheed Electra, VH-ECA. *Photo: Qantas*
255. VH-EAY, the first Otter to land at Karimui, 1960. *Photo: Qantas*
256. Qantas DC3, VH-EBU, flying in the Highlands, 1960. *Photo: Qantas*
257. Qantas Otter at Bulolo airstrip. *Photo: Qantas*
258. Flight deck of a Qantas DC4. *Photo: Qantas*

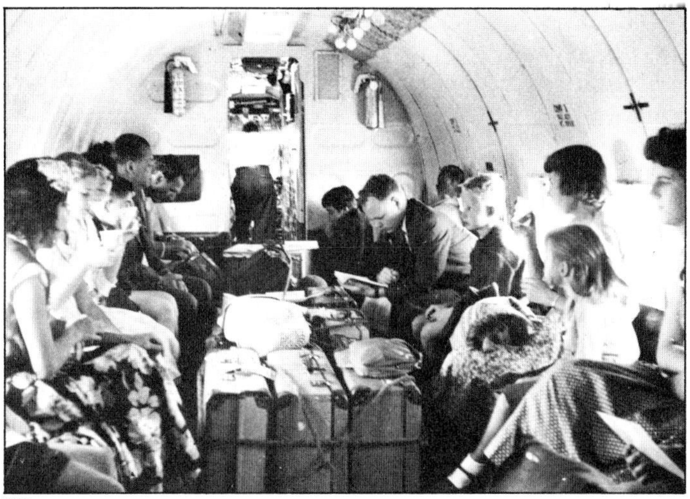
251

252

253

254

255

256

257

258

The Good Old Days

The swan song of the Skymaster seemed a fitting epitaph for Qantas to the large crowd who watched the departure from the historic Lae airport, where aviation in the Territory began. The roar of the big Pratt and Whitneys faded, and the long silver shape of the graceful old airliner slowly vanished into the blue over the Huon Gulf. Qantas was gone, but the memories remained.

Ask any group of veteran Qantas men what were the most rewarding years of their aviation careers, and most of them will answer: 'Our time in New Guinea.' Lionel Purkiss summed it up during an interview with the author in August, 1983: 'They were good days, great days. Probably the happiest days of our lives...'

Purkiss retired from Qantas as a Boeing 747 captain. He recalls: 'I was telling the second officer one day just before I retired that I had flown Fox Moths for Qantas commercially, and Jumbos for Qantas commercially. He looked at me, and said, "Gawd, did you know the Wright brothers very well?" '

Whenever Qantas pilots gather, they talk of airstrips, incidents and aeroplanes. Stories about Bill Forgan-Smith inevitably surface. Mal Shannon remembers being checked out by Forgan-Smith on a Dragon, soon after his arrival: 'A great gentleman, 'Forgie'. Funny man, very dry. Didn't actually inspire me with confidence the first day, because there's no dual control in a Dragon and he lifted up the leg of his pants and shows me a great scar on his shin, and said, "Don't run me off the bloody strip like the last bloke did — look what he did to me!" I'll never forget his advice: "Two things you don't do up here, son, don't fly in cloud, and don't forget to take your malaria pills." I've always remembered that...'

Forgan-Smith was a hard taskmaster, with old-fashioned ideas about duty and service. When chief pilot of New Guinea Internal, he constantly urged his pilots to keep flying, and was notoriously loath to grant time off. Roger Wilson recalls: 'It was Lionel Purkiss who finally got the better of 'Forgie'. Whenever someone asked 'Forgie' for a day off, he'd drive them into the ground like a tack. You were limited to 1,000 hours' flying a year, then, by DCA. Theoretically, by flying six or seven days a week for ten or eleven months, you could build up 1,000 hours, and you were not allowed to extend this. Lionel was based at Madang, and he was getting close to 1,000 hours and he said to 'Forgie', 'I'm getting close, you'll have to watch it'...'Forgie' just said, "You go ahead and fly, we'll take care of it..." '

'So Lionel continued to fly hard, and one day he went in to 'Forgie', put his logbook on the desk, and said, "I've now flown 1,000 hours in ten months. I'm going to take eight weeks off, and there's nothing you can do about it..."

' 'Forgie' was livid, but it was his own fault. He had to give Lionel the time off...'

Bill Forgan-Smith was a pilot of the old school. As noted earlier in this account, he learned to fly before the war, and was of a generation where pilots did not usually fly on instruments — 'on the gauges'. He was endorsed onto an astounding variety of aeroplanes during his long aviation career, including many de Havilland types: the DH50, DH60 Moth, DH 61 Giant Moth, DH75 Hornet Moth, DH82 Tiger Moth, DH83 Fox Moth, DH 84 Dragon, DH85 Leopard Moth, DH 86 four-engined Express airliner, DH 89 Rapide, DH90 Dragonfly, DH94 Moth Minor. He flew the Klemm Swallow, the Avro Avian, the Fairchild 24 and the Lockheed 10 and 14 — all before coming to PNG. He loved the old aeroplanes and the old flying ways. He retired as a pilot at age 60, with over 25,000 hours in his logbook, and still says today: 'The fun went out of flying with the biplane'.

As related earlier in this book, Forgan-Smith returned to Australia in 1953. He flew DC4's and Super Constellations on Qantas overseas routes, but he was too confirmed in his ways to be trained as a jet pilot. He was the operations manager at the old Qantas office at Breakfast Creek, Brisbane, when he retired. To this day,

'Forgie' says that Qantas is 'the best airline in the world, bar none', but he was treated less than generously when he retired. 'Forgie ' remembers: 'It was a bloody tough deal with Qantas — if you didn't make it, you didn't get it, no matter who your mates were. Qantas was going into jets, and I was too old. At that time it cost £16,000 plus a year of time to qualify onto jets. So I was told I'd better look around for a new job, and retire early if I found one. So I did. I got out in 1961, and joined Patair.

'Qantas didn't look after me as much as they should have, I reckon. I got what I was entitled to, but nothing like the pensions they get today. I got £9,000 — fair enough, you couldn't complain, but they had a surplus of funds in the superannuation scheme, and they said they'd look after the old blokes, but really they didn't. Still, when I joined, the retiring age was 45, and the pension was £1,500. I joined on £480 a year. Oh well, that's it, but it was a good life. You weren't worried about money, only about flying...'

Forgan-Smith is still proud of the Qantas record in PNG when he was chief pilot: 'Qantas was first in, all over the place. Gerry Toogood, the ADO at Kainantu, told me he had got the strip good enough for DC3 "ops". I said, "No way — I was in there with a Dragon the other day, and it's too soft." 'I'm the strip authority,' says Gerry, 'and it's ready.' "It's too soft," I told him, "but I'm going up to Goroka with a 'Doug' tomorrow, and I'll drop in on the way back."

'We had an empty 'Doug,' and Bruce Fawcett, the first officer, and I lobbed into Kainantu, and went up to our ears in mud. We had to use 'rice-power' to lift it out. When the dry season came, Gerry put in some herring-bone drains, and we were able to operate into Kainantu then...

'Same with Minj. Too soft — built on a swamp. Nondugl, too — Qantas was the first into all these places...'

Lionel Purkiss, who went to Lae with Bill Forgan-Smith in June, 1948, to start New Guinea Internal, is grateful to another pre-war New Guinea pilot for good advice when he first started flying in the Territory: 'I was advised very wisely very early in the piece, on my first trip. I flew with old Fred Bryce. We were flying out of Lae, and Fred handed me a map. He said, "What's that place down there?" I looked at the map, and I looked at the place, and I took my courage in my hand and said, 'It's not on the map.' He said, "Good. It's not on the map. That's the old Black Cat mine. I'll tell you something about these maps. The only use they are is to tell you the place names, and some of them are not on the map, as you've just seen. Ignore the spot heights, most of them are way out, the best thing you can do with that map is throw it away. Spend your time looking at the country..." '

Pilots new to PNG in the early post-war years very quickly came to realise that the maps then available were often highly inaccurate. The heights of mountain ranges and individual peaks were often hundreds of metres out. It was many years before better maps became available, and even then the experienced pilot used them with caution.

Roger Wilson tells a story that illustrates the attitude of the old hand towards maps: 'When Phil Miller first went to New Guinea, Jack Stammer was his check captain. Phil was a very experienced pilot, but he didn't know New Guinea. They took off from Moresby for Lae on the direct track, and Jack noticed that Phil had this map. He said, "What are you doing?" Phil said, 'I'm map-reading my way from Port Moresby to Lae.' "Is that so?" said Jack. He opened his window and leaned over and grabbed the map, and threw it out. He said, "This is the way you go, don't worry about a map..."

'Local knowledge is everything in New Guinea. You could get caught, though. Sometimes you had to assume that a cloud level would hold, and not get any lower. Once you were in it, you were committed. Weather reporting could be crook, too. One afternoon, I took a load of golfers from Moresby to Madang. By the time I got to Madang and got rid of this mob and refuelled to fly back to Moresby, the usual afternoon storms had stacked in along the range. Only the two of us and the purser, we were empty, but I had to go to 22,000 feet [6705 metres] and even then it wasn't any help, but at least I knew I was well above any of the mountains. Finally, I went IFR. This was where you had to break the regulations, either do that or go to Lae, or back to Madang.

'We sucked all the little oxygen bottles we had in the DC3 in those days, and they weren't much good to us. When I knew I had gone far south enough, I let down into it, and went along the south coast. I called up the Moresby bloke and said,

"What's the weather like?" 'Visibility forty miles [sixty-four kilometres], fine and beaut.' I said, "Terrific..." We were getting a pounding, all the afternoon storms had drifted down to the coast. We were at 20,000 feet [6096 metres] with 110 miles [177 kilometres] to go, and I was letting down slowly, but we were getting a real pasting. Raining like you wouldn't believe...

'So I got down to about 100 DME and asked for the weather again...still OK...sixty miles [ninety-six kilometres], same weather...I got to forty miles, and it was just as bad as it had been back at 100. Hailstones and Christ knows what. We went on and on...I broke out at Galley Reach, and it was OK there, twenty miles [thirty-two kilometres] from Moresby. I said to the tower, "You were giving me forty miles visibility, I only broke out at Galley Reach, how come?" And he said, 'Oh, I was looking down towards Cairns!'

'There was nothing to replace local knowledge in New Guinea. Nothing at all. I can remember flying with 'Two Heads' Taylor down from Madang to Lae one morning in the wet season, really claggy. He was checking me, and if I'd not known exactly where I was...we could have been anywhere...you'd look at a river and say, "Yeah, Kaiapit is just about there," or, "This is the Asaloka Gap." You had to know where you were — no question about it.'

Despite enormous local knowledge a pilot could still come to grief, through over-confidence. John Simler recalls: 'I used to do Moresby-Goroka-Madang on scheduled services. One day I was taking off from Madang, and a pilot had taken off ahead of me in a little Cessna. I dropped everything, flaps...tucked my wing right behind him, and he's sitting there, reading a book. I said on VHF, "Hope that's a good book, Father." It was Father Joe. You could always recognise Father Joe in the bar, he had a green tongue — used to drink creme- de-menthe. I used to call him 'Father Christmas'. He asked me why, and I said, "You're always hanging things on trees, like aeroplanes..." '

Father Joe Walachy killed himself in his Cessna in 1965, hanging up his last aircraft.

Some funny things happened in the early days. Len McNeil remembers one incident in a Fox Moth: 'I got back to Lae from somewhere in a DC3. It was early afternoon and I thought, that's it for the day. Girl in the office, Pat Murdoch, had flaming red hair, said, "Oh Len, would you do a run to Finschhafen in the Fox Moth. There's a bloke here who wants to go over there in a hurry." I said, 'I haven't flown it for a while, is it serviceable?' She said, "Yes, all set to go."

'So I signed the load sheet, didn't look at it, just signed it, and away I went. Half way across to Finschhafen this bloke started putting on an act, screaming and yelling, carrying on like a mad thing. There was a little hole where you could pass a note back to the pilot — the pilot was out in the fresh air, and the passenger was in the cabin — and he passes a note saying, "Where the hell are we going? I want to go to Bulolo!" So I went back to Lae, and this girl Pat saw the thing coming, and I yelled, 'Where is she!' and she'd gone. So then I looked at the load sheet, and it's got 'Bulolo' on it. I put in a bit more gas, and took him up to Bulolo. I thought he'd gone stark, raving mad. Pat had said Finschhafen when she meant Bulolo...'

Bill Forgan-Smith recalls the time when the Qantas office girls at Lae were quartered at the old Hotel Cecil, of infamous memory, and the single men at the Mess: 'We were getting good meals, Qantas provided well for us, flew up fresh stuff from South, fruit, vegetables and so on. At the pub, the girls had to live on tinned stuff and they started breaking out in sores. One of the girls wrote home to her parents, said all the girls were breaking out in sores, but nothing to worry about — it was just V.D. Weren't the parents upset! She told us, "They're saying, 'what sort of a girl am I, what sort of a life am I leading, what am I doing in New Guinea!' I'd better write back and tell them it's Vitamin Deficiency!"

'Qantas decided they'd better have the girls up at the Mess after that fuss. Built an extra hut, washroom and all. The girls all had a good laugh over the 'VD' thing, and Pat Murdoch said to the boys one night, "Come on down"...and they'd put a sign on the wall — DC3 rates — £1 2s 6d a minute! That was the DC3 charter rate, £63 an hour...'

The Dragon was a favourite subject at pilot gatherings. Jack Stammer remembers an incident when he was flying for MAL, before joining Qantas: 'One day I was landing at Yangoru. Very short strip, and you attempted to land very short. I got the main wheels onto the strip OK, but the tailwheel hit the bank of the river at the end of the strip, and ripped the stern-post out of the Dragon. With no radio, there was no way to let anyone know what had happened. The patrol officer there didn't even have a radio then, but he had a few tools.

'So we got a *sarif*, and tied that round the sternpost. The top longerons were unbroken, and we drilled a quarter-inch [about six millimetres] hole in the *sarif* at each end, wrapped it round the back and took it forward to where the bottom longerons were broken, drilled another quarter-inch hole, and tied the thing with a couple of quarter-inch bolts he had. I flew the aircraft back to Wewak where a proper repair was carried out. In those days, we could get away with that sort of thing. Couldn't do it now...'

The Dragon was flown by a single pilot in a cramped cockpit. There were no dual controls or auto-pilot, which created problems when the pilot had to leave the seat in flight for any reason. Ross Biddulph, now living in England, did not enjoy his flying years in PNG: in a letter to the author, he wrote, 'It was a tiresome and boring existence, punctuated with occasional moments of intense fear — like all flying. And it was bloody hot in Lae...'

Roger Wilson tells of a time that probably induced one of those moments of fear in Ross Biddulph: 'The Dragon had a fairly critical centre of gravity. You couldn't load much either side, and if you did you had to be careful what you were doing. Ross was then a keen cigarette smoker. When he got in the aeroplane this day, he hung his jacket in the back of the cabin, which wasn't very long, eight or ten feet [about two-and-a-half or three metres]. He took off, and soon was dying for a smoke, and his cigarettes were in his coat pocket. So he got out of the seat and took one step back over the partition divider, and the nose came up. He went forward and trimmed it, then took another step back, and was in balance. Then Ross took another step back, and the nose came up again, and he kept on with this juggling business, trimming backwards and forwards, and finally got to the point where he couldn't do anything more, so he rushed back to the tail of the aeroplane and grabbed his cigarettes, and as he did so the nose came up and the aeroplane stalled...at least he got his cigarettes. Had no passengers of course, and he was able to get back to the seat and recover. It had him really worried for a time...

'Most of us had prangs in Dragons — I think the only one of us in those 1949-50 years who didn't was Ron Ballard. I had mine at Gusap. Ross Biddulph had one at Chimbu. He got it started, had too much throttle cracked, and the brakes were only air-brakes in the DH84, not much good. He began to run forward, and before he could get into the cockpit, it hit a pawpaw tree and smashed the prop. No radio, of course. Once you left Lae you were on your own, nobody knew where you were.

'I had a couple of engine failures in the Dragon. One gave up once on a take-off from Chimbu. Strip was very short then, and you dropped off one end towards the river. Because of the loss of power when I was running down the strip there was no way you could pull up of course, so I fell off the end of the strip. The port engine was only giving a certain percentage of power. I managed to climb up on what was left, enough to get me to Kerowagi, and there I sat...

'We used to fly the DH84 from Lae to Brisbane for C of A. That was a horrendous trip. Go over to Kerema, then Daru, then Horn Island where you checked with Customs and Immigration, then work your way down the coast...

'It was a good holiday, though, because it took about ten days to do a C of A. I remember one time on the way back I left Cairns and was heading for Cooktown and the weather was bad, and started to force me inland. I finally picked up the overland telegraph line, and I landed at Coen, among the ant-hills. I sent 'Forgie' a signal from there, and he said, "What the hell are you doing there, get back here!"

'I did the first Qantas landing at Mendi in a DH84. Some young PO came up to me at Hagen and asked to be taken to Mendi. I said, "OK, where is it?" He said, 'I've never been south of Hagen...'

'So we took off from Hagen, turned left, and went through the Gap. I said, "Where do we go now?" 'Over here, I think'...we went over there for a while, then he said, 'No, I think it's over here.' This went on for twenty minutes...he began to recognise the country, and we found Mendi.

'You should have seen the bloody place. A brand new strip, and it was like a ploughed field. I swear to God. One grass *kunai* hut at the top of the strip, and that was all it was. That was all Mendi consisted of. The strip pointed towards big cliffs, on the other side of the Mendi River. We got in OK, and I left them to it, and took off down this bloody ploughed field. Poor old Dragon. I got off, and I had nowhere to go. Everywhere was up. Interesting...'

Len McNeil had a different kind of C of G problem in a Dragon: 'I was flying old Jerry Pentland somewhere in a Dragon, with a sow and about eleven little ones. During the take-off, with the different forces all the pigs were pushed down the back. We staggered into the air, tail-down. Once we got going, I said to Jerry, "For Christ's sake, get her forward, I can't fly this thing tail- heavy like this all the way." He got her about half way along, but a bloody pig has a mind of its own, and once he started her off she went right up the front, and all the little ones followed her. Then it was nose down all the way...'

Animals were always a problem on aircraft. Says Dennis O'Malley: 'My first dead passenger was a goat. I was ferrying it in a Dragon from Nondugl to Baiyer River, where the Department of Agriculture had a farm. He was a billy-goat, and we lashed him down to a space between the seats. He bleated alarmingly all the way up the valley to Baiyer River, and suddenly he was quiet — permanently quiet. He'd had a heart attack. Died en route — DOA.

'Another time I had just about got airborne in a Dragon from Arona, nowhere to go except up, and a goat wandered across the strip. I waited until the last moment, rotated the old Dragon, and staggered into the air with a thump. I thought, I wonder how the tail-wheel is, but I couldn't do anything about it, so I circled the strip, and there, lying in the middle, was this goat. Next time into Arona, I asked how the goat got on. Fellow said, "He just lay there for five minutes, then got up and wandered away." He'd been zotted by the tail wheel, which had just knocked him out...'

Jock McKee had a spot of excitement in a Dragon in the mountains behind Madang: 'I went in with Ian Downs in November, '57, and searched for a small patrol of native police, held at bay by hostile tribesmen at Adelbert Mission. We buzzed the warriors with empty beer bottles — they screamed as the bottles fell. Then we dropped food and supplies to the beseiged police boys...'

Ross Christensen had his share of experiences with the Dragon in the late 1940s and early '50s, before ten years on the Bird of Paradise route: 'My memories are vividly coloured with the continual struggle against the weather and the mountainous terrain, flying under-powered aircraft with heavy loads in and out of fairly primitive airstrips. I had the usual frights, as all pilots did in those days when trapped with weather, but as time went by, learning the hard way, I survived those years and remember clearly the thrill and the beauty of early morning flights out of Lae and Madang, particularly up into the Goroka and Wahgi valleys, to Hagen and Wabag...

'I remember the airstrip at Chimbu, with its approach over the valley towards the mountain, and then the take-off down the hill, falling off the end of the strip, and having to circle to gain height to get back to the level of the strip. Kegsugl, at about 8,000 feet [2438 metres], and uphill, where one had to use full power just to taxi up the hill; Tapini, which I thought was the worst strip I've ever used in my life. The early morning flights were wonderful in the still clear air over the hills and into the valleys before the cloud build-ups in mid-morning and afternoon made flying difficult in Dragons, which were so under-powered that the clouds usually built up faster than the aircraft could climb...

'Other memories are of a minor prang on the Wabag strip in a Dragon, when a small willy-willy picked up my aircraft as I was taxying up the strip and lifted me back into the air, to come to grief with the aircraft's nose on the ground. Another was a forced landing on the almost disused Wapenamanda strip in the Wabag valley, followed by a five-mile [eight-kilometre] hike to the home of a Dutch missionary to report my whereabouts. Of course the aircraft did not carry any radio or navigational aids of any sort. It was all eyeball navigation, learned only by hard experience...'

Clearly, Ross Christensen had at least one bad fright early in his PNG days. Dick Davis recalls: 'Flying through cloud to get somewhere was called 'pressing on' and it was a no-no. Ross — a fine pilot — must have given himself something to think about in the region of the Bena Bena gap, because we all noticed a sketch on the front of his journey logbook — it was a heraldic coat- of-arms he had drawn, underneath which a scrolled motto said, "*MASKI* PRESS ON — BENA LESSON." *Maski* was Pidgin English for "don't"...'

Bill Taylor recalls a sticky landing at Wapenamanda in a DC3: 'Jack Stammer was the pilot, I was first officer. We arrived over Wapenamanda, and there was a DC3 already on the ground. Tommy Howes was the pilot, and he's on the radio to us, "Don't land the thing, I'm bogged." Jack says, 'He can't be, he doesn't know where the soft spots are.' So Jack puts the gear down and we go in to land, and we are immediately bogged up to the axle right alongside Tommy. "I told you so..." But we got the locals under the wing of the DC3, and everybody heave-hoed, and we eventually got away again...'

Stories of the legendary DC3 abound whenever pilots get together. It was everybody's favourite — 'a magic aeroplane', in Dick Davis's words. Most of the stories are about seemingly impossible overloads. Tommy Howes figured in one well-documented incident, where his DC3 carried exactly double the normal load all the way from Lae to Garaina, over high mountains. Says Howes: 'Qantas had the contract for flying in the prefabricated tea factory to Garaina, and through a mix-up they put on a double load of iron girders. It was immediately obvious after dragging it off the strip at Lae that all was not normal, and much later, after I had got it high enough to get into Garaina, I checked its stalling speed and approached accordingly. For months the pundits argued whether or not I shouldn't have returned immediately to Lae. As far as I was concerned, the longer I flew it the closer it got to the correct maximum landing weight. I don't think Qantas ever quite made up their mind whether getting two loads in for the price of one was a good idea.'

That was the DC3. As Bill Forgan-Smith once said, 'If you could get it in and close the door, she'd fly. She'd fly with things sticking *out* of the door. 'Forgie' remembers: 'One day we went to take a drill rig up from Madang to Goroka for the mining engineer, J.C. Coldham. They were going to take it on by road from Goroka to Bena Bena. There was a strip at Bena, but you weren't allowed to land a 'Three' there. Phil Chandler, our station manager at Madang, brought out this drill rig to the strip, and it was an enormous great thing. We couldn't get it in, had about eight feet [about two-and-a-half metres] of it hanging out the back of the aeroplane, so we left the door off. Couldn't take it to Goroka, where DCA could see us, so we took it into Bena, unloaded it, and flew back to Madang. No trouble with the 'Doug'...'

Lionel Purkiss had an unpleasant experience in a DC3 soon after New Guinea Internal began: 'No flight plans then, such as we have now. The only radio aids were NDB's at Moresby and Lae, one in Rabaul, and one in Madang. That was it. No DME or anything like that then. You had to use your skull to get out of problems.

'I was doing a weekend football charter in a DC3, from Lae up to Wau. Overnight stop. Before leaving Wau for Lae in the morning, I called up and asked what the weather was like. They said, "It's all clagged in down here," and I said I'd call back in a couple of hours. So I called back, and they said there was a break up the Markham. So I put my footballers on board, and away we went. Over the top of Bulolo, had a look down through the Wampit, and it was clagged up. So went back through the Watut, and it was clagged up. Had to go back to Wau, and the weather had moved in there and Wau was closed. So I went back to Bulolo, and it was closed, too. I said to the first officer, "From here on, it's survive..."

'I flew back down the Wampit as far as I could, then turned in a more southerly direction, climbing, until I reckoned I was higher than any of the mountains on the coastline, then over the Huon Gulf, fully in cloud, picked up a back- bearing on the NDB, and let down away from Lae until I could see the water, very, very low, then turned round and flew back towards Lae. Of course, the fuel I had on wasn't officially enough to do all this, and DCA called me up and said, "Can you make Finschhafen?" I said, 'Negative.' They said, "We reckon you only have about ten minutes fuel left." But I had a little better than that. We just kept coming in until we saw the colour of the water change, where the Markham came in, and put wheels and flaps down, and landed straight in down-wind in teaming rain. Lae was completely closed. The *Tenyo Maru* was the final flap-down stage...

'I was going in to Goroka late one afternoon. Lot of radio action — they wanted a doctor urgently up at Wabag, but it was closed to DC3's — too much rain. A silly mission woman had decided to stay at Wabag to have her baby, instead of going out to the coast, and the inevitable had happened — a fouled- up birth, retained placenta, and the nurse there tried to do something about it and made a bigger balls of it. She needed a doctor, badly. So I said, "To hell with the rules," put the doctor on board, and departed Hagen, presumably for Madang.

Came out where I could turn in to Wabag, switched the radio off, and landed. The co-pilot said, "You'll be a smash hit when you get back..." We finished up sliding up the strip sideways — like grease —, got to the end, and slid back down to the parking area. The poor woman's husband was there, hadn't shaved for a couple of days, ashen grey, down on his knees at the steps, saying, 'Thank God'...

'I took off again, and then called up Madang. I said, "Madang, Madang, I can't hear you, do you read me," over and over...so I got away with it.

'You flew the best that you could, and figured out what you were going to do if anything went wrong. I worked in with Bert Kienzle at Kokoda, to get his strip up so that he could take maximum loads off it. He asked me one day what I'd do if I lost a motor. So I said, "Which motor?" I had two separate plans prepared, depending on which motor went, where I was going to put it. No hope of flying out of Kokoda on one engine if you lost one on take-off, and I knew that. Luckily, with the Pratt and Whitney you just didn't lose them. The motor could be pretty sick, but it would get you into a safe area before you closed the thing down. You relied on your aircraft, and your own abilities. That's what made flying in New Guinea worth while. You knew what your aircraft cost per minute to charter, and you tried to do the best you could for the bloke who was chartering it. Partly from the personal point of view, because you invariably knew him, and also if you did a good job, the company would get the charter again...'

On at least one occasion, a DC3 was used as a taxi. The tale is told by Dennis O'Malley: 'As you know, in those days we didn't fly at night in New Guinea. Everything shut down when the sun went down, and wherever you were at that time, was where you stayed until next morning.

'One night, Clyde Johnson got stuck overnight at Minj, then the largest strip in terms of length in the Highlands. Just across the valley from Minj was the Hallstrom farm at Nondugl, where Nep Blood was in charge. At about 9.30 that night, Blood was about to go to bed when he heard Clyde Johnson's DC3 start up. That's strange, he thought, they must be doing some maintenance, or something. The lights of the aeroplane went on, and it taxied down the full length of Minj strip, where the engines stopped, and the lights went out.

'Nep thought, They're probably checking the brakes or something, ready for the morning...towards midnight, he was awakened by the sound of DC3 engines again, and he looked out his bedroom window, and there was the DC3 with all lights blazing, trundling back to the government end of the strip, and again things all shut down, and quiet prevailed.

'Clyde and his crew were staying overnight with the patrol officer. They had a few drinks, and when they had run out of grog, the PO suggested they visit the Catholic mission at the far end of the strip to get some more. They tottered out and onto the strip to walk up there, but it was a long way, and it was very dark, so Clyde had the brilliant idea of going up there in the DC3...'

Clyde Johnson was the first officer on a DC3 piloted by Dick Davis during one celebrated incident, when Colonel J.K. Murray was Administrator. Says Davis: 'We were flying an RAAF charter, to take Colonel Murray and his wife to Momote, on the occasion of a visit by the Royal Australian Navy to the RAAF station there. The schedule was to night-stop at Momote, and that night the Navy was to return the hospitality of the RAAF with a party on the quarter-deck of HMAS *Melbourne* the flagship. The Administrator and his wife were the guests of honour, and Clyde and I were also invited.

'A splendid evening. Clyde and I knew most of the RAAF people at Momote as we flew in there fairly regularly with mail and supplies from Townsville on the courier which Qantas operated for them. I also knew several of the Navy pilots, and the Flag Lieutenant of the Admiral...to cut a long story short, it was one of those nights, and somehow it was arranged that we would make a pseudo approach to land on the aircraft carrier as we departed next morning with the Administrator and his wife. I was quite dubious about this, as I knew it was unpardonable to fly in close proximity to HM ships. However, they convinced me...

'Luckily, I was sober enough to mention the plan to Mrs Murray, who adored aeroplanes and ships and anything to do with them. She was very excited, and fortunately for me, she told her husband. He thought it was tremendous, so off we went to Rabaul in the morning, via HMAS *Melbourne*.

'Took me about ten minutes to catch up with the Navy, which had left at dawn. From about ten or fifteen miles [about sixteen or twenty-four kilometres] out, lined up on the deck of the carrier and started a very slow descent. Mrs Murray had come up to the cockpit to get a better view...the ships were a beautiful sight, steaming along in the early morning sun, the sea as blue as it ever gets, and as smooth as a millpond...

'We lowered our flaps and undercarriage, and as we approached *Melbourne* the Navy batman had taken up his position, complete with bats and signals as though we were one of the carrier's own aircraft. We must have made a great sight as we slid down the approach path (all the Navy fellows I spoke to later said we looked just great!) but as we got to about 400 feet [about 122 metres] or so, I figured it was time to break off. Called to Clyde for gear up, and applied power to climb. Mrs Murray had her foot on the undercarriage release latch — poor Clyde, he dived his left hand down to release the latch, and grabbed Mrs Murray's ankle...rather embarrassing for Clyde, he said later he didn't know whether to lift her foot, or ask her to do so, or what...

'It took but a few seconds delay really, but the DC3 doesn't climb in a very spectacular fashion with the wheels and flaps down, and we lost a few more feet. We passed by *Melbourne* much, much closer than I had intended. I learned later that the Admiral was furious...he was about the only one on board who was not aware of the pseudo approach plan. Didn't look very good for me at all for a while, but the Administrator joined the fray on my behalf, and the matter was finally dropped...'

Qantas pilots who joined New Guinea Internal in the very early days had to put up with very poor accommodation and living standards. Fred Phillips remembers life at Lae in 1949: 'We lived in a sisal paper house with galvanised iron roof. There was no electricity, just kerosene lamps, kero fridge, kero primus stove, and even the iron was run on kerosene. Water was caught in forty-four gallon [200-litre] drums placed around the house, there being no guttering. The deafening nightly thunderstorms at least kept our water supply adequate. The cumulous clouds built up over the mountains during the day and some slipped down the Markham valley during the night on their way out to sea. The mountains on either side of the valley channelled the thunder right at Lae, making the crockery rattle...

'Huge tarantulas, the size of bread and butter plates, scared the daylights out of my wife. Even though they took all the prescribed precautions, both my wife and baby daughter contracted malaria, which took years to work out of their systems.'

Time brought an end to such spartan living conditions, as Qantas gradually expanded and improved the living and working conditions at the Lae base. The aeroplanes the pilots flew improved, too — except for the eternal DC3. The Dragon gave way to the Drover, which yielded to the Beaver, replaced in its turn by the Otter. Some liked the Drover; others didn't. As we have seen, most pilots liked the Dragon, despite its faults. The Beaver was acknowledged as a fine little aeroplane, but the Otter was generally considered to be a lesser machine.

The DHC-3 Otter was the last new aeroplane type to be introduced by Qantas in PNG. The Otter was essentially a scaled-up version of the Beaver, with an all-metal airframe and a Pratt and Whitney R1340-S1H1-G engine of 600 hp, the same as that fitted to Bobby Gibbes' Norsemans. Empty, the landplane version of the Otter weighed 2010 kilograms, and maximum take-off weight was 3628 kilograms. The amphibian version had an empty weight of 2398 kilograms, and was thus at a considerable weight disadvantage to the landplane.

The Otter had an amazing STOL performance, had a capacious fuselage capable of taking bulky loads, and had a reasonable turn of cruising speed — 222 and 201 kph respectively. On paper, it looked to be the ideal utility light aircraft for PNG conditions, apart from one overriding deficiency: engine power.

The Otters were the first Qantas aircraft in PNG to carry names, apart from the Catalinas. They were christened *Kikori*, *Kokopo*, *Kieta* and *Kerowagi*, in honour of the four principal regions in which they operated — Papua, New Britain, Bougainville and the Highlands. The type made an inauspicious beginning with the incident at Tapini, mentioned in the previous chapter, and Ken Montagu was not a fan of the Otter, particularly the amphibian: 'It was a very poor substitute for the Catalina. It was not a particularly good aeroplane, somewhat underpowered, rather fragile, and offered virtually nothing that the old GSA Norseman offered at a mere fraction of the cost, and the Norseman was far cheaper to repair and maintain. I didn't really think an awful lot of the Otter...

'May 30, 1959, was my very last flight in New Guinea. In the Otter amphibian, a check flight for first officer Jim Conway, prior to him taking over the aircraft and continuing operations for Qantas until TAA took over...'

Phil Oakley also thought poorly of the amphibious Otter: 'The Otter amphibian really was a hopeless exercise. When it flew, it always continued to fall behind in the load it carried — could never carry enough to satisfy demand. It had to go to Lae not only for its C of A, but even its 100-hourly and things like that. Inevitably, some problem would occur, and it used to be away far longer than it should have been. Good landplane, but not as good as the Beaver, pound for pound...

'The float Beaver had a comparatively better performance than the float Otter, partly because it had a better powerweight ratio, and partly because it didn't have wheels — just the floats. A much better proposition, performed better and take-off better. Pretty successful, and APC used it a lot, based at Middletown. Ken Montagu did a lot of work with the float Beaver...

John Simler agrees that the float Beaver was a much better aeroplane than the amphibious Otter: 'The total all-up weight of the Beaver was 5,100 pounds [2313 kilograms], with a 450 hp engine. Otter weighed 3,000 pounds [1361 kilograms] more, and only had 600 hp — and not as good an engine. Under- powered. Tough, though. TAA took over the Otter amphibian from Qantas, and Chris Cross used to do most of the flying out around the delta. The amphibian had wheels below the floats. Chris used to do half his landings on water, and half on land. He got out of sequence once. Landed at Vaimuru, which had marsden matting with grass growing up through it. He landed with the wheels up. Luckily for him, it was wet. He made a perfect landing, and didn't even know that he didn't have the wheels down until he opened the throttle, and the thing rocked. All he did was jack it up, put the wheels down, and away he went. No damage at all...

'I remember sitting up with Col Williamson in an Otter landplane, going in to Wau. First time I'd been in one. The thing went from about 45 degrees down to 45 degrees up on the approach. The angle was unbelievable. You could stop in no distance at all, especially on an uphill slope...'

Colin Williamson had a close shave with disaster, while flying an Otter. He was dropping supplies to a new government outpost at Wonenara, and lost much of his engine power at a critical moment. Says Mal Shannon: 'Colin staggered in towards Menyamya, tried to reduce power, he threw everything off, and it was holding height, but the vibration was so terrible that he had to increase power to stop it...got in to Menyamya eventually. Someone asked Colin later what aeroplane it was his ambition to fly, and he said, "An aeroplane with six engines. I'll say, 'Start Three!' and they'll say, 'Which side' "..'

Bill Taylor flew many hours in both the amphibian and land Otters. He was checked out on the amphibian by Ken Montagu, and then alternated between Beavers and land Otters out of Lae before moving to Port Moresby to take over the amphibian. In 1959 he was posted to Goroka on the Otter, doing charter flights throughout the Highlands at £45 per hour. Bill Taylor found the work interesting, but demanding: 'A busy time for me. Our agent would line up the daily work effort. "OK," he'd say, "the Catholic mission have got stuff to go to this place and that, when can you do it?" And I'd work out my programme for the day, and I'd charge off in the Otter with a load of stuff for Chimbu or Omkolai, then maybe to Hagen and pick up a load for Ialibu or somewhere else in the Southern Highlands, do three or four shuttles, then back to Goroka...

'In the meantime, if other business came along, the agent would have it ready. Passengers would come along too, all sorts of bits and pieces, and I would write it in the logbook, then have a reckoning at the end of the week. Passengers would present money, and I'd make out a manifest, tickets, collect money, do the load sheet, strap them in, get in, and fly them wherever it might be. A lot of paper-work involved, but very good training for anyone in the airline business. It doesn't happen today, with computers...

'This went on, six days a week, week in, week out. On the weekend, I'd get out the typewriter and do the reports — figure it all out, who got the charters, where the trips were, flying times, charter rates etc, then send down the report and the money I had collected to the area accountant at Lae — Bert Gould in those days, one of Nature's gentlemen. He'd go through all my stuff...never had a query back from him...gun fun, and great experience.

'I found the best people to do business with were the Catholic missions. They had an awful lot of work into the Southern Highlands, pre-fabricated houses and things like that...no suggestion of overloading the aeroplane...other missions were prone to do it on odd occasions, but never the Catholics.

'The Otter was not a bad plane from the lifting point of view for bulk — not that I agree with a single engine in New Guinea. But it had big doors, and you could get a fair bit of stuff into it. Did a little better than Bobby Gibbes' Norsemans.

'One of the other interesting aspects of the Otter operation was the late arrivals back at Goroka in the evenings. You'd be flat out in the Wahgi or somewhere during the day, and you had to get through the Daulo to get back into the Goroka valley before dark. You would quite often get to Chimbu, and it would look all right towards the Daulo, and it wasn't. I got caught one night, andd had to turn back, was getting a bit desperate, and found a little clearing on the side of a hill just to the south and east of the Daulo, had been an airstrip once but had been neglected, had stones all over it. I managed to get the Otter down in one piece.

'When you were doing ten or fifteen landings a day, and flying six days in a row, you got to know precisely where the ground was in the Otter, and you could put it onto a square foot of space very accurately, each wheel in its correct place...you could really manipulate this aeroplane on the ground. One did get pretty good at aircraft manipulation, people used to remark on this. With the practice you were getting, it was obvious that you would improve, or be dead. We did improve, and most of the people who flew the Otter up there got a lot of enjoyment out of being able to place the thing very accurately on the ground.

'Anyway, I got this thing safely on the ground, and called up Phil Graham in the Goroka tower, and asked him to ring my wife. I had just about had it, and I just sat there. Next thing I know, the big cockpit door opens, and this hand comes through my legs and grabs my family jewels from behind. I hadn't experienced this before, and I jumped about three feet, turned round, and it was an old hag about eighty! She was just saying "hello", Highlands- fashion...

'About March, 1960, we packed up from Goroka, flew down to Madang with Roger Wilson, and I commenced DC3 command training. Col Williamson relieved me at Goroka on the Otter...'

....

The end had come for Qantas in PNG. New Guinea Internal was formally handed over to TAA in a brief ceremony at Lae on 1 September, 1960. Fittingly, the honour of hauling down the Qantas house flag at the airport was given to Nanandai, the Papuan engineering foreman at the workshops, who had been with the company from the beginning. '*Sori tumas,*' said Nanandai, as the flag came down.

It was a sad time for the Qantas staff watching. Present were area manager Captain Phil Miller, assistant manager Doug Hilliger, and general services officer Fred Moger, and many others. It was a happy time for the TAA men, about to start on a new adventure. Newly-appointed TAA manager for PNG, Captain Lionel Thrift, was on hand, and Lae manager, Geoffrey Jones. The TAA general manager, John Ryland, was there for the occasion, and said: 'We of TAA are here to stay and succeed in the Territory'. He promised a service that would meet the demands of a Territory so dependent upon the aeroplane. He paid tribute to the 'magnificent job' that Qantas had done, and assured listeners that 'personalised service' would be the major policy for TAA in PNG. 'TAA is here to carry on the great work that Qantas have left behind them,' Ryland concluded.

A touch of humour enlightened the proceedings. As a gesture towards TAA — like Qantas, government-owned — a picture of one of their aeroplanes had occupied pride of place over the bar in the Qantas Social Club at Lae. Senior Qantas official Cedric Oliver, up from Sydney for the handover, said that now Qantas was moving out, the TAA photograph should be replaced by one featuring a Qantas aeroplane, and he presented John Ryland with a framed photograph of a Boeing 707.

The ceremony was soon over. The TAA house flag was raised, and the first 'Sunbird Service' flight took off for Rabaul.

Qantas was gone, but it was business as usual.

Abbreviations

ABC Australian Broadcasting Commission
ADO Assistant District Officer
ADT Australian Directorate of Air Transport
ALA Authorised Landing Ground
ANA Australian National Airways
ANG Air Niugini
ANGAU Australian New Guinea Administrative Unit
ANO's Air Navigation Orders
ANR's Air Navigation Regulations
APC Australasian Petroleum Company
ASIO Australian Security Intelligence Organisation
ATC Air Traffic Control
ATI Ansett Transport Industries
AWA Amalgamated Wireless (Australasia) Limited
AWAS Australian Women's Army Service
BAS Bougainville Air Services
BGD Bulolo Gold Dredging Limited
BOAC British Overseas Airways Corporation
BOUGAIR Bougainville Air Services
BSIP British Solomon Islands Protectorate
CAA Civil Aviation Agency
CFI Chief Flying Instructor
C of A Certificate of Airworthiness
C of G Centre of Gravity
DC District Commissioner
DCA Department of Civil Aviation
DF Direction Finding
DH de Havilland
DO District Officer
DOT Department of Transport
DME Distance Measuring Equipment
ELG Emergency Landing Ground
ETA Estimated Time of Arrival
ETD Estimated Time of Departure
GAS Goilala Air Services
GAT Guinea Air Traders
GSA Gibbes Sepik Airways
HF High Frequency
ICAO International Civil Aviation Organisation
IATA International Air Transport Association
IFR Instrument Flight Rules
ILS Instrument Landing System
JATO Jet Assisted Take-Off
KLM Koninklijke Nederlansche Indische Luchtraat Maatachappij
LAME Licenced Aeronautical Maintenance Engineer
MAF Missionary Aviation Fellowship
MAL Mandated Airlines
MAS Madang Air Services
NAC National Airways Commission
NAMASU Native Marketing and Supply Services Limited
NASA Nationwide Aviation Space Academy
NDB Non Directional Beacon
Notam Notice to Airmen (issued by Department of Civil Aviation, Department of Aviation)
OIC Officer-in-Charge
PATAIR Papuan Airlines
PIM Pacific Islands Monthly
PO Patrol Officer
PNG Papua New Guinea
QEA Qantas Empire Airways
RAAF Royal Australian Air Force
RAF Royal Air Force
RPM Revolutions Per Minute
RPT Regular Public Transport
SDA Seventh-Day Adventists
SIL Summer Institute of Linguistics
SPAC South Pacific Aero Clubs
STOL Short Take-Off and Landing
TAA Trans-Australia Airlines
TAC Town Advisory Council
TAL Territory Airlines
TAT Taylor's Air Transport
TOA Trans Oceanic Airways
UHF Ultra High Frequency
UN United Nations
UPNG University of Papua New Guinea
USAAF United States Army Air Force

Bibliography

1. Official Sources

Files, reports and papers of the Department of Civil Aviation, Canberra, Melbourne and Port Moresby.

Records of the Administration of the Territory of Papua and New Guinea.

Annual Reports

Papua 1945-1971

New Guinea 1945-1971

Papua New Guinea 1971-1973

Department of Civil Aviation 1944-1973

Department of Transport 1973-1975

Various district annual reports and patrol reports

Government Gazettes

Papua and New Guinea

Papua New Guinea

Hansard

Legislative Council Debates 1951-1963

First, Second and Third House of Assembly Debates, 1964-1975

Other Documents

Aircraft Accident Reports 1946-1982 — Department of Civil Aviation

Aviation Studies (Air Transport Policy Branch, Department of Civil Aviation)

Aviation Safety Digests

Registers of Civil Aircraft

United Nations Reports on Visiting Missions to the Trust Territory of New Guinea 1950-1971

2. Journals, Newspapers, Handbooks — various issues

Journals

The Aeroplane

Aircraft

Australian Air Log

Australian Territories

Aviation Heritage: Journal of the Aviation Historical Society of Australia

A.H.S.A. Journal

Airways — Qantas flight magazine

Australian Flying

Aviation Australia — Department of Aviation newsletter

Balus Toktok — Air Niugini Staff Magazine

Bulletin

Flight International

Islands Newsbeat, Rabaul

Pacific Islands Monthly

Paradise — Air Niugini flight magazine

Rag and Tube: Newsletter of the Antique Aeroplane Association of Australia

Whites Magazine

Newspapers

Australian

Financial Review

New Guinea Times-Courier

Niugini Nius

Papua New Guinea Post-Courier

Rabaul Times

South Pacific Post

Sydney Morning Herald

The Times of Papua New Guinea

Handbooks

Aircraft Bluebook

Australian Aviation Year Books

Pacific Islands Year Books

Papua New Guinea Handbooks

3. Books, Articles, Unpublished Papers, Selected Departmental Reports

Affleck, Arthur, **The Wandering Years**, Melbourne, 1964

Affleck, Arthur, and Tenenbaum, Icko, **Survey of Civil Aviation Requirements in New Guinea**, Melbourne, 1947

Airlines of Oceania, McDonnell Douglas, U.S.A., 1982

Air Niugini,An Interim Report by the Ombudsman Commission, Port Moresby, June, 1980

Ambrose, Vic, **History of Missionary Aviation Fellowship**, unpublished typescript, 232pp

Anderson, David, **Ansett Airlines of Papua New Guinea DC3's**, Aviation Heritage, Vol. 23, No.1

Aplin, Douglas, **Rabaul 1942**, Melbourne, 1980

Augenson, V., **Civil Air Services in New Guinea**, Department of Civil Aviation, typescript, November, 1945

Australasian Petroleum Company, photographic records

Behr, John, **Qantas Airways in New Guinea 1946-1960**, unpublished typescript, 349pp

Boughton, Terence, **The Story of the British Light Aeroplane**, John Murray, London, 1963

Boughton, Trevor W., **Some Impressions of Aviation in the Territories of Papua and New Guinea**, A.H.S.A. Journal, February, 1963

Bovelt, Al, **Catalinas in the South Pacific**, A.H.S.A. Journal, 1971

Bennett-Bremner, E, **Front-Line Airline: The War Story of Qantas Empire Airways Ltd**, Angus and Robertson, Sydney, 1944

Briggs, Tom, newspaper articles and unpublished papers

Brogden, Stanley, **The History of Australian Aviation**, Melbourne, 1960

Australia's Two Airline Policy, MUP, 1968

Civil Aviation, article in Encyclopaedia of Papua New Guinea, MUP, 1972

Civil Aviation in Papua New Guinea, DCA report, November, 1971

Collins, Rick and Marianne, **PNG Aviation Survey** in Aircraft, June, 1971

Cooke, John, **Working in Papua New Guinea 1931-1946**, Lara Publications, Brisbane, 1983

Croker, M.T., **Trans Oceanic Airways Pty. Ltd.**, Aviation Heritage Vol. 23, No.1

Ellison, Norman, **Flying Matilda**, Angus and Robertson, Sydney, 1957

Encyclopaedia of Papua New Guinea, Peter Ryan [Ed.], 3 volumes, MUP, 1972

Fairbairn, I.J., **Namasu: New Guinea's Largest Indigenous-Owned Company**, New Guinea Research Bulletin, No.28, 1969

Fysh, E.H., **Qantas At War**, Sydney, 1968

Wings to the World, Sydney, 1970

General Aviation Air Safety Symposium, Goroka, 1968. Unpublished papers

Goodall, Geoff, **Bristol 170 in Australia**, A.H.S.A. Journal, Vol. 20, No.3

The Lockheed Era, A.H.S.A. Journal, Sept-Oct 1975, Nov-Dec 1975, Jan-Feb 1976

Unpublished aircraft histories — various, but particularly DH84, Avro Anson, Lockheed Hudson

Griffin, James; Nelson, Hank; Firth, Stewart, **Papua New Guinea: A Political History**, Heinemann, Melbourne, 1979

Gunston, Bill [Ed.], **The Illustrated Encyclopaedia of Commercial Aircraft**, Windward, London, 1980

The Illustrated Encyclopaedia of Propeller Airliners, Windward, London, 1980

Gunther, Sir John, correspondence and papers on PNG aviation matters

Hall, Timothy, **Flying High**, Methuen, Sydney, 1979

Hasluck, Paul, **A Time for Building**, MUP, 1976

Imree, Alex, **Fressie's Aeroplane: Reflections on Flying the DH84 Dragon**, Aircraft Illustrated, February, 1971

Ingells, Douglas J., **The Plane That Changed The World**, Aero Publishers, California, 1966

Jackson, R., **Ok Tedi: The Pot Of Gold**, University of PNG, Port Moresby, undated
Mackenzie, Daryl, **TAA's Utilisation of the Douglas DC3 in New Guinea 1963**, A.H.S.A. Journal, October, 1963
Masters, Geoff, **Douglas DC2**, Australian Aviation, March, 1983
McKee, Jock, **My Flying During the Mt Lamington Eruption**, ms, 1985
Mondey, David [Ed.], **The Encyclopaedia of the World's Civil Aircraft**, Hamlyn-Aerospace, London, 1981
Morton, Harold, **Below Me The Mountains: Flying for God in New Guinea**, Jordan Books, Sydney, 1964
Moss, Peter W., **The de Havilland 84 Dragon**, Aircraft Illustrated, July, 1969
Nye, W.T.H., papers and correspondence
Pacific Islands Aviation Society. PNG aircraft fleet lists, aviation company directories, newsletters etc., — to 1970
Papua New Guinea Notes, in A.H.S.A. Journals
Pentland, Geoffrey, and Malone, Peter, **Aircraft of the RAAF 1921- 1971**, Kookaburra, Melbourne, 1971
Pattison, Barry, and Goodall, Geoff, **Qantas Empire Airways Indian Ocean Service 1943-1947**, A.H.S.A. Journal, 1979
Poulton, H.W., **New Factors in Papua and New Guinea Air Transport**, Lawrence Hargrave Memorial Lecture, 1969
Law, History and Politics of the Australian Two-Airline System, Melbourne, 1981
Report on Air Transport Operations in the Territory of Papua and New Guinea, DCA, typescript, March, 1962
Report on the Committee established to study Third Level and Charter Operations in PNG, B.L. Mitchell, chairman, unpublished, 1977
Rhys, Lloyd, **High Lights and Flights in New Guinea**, Hodder and Stoughton, London, 1942
Sabey, Ian, **Challenge in the Skies: The Founding of TAA**, Hyland, Melbourne, 1979
Schofield, J.E., **History of the Department of Aviation**, Canberra, 1984
Sinclair, James, **Sepik Pilot**, Lansdowne, Melbourne, 1971
Wings of Gold, Pacific Publications, Sydney, 1978
Uniting A Nation, Oxford, Melbourne, 1984
Smith, Vic, correspondence and papers on early days of Territory Airlines
Stedman, Phil, manuscripts of speeches, newspaper cuttings
Stuart, Ian, **Port Moresby: Yesterday and Today**, Pacific Publications, Sydney, 1970
Summer Institute of Linguistics, tapes, correspondence, reports and photographs on aviation operations in PNG
Taylor, G.A., **The 1951 Eruption of Mt Lamington, Papua**, Bureau of Mineral Resources, Geology and Geophysics, Bulletin No.38, Canberra, 1958
Taylor, John W.R. [Ed.], **Combat Aircraft of the World**, Edbury Press and Michael Joseph, London, 1969
Taylor, John W.R., and Munson, Kenneth, **History of Aviation**, Octopus, London, 1972
Territory Airlines and Talair Pty. Ltd., records, files, Board minutes and photographs
The First Sixty Years: Civil Aviation in Papua New Guinea, Australian Government Publishing Service, Canberra, 1982
Turnbull, Clive, **Wings of Tomorrow**, F.H. Johnston, Sydney, 1945
Vincent, David, **Catalina Chronicle**, Catalina National Committee, Adelaide, 1981
White, Leo, **Aviation in the Southwest Pacific**, Whites magazine, October, 1948
Williams, W., **A Survey of the Development of Communications in Papua and New Guinea**, typescript, 206pp, 1962
Air Communications in Papua and New Guinea, Australian territories, Vol. 4, No.5
Wilson, John, **TAA Fleet List**, Aviation Heritage, Vol. 23, No.4
Woodford, Don, **Papua New Guinea: Initiation and Independence**, University of Queensland Press, Brisbane, 1976

PAPUA NEW GUINEA

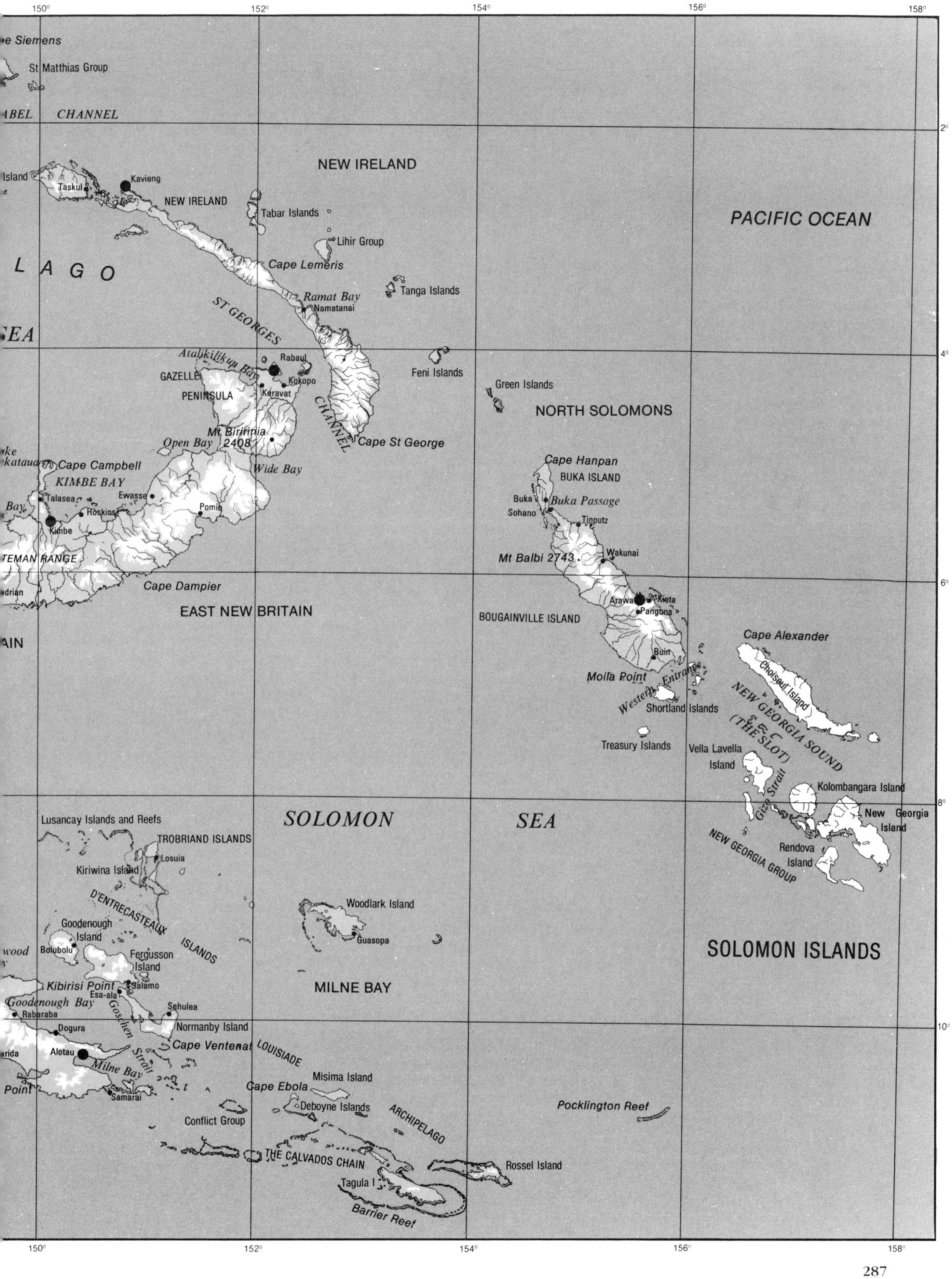

150°
152°
154°
156°
158°
2°
4°
6°
8°
10°
e Siemens
St Matthias Group
BEL CHANNEL
NEW IRELAND
Island
Kavieng
Taskul
NEW IRELAND
Tabar Islands
Lihir Group
PACIFIC OCEAN
L A G O
Cape Lemeris
Tanga Islands
Ramat Bay
Namatanai
ST GEORGES
EA
Atalikilikun Bay
Rabaul
GAZELLE
Kokopo
Keravat
PENINSULA
Feni Islands
Green Islands
CHANNEL
NORTH SOLOMONS
Mt Biririnia 2408
Open Bay
Cape St George
Cape Campbell
KIMBE BAY
Wide Bay
Cape Hanpan
BUKA ISLAND
Talasea
Ewasse
Buka
Buka Passage
Sohano
Roskins
Pomio
Tinputz
Kimbe
TEMAN RANGE
Mt Balbi 2743
Wakunai
Cape Dampier
EAST NEW BRITAIN
Arawa
Kieta
Panguna
BOUGAINVILLE ISLAND
Cape Alexander
Buin
Moila Point
Western Entrance
Choiseul Island
NEW GEORGIA SOUND (THE SLOT)
Shortland Islands
Treasury Islands
Vella Lavella Island
Gizo Strait
Kolombangara Island
SOLOMON
SEA
New Georgia Island
Lusancay Islands and Reefs
TROBRIAND ISLANDS
Losuia
Kiriwina Island
NEW GEORGIA GROUP
Rendova Island
D'ENTRECASTEAUX ISLANDS
Woodlark Island
Goodenough Island
Guasopa
Bolubolu
Fergusson Island
SOLOMON ISLANDS
Kibirisi Point
Salamo
Esa-ala
MILNE BAY
Goodenough Bay
Goschen Strait
Sehulea
Rabaraba
Dogura
Normanby Island
Cape Ventenat
LOUISIADE
Alotau
Milne Bay
Misima Island
Point
Cape Ebola
Samarai
Deboyne Islands
ARCHIPELAGO
Pocklington Reef
Conflict Group
THE CALVADOS CHAIN
Rossel Island
Tagula I
Barrier Reef

Index

D

E

F

G

H

I

Mc

N

O

P